Nanoparticle Technology for Drug Delivery

DRUGS AND THE PHARMACEUTICAL SCIENCES

Executive Editor

James Swarbrick

PharmaceuTech, Inc.
Pinehurst, North Carolina

Advisory Board

DRUGS AND THE PHARMACEUTICAL SCIENCES
A Series of Textbooks and Monographs

Nanoparticle Technology for Drug Delivery

edited by

Ram B. Gupta
Auburn University
Auburn, Alabama

Uday B. Kompella
University of Nebraska Medical Center
Omaha, Nebraska

Taylor & Francis
Taylor & Francis Group
New York   London

FIRST INDIAN REPRINT, 2014

Published in 2006 by
Taylor & Francis Group
270 Madison Avenue
New York, NY 10016

© 2006 by Taylor & Francis Group, LLC

Printed and bound in India by Bhavish Graphics.

International Standard Book Number-10: 1-57444-857-9 (Hardcover)
International Standard Book Number-13: 978-1-57444-857-3 (Hardcover)

Library of Congress Cataloging-in-Publication Data

Catalog record is available from the Library of Congress

Taylor & Francis Group
is the Academic Division of Informa plc.

Visit the Taylor & Francis Web site at
http://www.taylorandfrancis.com

FOR SALE IN SOUTH ASIA ONLY

Preface

Products of nanotechnology are expected to revolutionize modern medicine, as evidenced by recent scientific advances and global initiatives to support nanotechnology and nanomedicine research. The field of drug delivery is a direct beneficiary of these advancements. Due to their versatility in targeting tissues, accessing deep molecular targets, and controlling drug release, nanoparticles are helping address challenges to face the delivery of modern, as well as conventional drugs. Since the majority of drug products employ solids, nanoparticles are expected to have a broad impact on drug product development. The purpose of this book is to present practical issues in the manufacturing and biological application of nanoparticles. Drug delivery scientists in industry, academia, and regulatory agencies, as well as students in biomedical engineering, chemical engineering, pharmaceutical sciences, and other sciences with an interest in drug delivery, will find this book useful. It can also be used as a textbook for drug delivery courses focusing on nanoparticles.

This book is organized into four sections. The first section describes the distinguishing fundamental properties of nano-particles (Chap. 1) as well as technologies for nanoparticle manufacturing (Chaps. 2–4). Nanoparticles can be manufac-tured by either breaking macro-particles using technologies such as milling and homogenization (Chap. 2) or by building particles from molecules dissolved in a solution using super-critical fluid technology (Chap. 3). Nanoparticle manufacturing and properties can be further optimized by employing poly-mers or proteins as stabilizers (Chap. 4).

The second section describes the characterization of nano-particles at the material or physicochemical level (Chap. 5) and relates these properties to the delivery and effectiveness of nanoparticles (Chap. 6) as well as toxicological characteristics (Chap. 7).

The third section presents the various applications of nanoparticles in drug delivery. Depending on the route and purpose of drug delivery, the requirements for nanoparticulate systems can vary. These aspects are discussed in Chapter 8 for injectable delivery, Chapter 9 for oral delivery, Chapter 10 for brain delivery, Chapter 11 for ocular delivery, and Chapter 12 for gene delivery.

Finally, the fourth section provides an overview of the clinical, ethical, and regulatory issues of nanoparticle-based drug delivery. These are evolving areas and the drug product development experience with nanoparticles is limited. As more data is gathered on the safety and efficacy of nanoparti-culate systems, a clearer view will emerge.

Preparation of this book would not have been possible without the valuable contributions from various experts in the field. We deeply appreciate their timely contributions. Also, we are thankful to our colleagues at Auburn University and the University of Nebraska Medical Center for their support in preparing this book.

Ram B. Gupta
Uday B. Kompella

Contents

PART II: NANOPARTICLE CHARACTERIZATION AND PROPERTIES

PART IV: CLINICAL, ETHICAL, AND REGULATORY ISSUES

Contributors

Aniruddha C. Amrite   Department of Pharmaceutical
Sciences, University of Nebraska Medical Center, Omaha,
Nebraska, U.S.A.

Vivekanand Bhardwaj   Department of Pharmaceutics, National
Institute of Pharmaceutical Education and Research, Punjab,
India

Paul J. A. Borm   Centre of Expertise in Life Sciences,
Zuyd University, Heerlen, The Netherlands

Faris Nadiem Bushrab   Department of Pharmaceutical
Technology, Biotechnology and Quality Management, Freie
Universität, Berlin, Germany

Mahesh V. Chaubal   BioPharma Solutions, Baxter Healthcare,
Round Lake, Illinois, U.S.A.

Svetlana Gelperina   Institute of Molecular Medicine, Moscow Sechenov Medical Academy, Moscow, Russia

Ram B. Gupta   Department of Chemical Engineering, Auburn University, Auburn, Alabama, U.S.A.

Makena Hammond   College of Pharmacy, University of Nebraska Medical Center, Omaha, Nebraska, and Virginia State University, Petersburg, Virginia, U.S.A.

Roy J. Haskell   Pfizer Corporation, Michigan Pharmaceutical Sciences, Kalamazoo, Michigan, U.S.A.

Uday B. Kompella   Department of Pharmaceutical Sciences, College of Pharmacy, University of Nebraska Medical Center, Omaha, Nebraska, U.S.A.

Majeti Naga Venkata Ravi Kumar   Department of Pharmaceutics, National Institute of Pharmaceutical Education and Research, Punjab, India

Vinod Labhasetwar   Department of Pharmaceutical Sciences, University of Nebraska Medical Center, Omaha, Nebraska, U.S.A.

Jan Möschwitzer   Department of Pharmaceutical Technology, Biotechnology and Quality Management, Freie Universität, Berlin, Germany

Rainer H. Müller   Department of Pharmaceutical Technology, Biotechnology and Quality Management, Freie Universität, Berlin, Germany

Moses O. Oyewumi   Division of Medicinal and Natural Products Chemistry, College of Pharmacy, University of Iowa, Iowa City, Iowa, U.S.A.

Barrett Rabinow   BioPharma Solutions, Baxter Healthcare, Round Lake, Illinois, U.S.A.

Kevin G. Rice   Division of Medicinal and Natural Products Chemistry, College of Pharmacy, University of Iowa, Iowa City, Iowa, U.S.A.

Sanjeeb K. Sahoo   Department of Pharmaceutical Sciences, University of Nebraska Medical Center, Omaha, Nebraska, U.S.A.

Roel P. F. Schins   Institut fur Umweltmedizinische Forschung (IUF), University of Düsseldorf, Düsseldorf, Germany

1

Fundamentals of Drug Nanoparticles

RAM B. GUPTA

Department of Chemical Engineering,
Auburn University, Auburn, Alabama, U.S.A.

INTRODUCTION

In pharmaceutics, $\approx 90\%$ of all medicines, the active ingredient is in the form of solid particles. With the development in nanotechnology, it is now possible to produce drug nanoparticles that can be utilized in a variety of innovative ways. New drug delivery pathways can now be used that can increase drug efficacy and reduce side effects. For example, in 2005, the U.S. Food and Drug Administration approved intravenously administered 130-nm albumin nanoparticles loaded with paclitaxel (Abraxane^{TM}) for cancer therapy, which epitomizes the new products anticipated based on nanoparticulate systems. The new albumin/paclitaxel–nanoparticle formulation offers several advantages including elimination of toxicity because of cremophor, a solvent used in the previous formulation, and improved efficacy due to the greater dose of the drug that can be administered and delivered. For better development of the nanoparticulate systems, it is essential to understand the pharmaceutically relevant properties of nanoparticles, which is the purpose of this chapter and this book in general. In the following narrative, some fundamental properties of nanoparticles including their size, surface area, settling velocity, magnetic and optical properties, and biological transport are brought into the perspective of drug delivery.

NANOPARTICLE SIZE

To put the size of nanoparticles in perspective, Table 1 compares sizes of various objects. Because of the comparable size of the components in the human cells, nanoparticles are of great interest in drug delivery. It appears that nature, in making the biological systems, has extensively used nanometer scale. If one has to go hand in hand with nature in treating the diseases one needs to use the same scale, whether it is correcting a faulty gene, killing leprosy bacteria sitting inside the body cells, blocking the multiplication of viral genome, killing a cancer cell, repairing the cellular metabolism, or preventing wrinkles or other signs of aging. One cannot use a human arm to massage the hurt leg of an ant. Size matching is important in carrying out any activity. Drug delivery is aimed at influencing the biochemistry of the body.

Table 1 Typical Size of Various Objects

Object	Size (nm)
Carbon atom	0.1
DNA double helix (diameter)	3
Ribosome	10
Virus	100
Bacterium	1,000
Red blood cell	5,000
Human hair (diameter)	50,000
Resolution of unaided human eyes	100,000

The basic unit of the biological processes is the cell and the biochemical reactions inside it. With the advent of nanoparticles it is now possible to selectively influence the cellular processes at their natural scales.

NANOPARTICLE SURFACE

We can generally see and discern objects as small as 100,000 nm (100 µm). It is only in the past 300 years, with the invention of microscopes, that we can see smaller objects. Today, one can see objects as small as individual atoms (about 0.1 nm) using the scanning probe microscope. Owing to their small size, nanoparticles exhibit interesting properties, making them suitable for a variety of drug delivery applications. The number of molecules present on a particle surface increases as the particle size decreases. For a spherical solid particle of diameter d, surface area per unit mass, S_g, is given as

$$S_g = \left(\frac{\pi d^2}{4}\right)\left(\frac{\pi d^3 \rho_s}{6}\right)^{-1} = \frac{3}{2d\rho_s} \tag{1}$$

where ρ_s is the solid density. If the molecular diameter is σ, then the percentage of molecules on the surface monolayer is given as

$$\%\text{Surface molecules} = \frac{(4/3)\pi[d^3 - (d-\sigma)^3]}{(4/3)\pi[d^3]}100$$

$$= 100\left[\left(\frac{\sigma}{d}\right)^3 - 3\left(\frac{\sigma}{d}\right)^2 + 3\left(\frac{\sigma}{d}\right)\right] \tag{2}$$

For a typical low–molecular weight drug molecule of 1-nm diameter, %surface molecules are calculated in Table 2. It is interesting to see that for a 10,000-nm (or 10-µm) particle, a very small percentage of the molecules are present on the surface. Hence, the dissolution rate is much lower for the microparticles when compared to nanoparticles.

When the particles are of nanometer length scale, surface irregularities can play an important role in adhesion, as the irregularities may be of the same order as the particles (1).

Table 2 % Surface Molecules in Particles

Particle size (nm)	Surface molecules (%)
1	100.00
10	27.10
100	2.97
1,000	0.30
10,000	0.03

Nanoparticles can show a strong adhesion because of the increased contact area for van der Waals attraction. For example, Lamprecht et al. (2) observed differential uptake/ adhesion of polystyrene particle to inflamed colonic mucosa, with the deposition 5.2%, 9.1%, and 14.5% for 10-μm, 1000-nm, and 100-nm particles, respectively.

NANOPARTICLE SUSPENSION AND SETTLING

Because of the small size of the nanoparticles, it is easy to keep them suspended in a liquid. Large microparticles precipitate out more easily because of gravitational force, whereas the gravitational force is much smaller on a nanoparticle. Particle settling velocity, v, is given by Stokes' law as

$$v = \frac{d^2 g (\rho_s - \rho_l)}{18 \mu_l} \tag{3}$$

where g is gravitation acceleration (9.8 m/sec at sea level), ρ_l is liquid density (997 kg/m^3 for water at 25°C), μ_l is viscosity (0.00089 Pa/sec for water at 25°C). For various particles sizes, settling velocities are calculated in Table 3 for a solid density (ρ_s) of 1700 kg/m^3.

Thermal (Brownian) fluctuations resist the particle settlement. According to Einstein's fluctuation–dissipation theory, average Brownian displacement x in time t is given as

$$x = \sqrt{\frac{2 k_B T t}{\pi \mu d}} \tag{4}$$

Table 3 Particle Settling Velocities

Particle size (nm)	Settling velocity (nm/sec)
1	0.00043
10	0.043
100	4.30
1,000	430
10,000	43,005

where k_B is the Boltzman constant (1.38×10^{-23} J/K), and T is temperature in Kelvin. Table 4 shows displacements for particles of varying sizes in water at 25°C. The Brownian motion of a 1000-nm particle due to thermal fluctuation in water is 1716 nm/sec, which is greater than the settling velocity of 430 nm/sec. Hence, particles below 1000 nm in size will not settle merely because of Brownian motion. This imparts an important property to nanoparticles, that they can be easily kept suspended despite high solid density. Large microparticles easily settle out from suspension because of gravity, hence such suspensions need to carry a "shake well before use" label. Also, a microparticle suspension cannot be used for injection. For the nanoparticles, the gravitational pull is not stronger than the random thermal motion of the particles. Hence, nanopaticle suspensions do not settle, which provides a long self-life.

However, settling can be induced using centrifugation if needed for particle separation. Particle velocity under centrifugation is given as:

$$v = \frac{\pi x d^2 (\rho_s - \rho_l)}{9 \quad \mu} \left(\frac{\mathrm{rpm}}{60}\right)^2 \tag{5}$$

Table 4 Brownian Motion of the Particles

Particle size (nm)	Brownian displacement (nm in 1 sec)
1	54,250
10	17,155
100	5,425
1,000	1,716
10,000	543

where the centrifugal rotation is rotations per minute (rpm). For various particle sizes, centrifugal velocities calculated for a solid density (ρ_s) of 1700 kg/m^3 in water at 0.1 m from the axis of rotation are presented in Table 5.

MAGNETIC AND OPTICAL PROPERTIES

Small nanoparticles also exhibit unique magnetic and optical properties. For example, ferromagnetic materials become superparamagnetic below about 20 nm, i.e., the particles do not retain the magnetization because of the lack of magnetic domains; however, they do experience force in the magnetic field. Such materials are useful for targeted delivery of drugs and heat. For example, interaction of electromagnetic pulses with nanoparticles can be utilized for enhancement of drug delivery in solid tumors (3). The particles can be attached to antibodies directed against antigens in tumor vasculature and selectively delivered to tumor blood vessel walls. The local heating of the particles by pulsed electromagnetic radiation results in perforation of tumor blood vessels, microconvection in the interstitium, and perforation of cancer cell membrane, and therefore provides enhanced delivery of drugs from the blood into cancer cells with minimal thermal and mechanical damage to the normal tissues.

Gold and silver nanoparticles show size-dependent optical properties (4). The intrinsic color of nanoparticles changes with size because of surface plasmon resonance. Such nanoparticles are useful for molecular sensing, diagnostic, and imaging applications. For example, gold nanoparticles can exhibit different colors based on size (Table 6).

PRODUCTION OF NANOPARTICLES

Although any particle of a size <1-µm diameter is a nanoparticle, several national initiatives are encouraging the development of particles <100 nm as they might exhibit some unique physical properties, and hence potentially different and useful biological properties. However, achieving sizes <100 nm

Table 5 Centrifugal Velocities of the Particles

Particle size (nm)	Centrifugal velocity (m/sec) of the particles for different rotation speeds				
	1,000 rpm	2,000 rpm	5,000 rpm	20,000 rpm	50,000 rpm
1	7.66×10^{-12}	3.06×10^{-11}	1.91×10^{-10}	3.06×10^{-9}	1.91×10^{-8}
10	7.66×10^{-10}	3.06×10^{-9}	1.91×10^{-8}	3.06×10^{-7}	1.91×10^{-6}
100	7.66×10^{-8}	3.06×10^{-7}	1.91×10^{-6}	3.06×10^{-5}	1.91×10^{-4}
1,000	7.66×10^{-6}	3.06×10^{-5}	1.91×10^{-4}	3.06×10^{-3}	1.91×10^{-2}
10,000	7.66×10^{-4}	3.06×10^{-3}	1.91×10^{-2}	3.06×10^{-1}	1.91

Table 6 Size-Dependent Color Variation of Gold Nanoparticles

	Wavelength for maximum absorption (nm)	
Nanoparticle size (nm)	In water	In AOT/water/isooctane ($w_0 = 10$)
9	519	535
20	523	531
30	525	535
40	526	545
52	528	543
59	535	546
79	550	560
100	567	583

Abbreviation: AOT, sodium bis(2-ethyl hexyl)sulfosuccinate.
Source: From Ref. 5.

is more readily feasible with hard materials compared to drug and polymer molecules, which are soft materials. For hard materials, such as silica, metal oxides, and diamonds with melting points above 1000°C, nanoparticles in the 1–100 nm size range have been prepared. However, for drugs that are usually soft materials with melting point below 300°C particles in the 1–100 nm size range are more difficult to prepare. For this reason, it is a reasonable goal to aim at <300 nm particles for drug and polymer materials. There are several success stories for pharmaceutical materials in this size range.

Table 7 Number of Molecules in a Spherical Particle

Particle diameter	Particle volume (mL)	Number of molecules
0.58 nm	8.18×10^{-22}	1
1 nm	4.19×10^{-21}	5.05
10 nm	4.19×10^{-18}	5.05×10^3
100 nm	4.19×10^{-15}	5.05×10^6
500 nm	5.24×10^{-13}	6.31×10^8
1 μm	4.19×10^{-12}	5.05×10^9
5 μm	5.24×10^{-10}	6.31×10^{11}
1 mm	4.19×10^{-3}	5.05×10^{18}

Note: Drug molecular weight = 500 and solid density = 1 g/mL.

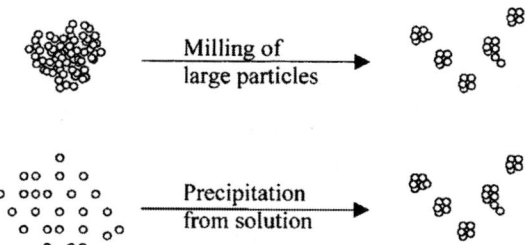

Figure 1 Schematic of the two general nanoparticle production techniques.

Production of nanoparticles of soft materials is much more challenging than that of hard materials because of the high stickiness of the former. The bulk pharmaceuticals are available in solids of large sizes (e.g., 1-mm-diameter powder), which can be often easily solubilized in solvent to obtain molecular size. Hence, there are two extremes of sizes: molecular size (each particle containing one molecule) and large size (e.g., each particle containing of the order of 10^{18} molecules). For a drug of 500 molecular weight and 1 g/mL solid density, the numbers of molecules in different size particles are given in Table 7.

Hence, to obtain nanoparticles in the 50–300 nm range for drug delivery, one requires of the order of 10^4–10^8 molecules in each particle. This size has to be achieved from either solution phase (single molecule) or millimeter-size particle (10^{18} molecules). The first approach is where the particle is broken down to nano size, whereas in the second approach, the particle will be built up from molecules. The two general approaches for the production of drug nanoparticles are sketched in Figure 1.

Milling of Large Particles or Breaking-Down Process

Comminuting or grinding or milling is the oldest mechanical unit operation for size reduction of solids and for producing large quantities of particulate materials. Here, the material is subjected to stress, which results in the breakage of the particle. Usually, the applied stress is more concentrated on the already present cracks in the material, which causes crack propagation leading to fracture. With the decreasing particle size, materials exhibit increasing plastic behavior making it more difficult to break small particles than large particles. For many materials, a limit in the grindability can be reached where subject to further grinding, no decrease in the particle size is observed (6). An empirical index, known as Bond work index (W_i) has been developed, which represents energy required for grinding (7).

$$W_i = 10\left(d_{\text{product}}^{-1/2} - d_{\text{feed}}^{-1/2}\right) \tag{6}$$

To reduce the size of a 1-mm particle, the energy required in terms of Bond index is given in Table 8 for various product sizes.

Hence, it is very energy intensive to go down to nanoparticles-size range. Other than size, parameters of importance are: (i) toughness/brittleness (in tough materials, stress causes plastic deformation, whereas in brittle materials cracks are propagated; hence, size reduction of brittle materials is easier than for tough materials; sometimes, a material can be cooled to embrittle), and (ii) hardness, abrasiveness, particle shape and structure, heat sensitivity (only about 2% of the applied energy goes to size reduction, the rest is converted to heat; hence, heat-sensitive drugs require cooling), and explodability (most pharmaceuticals are organic materials; as the size is reduced air combustibility of the material increases, hence proper inerting is needed).

Table 8 Energy Need (Bond Work Index) for Reducing Size of 1-mm-Diameter Particles

Product diameter (μm)	Energy required (Bond work index)
100	0.68
10	2.85
1	9.68
0.5	13.83
0.1	31.31

Most of the pharmaceutical size reduction operations utilize high-shear wet milling for the production of nanoparticles. Milling is explained in detail in chapter 2. Typical operation time for the wet milling may be hours to days, hence the drug has to show stability in that time period, otherwise milling cannot be used for unstable drugs. In addition, one has to be aware of contamination due to milling media.

Precipitation from Solution or Building-Up Process

In this process, a drug is dissolved in a solvent to achieve molecular solution. Then, the nanoparticle precipitate is obtained either by removing the solvent rapidly or by mixing an antisolvent (nonsolvent) to the solution, reducing its solubilizing strength. Initially, nuclei are formed, which grow because of condensation and coagulation giving the final particles. If the rate of desolubilization is slow, then sticky nuclei/particles are formed that have a higher tendency of agglomeration, giving large-size final particles. For example, if a drug is dissolved in a solvent (e.g., toluene) and then an

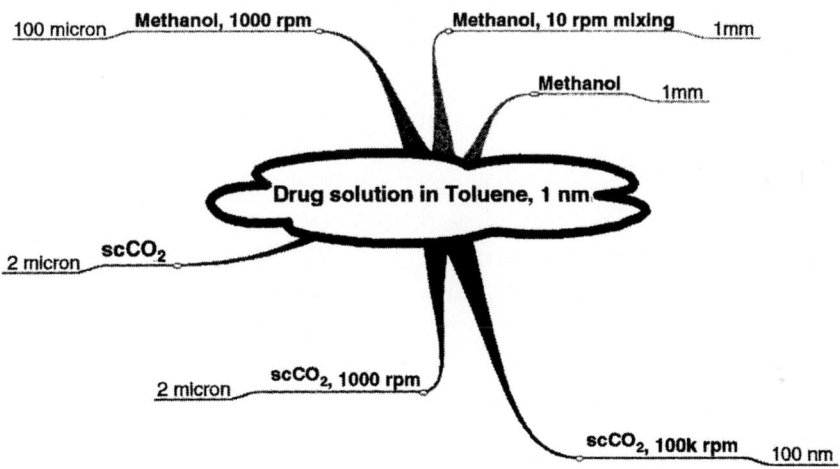

Figure 2 Variation of the particle size as the antisolvent and its mixing are varied in the solvent–antisolvent precipitation process.

antisolvent (e.g., methanol) is added with mild mixing, one will obtain drug precipitate of typically 1 mm particle size (Fig. 2). To obtain nanoparticles, a high desolubilization rate is needed, or use of a surfactant is required, which can isolate the particles until they are completely dry. Based on these requirements, two general methods for nanoparticle production are available: (i) supercritical fluid process, and (ii) emulsification–diffusion process. In the precipitation process, one can add compounds (e.g., polymers for controlled release) that will coprecipitate with the drug for smart drug delivery applications.

The key aspect of getting nanoparticles of the desired size and size distribution is to control both the rate of antisolvent action and the particle coagulation. Precipitation-based processes are described in chapters 3 and 4.

BIOLOGICAL TRANSPORT OF NANOPARTICLES

For drug delivery, most of the sites are accessible through either microcirculation by blood capillaries or pores present at various surfaces and membranes. Most of the apertures, openings, and gates at cellular or subcellular levels are of nanometer size (Table 9); hence, nanoparticles are the most suited to reach the subcellular level. One of the prime requirements of any delivery system is its ability to move around freely in available avenues and by crossing various barriers that may come in the way. Regarding the human body, the major passages are the blood vessels through which materials are transported in the body. The blood vessels are not left in any organ as an open outlet of the pipe, rather they become thinner and thinner and are finally converted to capillaries through branching and narrowing. These capillaries go to the close vicinity of the individual cells. After reaching their thinnest sizes, the capillaries start merging with each other to form the veins. These veins then take the contents back to the heart for recirculation. Hence, the supply chain in the body is not in the form of a pipe having an open inlet to the organ and outlet away from the organ. Consequently, for

Table 9 Approximate Sizes of Components in a Typical 20-μm Human Tissue Cell

Component	Size (nm)
Ribosomes	25
Golgi vesicles	30–80
Secretary vesicles	100–1000
Glycogen granules	10–40
Lipid droplets	200–5000
Vaults	55
Lysosomes	500–1000
Proteasomes	11
Peroxisomes	500–1000
Mitochondria	500–1000
Superfine filaments	2–4
Microfilaments	5–7
Thick filaments	15
Microtubules	25
Centrioles	150
Nuclear pores	70–90
Nucleosomes	10
Chromatin	1.9

Source: From Refs. 8, 9.

any moiety to remain in the vasculature, it needs to have its one dimension narrower than the cross-sectional diameter of the narrowest capillaries, which is about 2000 nm. Actually, for efficient transport the nanoparticle should be smaller than 300 nm.

But, just moving in the vessels does not serve the drug delivery purpose. The delivery system must reach the site at the destination level. This requires crossing of the blood capillary wall to reach the extracellular fluid of the tissue and then again crossing of other cells, if they are in the way, and entering the target cell. These are the major barriers in the transit. A nanoparticle has to do a lot during this sojourn of the carrier through the vessels (capillaries) and across the barriers.

There are two routes for crossing the blood capillaries and other cell layers, i.e., transcellular and paracellular. In the transcellular route, the particulate system has to enter the cell from one side and exit the cell from the other side to reach the tissue. The particulate system has to survive the intracellular environment to reach the target tissue.

The other route is paracellular. In this, the particlulate system is not required to enter the cell; instead, it moves between the cells. These intercellular areas are known as the junctions. Passing through the junctions would obviate destruction of the carrier by the cell system. Paracellular movement of moieties including ions, larger molecules, and leukocytes is controlled by the cytoskeletal association of tight junctions and the adherence junctions called apical junction complex. While tight junctions act as a regulated barrier, the adherence junctions are responsible for the development and stabilization of the tight junctions. Different epithelial and endothelial barriers have different permeabilities mainly because of the differences in the structure and the presence of tight junctions. While epithelia and brain capillary endothelium exhibit a high degree of barrier function, the vascular endothelium in other tissues has greater permeability. The tight junctions control the paracellular transport. For example, diffusion of large molecules may not be feasible, but migration of white cells is allowed. Understanding of this regulation mechanism is important as this might enable us to pave the way for the movement of nanoparticles in the body without actually entering into the unintended cells.

As the nanoparticle-based drug delivery is achieved by particle transport, it is important to understand the blood flow rates and volumes of various organs and tissues. Considering the body's distribution network, the blood vascular system, the body could be divided into several compartments based on the distributional sequencing and differentiation by the blood vascular system (Table 10).

Nanoparticles can have deep access to the human body because of the particle size and control of surface properties. Experiments by Jani et al. (13,14) have elegantly demonstrated the size effect. Polystyrene particles in the size range 50–3000 nm were fed to rats daily for 10 days at a dose of 1.25 mg/kg. The extent of absorption of the 50-nm particles

Table 10 Volumes and Blood Supply of Different Body Regions

Tissue	Volume (mL)	Volume of blood in equilibrium with tissue (mL)	Blood flow (mL/min)	Blood flow (mL/dL in an organ/min)
Viscera (well perfused)				
Adrenals	20	62	100	500
Kidneys	300	765	1240	410
Thyroid	20	49	80	400
Gray matter	750	371	600	80
Heart	300	148	240	80
Other small glands and organs	160	50	80	50
Liver with portal system	3,900	979	1580	41
White matter	750	100	160	21
Lean tissues (moderately perfused)				
Red marrow	1,400	74	120	9
Muscle	30,000		980	3.25
Skin (nutritive)	3,000		98	3.25
Nonfat subcutaneous	4,800	43	70	1.5
Adipose tissues (poorly perfused)				
Fatty marrow	2,200	37	60	2.7
Fat	10,000	123	200	2
Others				
Bone cortex	6,400	≈0	≈0	≈0
Arterial blood	1,400			
Venous blood	4,000			
Lung parenchymal tissue	600			1400
Air in lung	2.5 + half the tidal volume			

Note: Cardiac output, 6480 mL/min; data are of standard men: age 30–39 years, body weight 70 kg, and surface area 1.83 m^{-2}.
Source: From Refs. 10–12.

was 34% and that of the 100-nm particles was 26%. Of the total absorption, about 7% (50 nm) and 4% (100 nm) were accounted for in the liver, spleen, blood, and bone marrow. Particles >100 nm did not reach the bone marrow, and those >300 nm were absent from the blood. Particles were absent in the heart or the lung tissue. The rapid clearance of circulating particles from the bloodstream coupled with their high uptake by liver and spleen can be overcome by reducing the particle size, and by making the particle surface hydrophilic with coatings, such as poloxamers or poloxamines (15).

Gaur et al. (16) observed that 100-nm nanoparticles of polyvinylpyrrolidone had a negligible uptake by the macrophages in liver and spleen, and 5–10% of these nanoparticles remain in circulation even eight hours after intravenous injection. Because of longer residence in the blood, nanoparticles have potential therapeutic applications, particularly in cancer; the cytotoxic agents encapsulated in these particles can be targeted to tumors while minimizing the toxicity to the reticuloendothelial system.

Desai et al. (17) studied the effect of poly(d,l-lactide-co-glycolide) (PLGA). particle size (100 nm, 500 nm, 1 μm, and 10 μm) on uptake in rat gastrointestinal tissue. The uptake of 100-nm-size particles by the intestinal tissue was 15–250-fold higher compared to the larger-size microparticles. The uptake also depends on the type of tissue (i.e., Peyer's patch and nonpatch) and the location (i.e., duodenum or ileum). Depending on the particle size, Peyer's patch tissue had a 2–200-fold higher uptake of particles than the nonpatch tissue. The 100-nm particles were diffused throughout the submucosal layers, while the larger-size particles were predominantly localized in the epithelial lining of the tissue, because of the microparticle exclusion phenomena in the gastrointestinal mucosal tissue.

Hillyer and Albrecht (18) studied the gastrointestinal uptake and subsequent tissue/organ distribution of 4-, 10-, 28-, and 58-nm-diameter metallic colloidal gold particles following oral administration to mice. It was found that colloidal gold uptake is dependent on the particle size: smaller particles cross the gastrointestinal tract more readily. Interestingly, they observed that the particle uptake occurs in the small intestine by persorption through single, degrading enterocytes in the process of being extruded from a villus.

Cellular uptake is greater for nanoparticles compared to microparticles. In cultured human retinal pigment epithelial cells, an increase in the mass uptake of particles was observed with decreasing particle size in the range of 20–2000 nm polystyrene particles (19). Further, no saturable uptake was observed for these particles up to a concentration of 500 µg/ mL. With 20-nm nanoparticles, the uptake by the 1-cm^2 cell monolayer was as high as ~20%.

Because of possible differences in particle uptake, gene expression efficiencies can also be improved with smaller particles. Prabha et al. (20) studied relative transfectivity of 70- and 202-nm-PLGA nanoparticles in cell culture. The smaller particles showed a 27-fold higher transfection than the larger nanoparticles in COS-7 cell line and a fourfold higher transfection in HEK-293 cell line.

CONCLUSIONS

Nanoparticles offer unique properties as compared to micro- or macroparticles. Salient features include the following:

- Small size.
- High surface area.
- Easy to suspend in liquids.
- Deep access to cells and organelles.
- Variable optical and magnetic properties.
- Particles smaller than 200 nm can be easily sterilized by filtration with a 0.22-µm filter.

Drugs, being mostly organic compounds, are more sticky in nature as compared to inorganic materials, such as silica or metal oxides. Hence, it is harder to make smaller nanoparticles of drugs compared with hard materials. Drug nanoparticles can be produced either by milling of macroparticles or by fast precipitation from solutions, as described in the following chapters.

REFERENCES

1. Chow TS. Size-dependent adhesion of nanoparticles on rough substrates. J Phys: Condens Matter 2003; 15(suppl 2):L83–L87.

2. Lamprecht A, Schafer U, Lehr C-M. Size-dependent bioadhesion of micro- and nanoparticulate carriers to the inflamed colonic mucosa. Pharm Res 2001; 18(suppl 6):788–793.

3. Barros APH. Synthesis and Agglomeration of Gold Nanoparticles in Reverse Micelles. M.S. thesis, University of Puerto Rico, Mayaguez, PR, 2005.

4. Esenaliev RO. Radiation and nanoparticles for enhancement of drug delivery in solid tumors. PCT Int Appl 2000, WO 2000002590.

5. Kelly L, Coronado E, Zhao LL, Schatz GC. The optical properties of metal nanoparticles: the influence of size, shape, and dielectric environment. J Phys Chem B 2003; 107:668–677.

6. Prior M. Size reduction. In: Kirk-Othmer, Othmer DF, eds. Encyclopedia of Chemical Technology. John Wiley & Sons, Inc., 2000.

7. Bond FC. Crushing and grinding calculations. Can Min Metal Trans 1954; 57:466–472.

8. Alberts B, Bray D, Lewis J, Raff M, Roberts K, Watson JD. The Molecular Biology of the Cell. 2nd ed. New York: Garland Publishing, Inc., 1989.

9. Benjamin L. Genes V. New York: Oxford University Press, 1995.

10. Oie S, Benet LZ. The effect of route of administration and distribution on drug action. In: Banker GS, Rhodes CT, eds. Modern Pharmaceutics. New York: Marcel Dekker, 1996.

11. Dedrick RL, Bischoff KB. Pharmacokinetics in applications of the artificial kidney. Chem Eng Progr Symp Ser 1968; 64:32–44.

12. Mapleson WW. An electric analogue for uptake and exchange of inert gases and other agents. J Appl Physiol 1963; 18:197–204.

13. Jani P, Halbert GW, Langridge J, Florence AT. Nanoparticle uptake by the rat gastrointestinal mucosa: quantitation and

particle size dependency. J Pharm Pharmacol 1990; 42(suppl 12): 821–826.

14. Jani P, Halbert GW, Langridge J, Florence AT. The uptake and translocation of latex nanospheres and microspheres after oral administration to rats. J Pharm Pharmacol 1989; 41(suppl 12): 809–812.

15. Rudt S, Muller RH. In vitro phagocytosis assay of nano- and microparticles by chemiluminescence. III. Uptake of differently sized surface-modified particles, and its correlation to particles properties and in vivo distribution. Eur J Pharm Sci 1993; 1:31–39.

16. Gaur U, Sahoo SK, De TK, Ghosh PC, Maitra A, Ghosh PK. Biodistribution of fluoresceinated dextran using novel nanoparticles evading reticuloendothelial system. Int J Pharm 2000; 202(suppl 1–2):1–10.

17. Desai MP, Labhasetwar V, Amidon GL, Levy RJ. Gastrointestinal uptake of biodegradable microparticles: effect of particle size. Pharm Res 1996; 13(suppl 12):1838–1845.

18. Hillyer, JF, Albrecht RM. Gastrointestinal persorption and tissue distribution of differently sized colloidal gold nanoparticles. J Pharm Sci 2001; 90(suppl 12):1927–1936.

19. Aukunuru JV, Kompella UB. In vitro delivery of nano- and micro-particles to retinal pigment epithelial (RPE) cells. Drug Deliv Technol 2002; 2(suppl 2):50–57.

20. Prabha S, Zhou W-Z, Panyam J, Labhasetwar V. Size-dependency of nanoparticle-mediated gene transfection: studies with fractionated nanoparticles. Int J Pharm 2002; 244(suppl 1–2):105–115.

2

Manufacturing of Nanoparticles by Milling and Homogenization Techniques

RAINER H. MÜLLER, JAN MÖSCHWITZER, and
FARIS NADIEM BUSHRAB

Department of Pharmaceutical Technology,
Biotechnology and Quality Management,
Freie Universität, Berlin, Germany

INTRODUCTION

The number of newly developed drugs having a poor solubility and thus exhibiting bioavailability problems after oral administration is steadily increasing. Estimates by the pharmaceutical companies are that about 40% of the drugs in the pipelines are poorly soluble, and as high as 60% of the compounds come directly from the synthesis route (1). Therefore, since a number of years the pharmaceutical development is focused on formulation approaches to overcome solubility and related bioavailability problems, so that these new compounds are available for clinical use. Often forgotten, the problem of poor solubility arises even before the preclinical phase, which means that when screening new compounds for pharmacological activity a test formulation needs to be able to lead to sufficiently high blood levels. Therefore, there is an urgent need to come up with a smart formulation approach.

One has to differentiate between specific and nonspecific formulations for increasing solubility and, subsequently, bioavailability. Specific approaches can only be applied to certain drug molecules, e.g., in case of cyclodextrins (CDs) to molecules that fit into the respective CD ring. In the area of CDs, research is focused on CD derivates with higher solubility of the CD itself and simultaneously reduced side effects of these excipients; for example, the recent development of Captisol® CDs (2,3). On the other hand, the nonspecific formulation approaches are applicable to almost any drug molecule (apart from a few exceptions). Such a nonspecific formulation approach since many years is micronization, which means converting relatively coarse drug particles to micrometer crystals with a mean diameter in the range of approximately 2–5 µm, and a corresponding size distribution approximately between 0.1 and 20 µm (4). Here, the increase in the surface area leads to an increase in the dissolution velocity. That means micronization is a formulation approach to overcome the bioavailability problems of drugs of the biopharmaceutical specification class II (BSC II). Drugs of class II are sufficiently permeable but the rate limiting step is a too low dissolution velocity (i.e., low solubility in general is correlated with low dissolution velocity, law by Noyes–Whitney). Nowadays however, many of the new compounds are so poorly soluble that micronization is not sufficient to overcome a too low oral bioavailability. Consequently, the next step taken was to move from micronization to nanonization.

By going down one more dimension from the microrange to the nanoworld there is a distinct increase in the surface area and related dissolution velocity. For example, when moving from a spherical 50 µm particle to micronized 5 µm particles, the total surface area enlarges by a factor of 10, moving to 500 nm nanocrystals by a factor of 100. However, there is an additional—but often forgotten—effect further increasing the dissolution velocity, that is, the increase in saturation solubility c_s when moving to sizes below 1 μm. Because of the strong curvature of the particles, they possess an increased dissolution pressure comparable to the increased vapor pressure of ultrafine aerosol droplets. The theoretical background is provided by the Kelvin equation and the Ostwald–Freundlich equation, which will not be discussed here in detail (5). According to the Noyes–Whitney equation the dissolution velocity dc/dt is proportional to the concentration gradient $c_s - c_x$ (c_s—saturation solubility, c_x—concentration in surrounding medium, bulk concentration). The increases in saturation solubility of nanocrystals reported are by a factor of about 2 to 4–6 [(6,7) and unpublished data]. The increase is even more pronounced when the nanosized drug material is not crystalline but amorphous. Preparation of amorphous oleanolic acid nanoparticles increased the saturation solubility up to 10-fold in relation to the coarse drug powder (8). Nanonization has the advantage that it practically can be applied to more or less any drug material. In general, even highly water-sensitive drugs can be reduced to drug nanocrystals, even stored in the form of an aqueous nanosuspension (drug nanocrystals dispersed in aqueous surfactant/stabilizer medium). For example, aqueous Paclitaxel nanosuspension proved to be stable over a period of four years stored at 4°C, i.e., more than 99% of the drug was recovered intact (9). On the other hand, aqueous Paclitaxel solution degrades to an extent of 80% within 25 minutes (10).

Drug nanocrystals can be produced by bottom-up or top-down technologies. In the case of bottom-up technologies, one starts with the molecules in solution and moves via association of these molecules to the formation of solid particles, i.e., it is a classical precipitation process (11). To our knowledge, there is presently no pharmaceutical product on the market based on precipitation technology. There are a number of reasons, discussed in detail in Ref. (12). Briefly, the use of solvents creates additional costs. In addition, a prerequisite for precipitation is that the drug is at least soluble in one solvent, and this solvent needs to be miscible with a nonsolvent. Many of the newly developed compounds; however, are poorly soluble in aqueous and simultaneously in nonaqueous media, thus excluding this formulation approach. In the case of top-down technologies, one starts with a coarse material and applies forces to disintegrate into the nanosize range. The diminution technologies can be categorized into two principal classes:

1. Pearl/ball milling.
2. High-pressure homogenization, and other processes.

There are two products on the market based on the pearl/ball-milling technology by the company NANOSYSTEMS ÉLAN (13). Rapamune® coated tablet is the more convenient formulation for the patient compared to the drug solution (Rapamune solution). The bioavailability of the tablet is 27% higher than the solution form (14). Rapamune® was introduced in the market in 2002 by the company Wyeth. The second is Emend®, introduced in the market in 2003 by the company MSD®, Sharp & Dohme Gmbh. It is a capsule composed of sucrose, microcrystalline cellulose (MCC), hyprolose, and sodium dodecylsulfate (SDS) (15).

Also, the products based on drug nanocrystals produced with high-pressure homogenization are in clinical phases. Therefore, these two technologies are reviewed in this chapter because of their relevance for the pharmaceutical market.

Drug nanocrystals are of high relevance to pharmaceutical products; therefore, it is not surprising that most of the research and development are being done in pharmaceutical companies, especially looking at the production process itself. Of course as a consequence, articles published by companies are very low in number to protect internal knowledge; primarily, only published patents are accessible. Even less literature is available on how to transfer the drug nanosuspensions to the final products, i.e., tablets, capsules, and pellets for oral administration or aqueous/lyophilized nanosuspensions for intravenous injection. Producing drug nanocrystals is relatively easy compared to the much more sophisticated technology to formulate a final drug dosage form. A final traditional drug dosage form has to be based on patient convenience. However, to fully benefit from the special properties of nanocrystals, they need to be released as ultrafine, nonaggregated suspension from the final dosage form. It could be shown that in the case of strong nanocrsytal aggregation, the dissolution velocity is reduced (16). Therefore, the tricky business is how to transfer the drug nanosuspension to dosage forms with optimized release properties. This chapter also describes the production of tablets, capsules, and pellets.

PEARL/BALL-MILLING TECHNOLOGY FOR THE PRODUCTION OF DRUG NANOCRYSTALS

Traditional equipment used for micronization of drug powders such as rotor–stator colloid mills (Netzsch) or jet mills (Retsch) are of limited use for the production of nanocrystals. For example, jet milling leads to a drug powder with a size range of roughly between 0.1 and 20 μm, containing only a very small fraction of about 10% in the nanometer range (4). However, it could be shown when running a pearl mill over a sufficiently long milling time, that drug nanosuspensions can be obtained (13,17,18). These mills consist of a milling container filled with fine milling pearls or larger-sized balls. The container can be static and the milling material is moved by a stirrer; alternatively, the complete container is moved in a complex movement leading consequently to movement of the milling pearls.

There are different milling materials available, traditionally steel, glass, and zircon oxide are used. New materials are special polymers, i.e., hard polystyrene. A problem associated with the pearl milling technology is the erosion from the milling material during the milling process. Buchmann et al. (19) reported about the formation of glass microparticles when using glass as milling material. In general, very few data have been published on contamination of pharmaceutical drug nanosuspensions by erosion from the milling material. Most data have been given in the discussions after oral presentations; figures from such discussions range from 0.1 to 70 ppm contamination. Of course it should be noted that the extent of erosion depends on the solid concentration of the macrosuspension to be processed, the hardness of the drug, and based on this, the required milling time and milling material. Apart from the milling material, the erosion from the container also needs to be considered. For the parts in contact with the product various materials are offered by companies producing pearl mills depending on the area of application, such as technical purposes, food, or the pharmaceutical industry. Normally, product containers are made of steel and can be covered with various materials to fulfill the required quality specifications of the formulation.

Surfactants or stabilizers have to be added for the physical stability of the produced nanosuspensions. In the production process the coarse drug powder is dispersed by high-speed stirring in a surfactant/stabilizer solution to yield a macrosuspension. The choice of surfactants and stabilizers depends not only on the properties of the particles to be suspended (e.g., affinity of surfactant/stabilizer to the crystal surface) but also on the physical principles (electrostatic vs. steric stabilization) and the route of administration. In general, steric stabilization is recommended as the first choice because it is less susceptible to electrolytes in the gut or blood. Electrolytes reduce the zeta potential and subsequently impair the physical stability, especially of ionic surfactants. In many cases an optimal approach is the combination of a steric stabilizer with an ionic surfactant, i.e., the combination of steric and electrostatic stabilization. There is a wide choice of various charged surfactants in case of drug nanocrystals for oral administration. Even relatively "nasty" surfactants, such as the membrane damaging SDS, can be used, of course within the concentration accepted for oral administration, e.g., the formulation of Emend (15). SDS as a low molecular weight surfactant diffuses fast to particle surfaces; it has excellent dispersion properties. Adsorption onto the particle surface leads to high zeta potential values providing good physical stabilities. In case of parenteral drug nanocrystals, the choice is limited; e.g., for intravenous injection, accepted are lecithins, Poloxamer 188, Tween® 80, low molecular weight polyvinylpyrrolidone (PVP), sodium glycocholate (in combination with lecithin). Drug nanocrystal suspensions for parenteral administrations need to be sterile, depending on the administration route and the volume they need to be pyrogene-free. Production of parenteral drug nanosuspension using pearl mills is much more tedious compared to producing oral drug nanosuspensions. The equipment needs to be sterilized and the product needs to be separated from the milling pearls by a preferentially aseptic separation process. A terminal sterilization by autoclaving is only possible with a number of products (20). The use of an ionic stabilizer such as lecithin is recommended when autoclaving nanosuspensions. The autoclaving temperature of 121°C leads to dehydration of steric stabilizers, which reduces their ability to stabilize the suspensions. Gamma irradiation of nanosuspensions is an alternative, but is less favoured by the pharmaceutical industry due to regulatory requirements (e.g., proof of absence of toxic radicals, etc.). From the industrial point of view, in many cases a well-documented aseptic production is easier for the production of formulations for parenteral administration than gamma irradiation.

There are a number of pearl mills available on the market, ranging from laboratory-scale to industrial-scale volumes. The ability for large-scale production is an essential prerequisite for the introduction of a product to the market. One advantage of the pearl mills, apart from being low-cost products, is their ability for scaling up. Assuming, for reasons of simplicity, hexagonal packaging of the milling pearls, 76% of the milling chamber volume will be filled by the pearls. In case of a 1000 L mill this corresponds to 760 L milling material; based on the apparent density of zircon oxide pearls being 3.69 kg/L, this corresponds to 2.8 tons of milling material. Figure 1 shows the solution for this problem, a pearl mill with an external suspension container. The suspension is continuously pumped through the pearl mill. This approach reduces the weight of the pearl mill itself, but it prolongs the milling times.

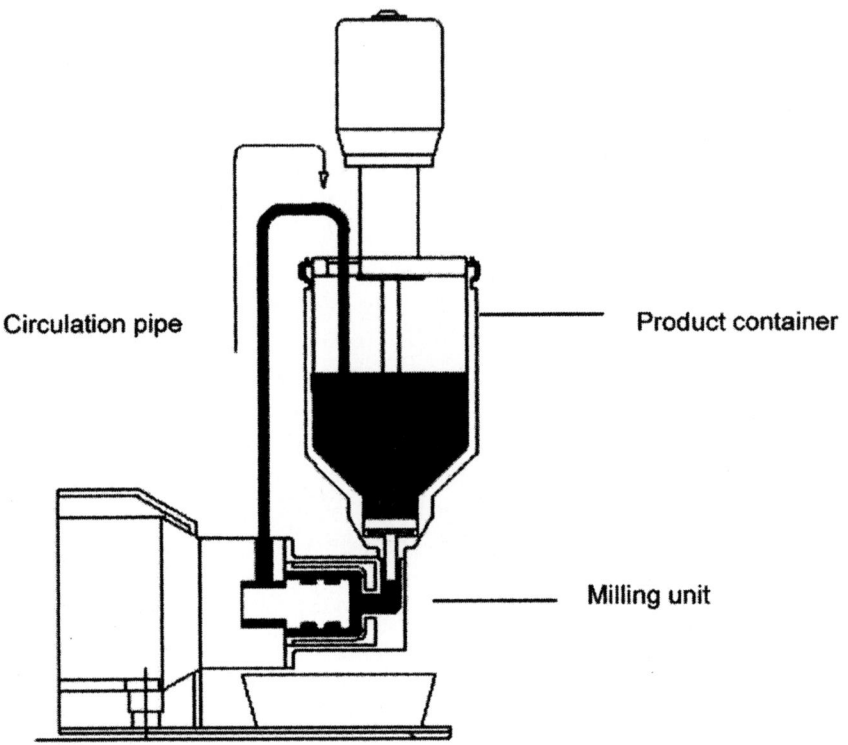

Circulation pipe   Product container

Milling unit

Figure 1   DISPERMAT® SL: schematic view of a bead mill using recirculation method. *Source*: From Ref. 21.

DRUG NANOCRYSTALS PRODUCED BY HIGH-PRESSURE HOMOGENIZATION

Theoretical Aspects

High-pressure homogenization is a technology that has been applied for many years in various areas for the production of emulsions and suspensions. A distinct advantage of this technology is its ease for scaling up, even to very large volumes. High-pressure homogenization is currently used in the food industry, e.g., homogenization of milk. In the pharmaceutical industry parenteral emulsions are produced by this technology. Commercial products such as Intralipid® and Lipofundin® possess a mean droplet diameter in the range of 200–400 nm (photon correlation spectroscopy data) (22). In the mid-1990s of the last century drug nanosuspensions produced with high-pressure homogenization were developed (23–27). Typical pressures for the production of drug nanosuspensions are 1000–1500 bar (corresponding to 100–150 Mpa, 14504–21756 psi); the number of required homogenization cycles vary from 10 to 20 depending on the properties of the drug. Most of the homogenizers used are based on the piston-gap principle, an alternative is the jet-stream technology (Fig. 2).

The Microfluidizer® (Microfluidics™ Inc., U.S.A.) is based on the jet-stream principle. Two streams of liquid collide, diminution of droplets or crystals is achieved mainly by particle collision, but occurrence of cavitation is also considered. The Microfluidizer has also been described for the production of drug nanosuspensions; however, according to the patent 10–50 cycle passes were required (28). Such a high

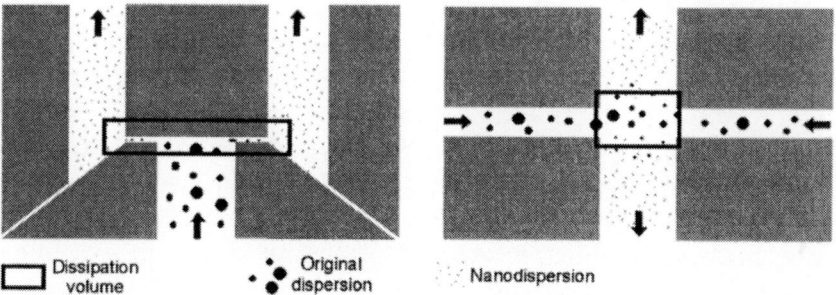

Dissipation volume   Original dispersion   Nanodispersion

Figure 2 Basic homogenization principles: piston-gap (*left*) and jet-stream arrangement (*right*). In the piston-gap homogenizer the macrosuspension coming from the sample container is forced to pass through a tiny gap (e.g., 10 μm), particle diminution is affected by shear force, cavitation, and impaction. In jet-stream homogenizers the collision of two high-velocity streams leads to particle diminution mainly by impact forces.

cycle number is not convenient for the production scale. Based on our own experiences, the Microfluidizer can be used for the production of drug nanosuspensions in the case of soft drugs. In the case of harder drugs, a larger fraction of particles in the micrometer range remain, which do not exhibit the increase in saturation solubility because of their too large size.

For many years cavitation was considered as the major force leading to particle diminution in the high-pressure homogenization process. Consequently, most high-pressure homogenization patents in various application areas focus on water as a dispersion medium. In the piston-gap homogenizer the liquid is forced through a tiny homogenization gap, typically in the size range of 5–20 µm (depending on the pressure applied and the viscosity of the dispersion medium). Using a Micron Lab 40 the suspension is supplied from a metal cylinder by a piston, the cylinder diameter is approximately 3 cm. The suspension is moved by the piston having an applied pressure between 100 and 1500 bar. In principle the piston-gap homogenizer corresponds to a tube system in which the tube diameter narrows from 3 to 5–20 µm. According to the Bernoulli equation, the streaming velocity and dynamic pressure increase extremely, the static pressure in the gap falls below the vapor pressure of water at room temperature. A liquid boils when its vapor pressure is equal to the static pressure, which means water starts boiling in the gap at room temperature leading to the formation of gas bubbles. The formation of gas bubbles leads to pressure waves disrupting oil droplets or disintegrating crystals. When leaving the homogenization gap, the static pressure increases to normal air pressure, which means the water does not boil any more and the gas bubbles collapse. Collapsing of the gas bubbles (implosion) leads again to shock waves contributing to diminution. There are different definitions of cavitation in the literature, describing cavitation either as the formation of gas bubbles in high streaming liquids or as the formation and subsequent implosion of these gas bubbles.

At the end of the 1990s it was found that similar efficient particle diminution can be achieved by homogenization in nonaqueous media such as oils and liquid polyethylene glycols

(PEGs), which means media with low vapor pressure. In the case of low vapor pressure liquids, the cavitation in the homogenization gap is distinctly reduced or does not exist at all. Figure 3 shows the change in static pressure when homogenizing in water as dispersion medium (left) and in a low-vapor liquid, whereas the static pressure does not fall below the vapor pressure (right).

Based on the aforementioned, cavitation does not seem to be essential for a diminution effect. Major forces are droplet or particle collision and the shear forces occurring in this highly turbulent fluid in the gap possessing a high kinetic energy. Homogenization in nonaqueous liquids has advantages for certain pharmaceutical final dosage forms. Preparation of drug nanocrystals in PEGs or oils (e.g., Miglyol 812 or 829) leads to nanosuspensions that can directly be filled into capsules (see the following) (29,30). It is also possible to homogenize in melted nonaqueous matrices, which are solid at room temperature. Solidification of such a matrix leads to a fixation of drug nanocrystals in the solid matrix, thus minimizing or avoiding crystal contact and subsequent crystal fusion/growth.

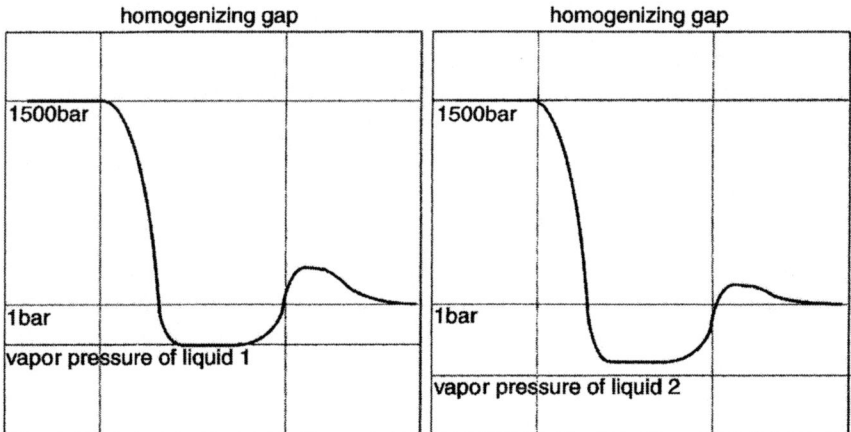

Figure 3 Variation of the static pressure (—) within the homogenizing gap. In the case of water the static pressure falls below the vapor pressure (*left*), whereas in the case of low-vapor media (*right*) the static pressure stays above the vapor pressure.

As a consequent next step, after homogenization in water (100% water) and homogenization in nonaqueous media (0% water), homogenization was performed in mixtures containing different percentages of water (1–99% water). The dispersion media were water mixed with water-miscible liquids (e.g., alcohols, glycerol). Preparation of drug nanosuspensions in water–ethanol mixtures is favorable for producing dry products, because later the spray drying can be performed under milder conditions when using such a mixture. Homogenization in water–glycerol mixtures (2.25% of water-free glycerol) leads to isotonic drug nanosuspensions for parenteral administration.

The laboratory scale homogenizers used by our group are the continuous and batch Micron Lab 40 (APV Systems, Unna, Germany). In the batch version, the batch size is a minimum of 20 mL and a maximum of 40 mL. A minimum of 20 mL is required for the machine to maintain the homogenization pressure because smaller volumes cannot be processed. In the batch Micron Lab 40, the homogenizer is equipped with two product containers having a maximum volume of 1000 mL. Considering the dead volume in the machine, a minimum batch size of about 200 mL is recommended. The advantage of the batch Micron Lab 40 is the relatively small batch volume, but unfortunately it is not produced any more. The successor model by the company APV is the APV-1000; however, the minimum batch size for this homogenizer is 150 mL (31). Scaling up to a size suitable for the production of clinical batches was performed using a Lab 60 unit. This homogenizer has a homogenization capacity of 60 L/hr, and can be qualified and validated (32). The commercially available Lab 60 was modified by equipping it with 10 L double-walled product containers; processing is possible in a continuous loop mode (2 kg batch) or alternatively in a batch mode (5–10 kg batch). Because of the termination of the production of the Lab 60 an APV 1000 or APV 2000 is recommended for a batch size in this range. Larger-scale machines from APV are the Gaulin 5.5 (160 L homogenization capacity per hour) or the Rannie 55 (600 L homogenization capacity per hour) at a pressure of 1500 bar.

Alternative suppliers of piston-gap homogenizers are the companies Avestin® (33) and GEA Niro Soavi (34).

PRODUCTION OF DRUG NANOCRYSTAL COMPOUNDS BY SPRAY-DRYING

For the production of tablets, an aqueous nanosuspension can be used as granulation fluid or a dry form of the nanosuspension, powder, or granulate can be employed. Starting from an aqueous macrosuspension containing the original coarse drug powder, surfactant, and water-soluble excipient, the homogenization process can be performed in an easy one step yielding a fine aqueous nanosuspension. In a subsequent step the water has to be removed from the suspension to obtain a dry powder. One method of removing the water from the formulation is freeze drying, but it is complex and cost-intensive leading to a highly sensitive product (35,36). Another simple and most suitable method for the industrial production is spray drying. The drug nanosuspension can directly be produced by high-pressure homogenization in aqueous solutions of water-soluble matrix materials, e.g., polymers [PVP, polyvinylalcohol or long chained PEG, sugars (saccharose, lactose), or sugar alcohols (mannitol, sorbitol)]. Afterward the aqueous drug nanosuspension can be spray dried under adequate conditions; the resulting dry powder is composed of drug nanocrystals embedded in a water-soluble matrix (37). Figure 4 schematically represents the whole production process of drug nanocrystal-loaded spray-dried compounds.

The loading capacity of the solid powder with drug nanocrystals can be adjusted by varying the concentrations of excipient and surfactant in the original aqueous nanosuspension. One aim of a solid nanoparticulate system is releasing the drug nanocrystals after administration in the gastrointestinal tract (GI) as a fine nonaggregated suspension; the other is to increase the physical stability for long-term storage. Contact of the drug nanocrystals is averted by fixation within the matrix. Thereby, the probability of physical instabilities as, e.g., aggregation and ripening are in principle clearly avoided

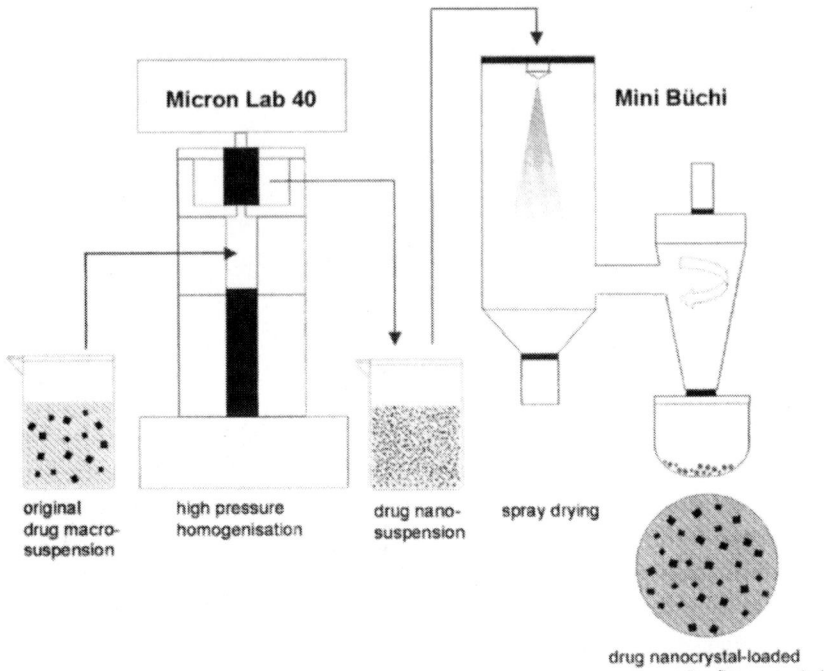

Figure 4 Two-step process of the production of drug nanocrystal-loaded compounds: the drug nanosuspension obtained by high-pressure homogenization (Micron Lab 40) is further processed by spray drying using a Mini Büchi. Drug nanocrystals embedded in the matrix are obtained.

or minimized to a negligible extent. However, appropriate investigations have shown a relation between the loading capacity of the compounds and the releasing behavior, as well as the storage stability. Exceeding a certain maximum loading capacity of the matrix with drug nanocrystals has an increasing negative effect on particle crystal growth and on release as fine dispersion (38).

Figure 5 shows the volume distribution of two spray-dried formulations A and B with increasing drug nanocrystal loadings after release in water. The formulation with the highest drug concentration, formulation B, clearly shows a negligible low but detectable aggregated volume fraction

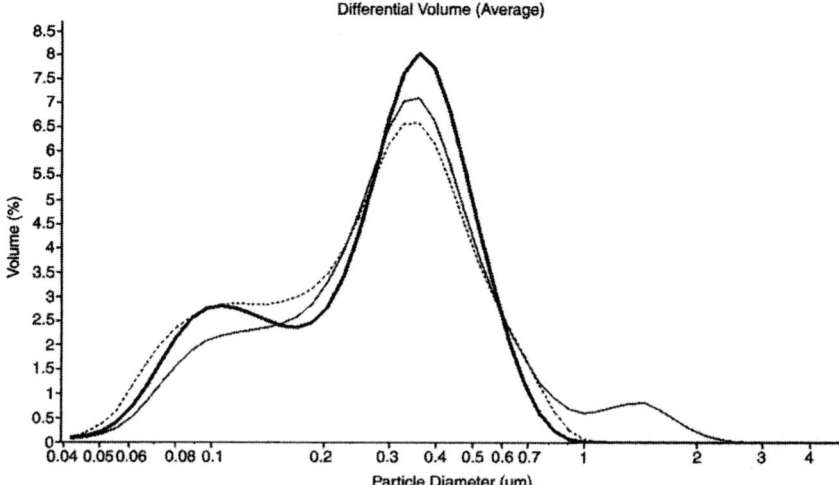

Figure 5 LD volume distributions of spray-dried formulations of Tween® 80 stabilized amphotericin B nanocrystals in PVP matrix after redispersion in water. An adequate volume of distilled water (22–23°C) was added to the compounds to obtain a 1% drug nano-suspension after release from the matrix. LD measurements were started after gently stirring until the matrix material was comple-tely dissolved. (····): Formulation A: 15.4% AmphoB, 7.7% Tween 80, 76.9% PVP (Kollidon 25); (—): formulation B: 50.0% AmphoB, 25.0% Tween 80, 25.0% PVP (Kollidon 25); (——): original nanosus-pension, 1% AmB, 0.5% Tween 80. *Abbreviation*: LD, laser diffrac-tion; PVP, polyvinylpyrrolidone.

and consequently, a reduced percentage within the lower nanometer range.

PRODUCTION IN NONAQUEOUS LIQUIDS

To avoid the removal of water after high-pressure homogeni-zation in aqueous media, homogenization can be performed directly in nonaqueous media. A number of nonaqueous media are suitable as dispersion media for drug nanocrystals. For example, PEG and triglycerides or self-emulsifying drug delivery systems are ideal liquid candidates and are suited for direct filling of hard or soft gelatine capsules (39). The production process can be easily performed similar to the process in water.

Influence of the Dispersion Media

Forces caused by shearing, cavitation, and impaction are dominant for the diminution of drug particles during the homogenizing process. However, the physical properties of the dispersion media/suspension as well as the type of homogenizer and geometry of its dissipation zone highly influence these forces and consequently, the diminution. Especially, the viscosity of the suspension shows significant effect on the properties of the homogenized products.

According to the law by Hagen–Poiseuille, flow rate, pressure, tube diameter, as well as the viscosity of the streaming suspension are interdependent in laminar flow systems. The Micron Lab 40 works with a constant flow rate. According to the input requirements the homogenization pressure is automatically adjusted by the width of the homogenizing gap (40). In this correlation the viscosity of the fluid/suspension can be considered as a determining factor for the width of the gap. Using a Micron Lab 40 simplified a doubling of the width is observed when decoupling the viscosity. Thus, increasing viscosities alter the flow conditions within the homogenizing region. A decrease in flow velocity and an increase in gap volume clearly influence the homogenization results. Figure 6 shows the calculated width of the homogenization gap of the Micron Lab 40 in dependency on the fluid viscosity.

Broadening of the homogenization gap leads to decreased shear forces and kinetic energies of the nanocrystals; consequently, the lower forces affect the breaking of the particles.

In summary, the grade of particle diminution is determined by the forces acting on each drug particle during the homogenization process and the drug properties (e.g., hardness of crystal, number of imperfections in crystal, and percentage of amorphous fraction) (41). A particle breaks if the acting force is higher than the breakage stress. The maximal dispersivity of a nanosuspension is reached if further

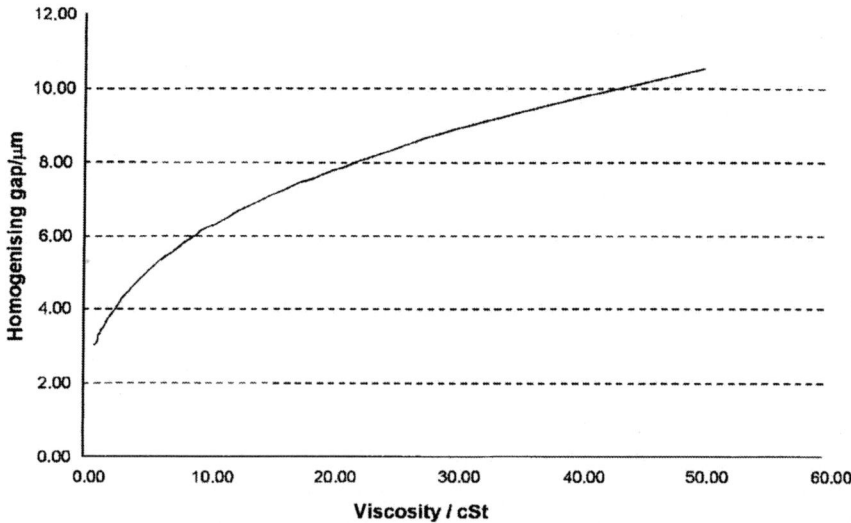

Figure 6 Width of the homogenizing gap as a function of the kinematic viscosity of the fluid (Micron Lab 40) at a homogenization pressure of 1500 bar.

homogenization cycles show no more effect on particle size distribution. In this case, the acting forces are not high enough for inducing particle breakage and further diminution.

For the production of drug nanosuspensions in nonaqueous media the maximal dispersivity has to be investigated for each drug and dispersion medium. In principle, the media with considerably higher viscosities than water require a higher number of homogenization cycles to achieve identical or similar particle sizes and distributions to lower-viscosity media. Figure 7 shows the particle size distribution of Amphotericin B nanoparticles, using laser diffraction (LD), after high-pressure homogenization in different dispersion media.

PRODUCTION IN HOT-MELTED MATRICES

A further possibility for the production of drug nanocrystals in solid matrices is high-pressure homogenization in hot

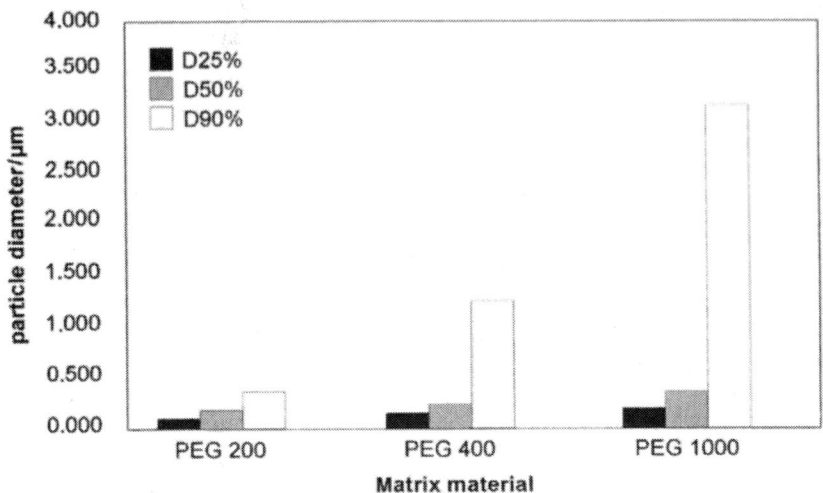

Figure 7 LD particle diameters 25–90% of Amphotericin-B nanocrystals after high-pressure homogenization in PEGs of various chain lengths and thus viscosities. *Abbreviations*: LD, laser diffraction; PEG, polyethylene glycol.

melts. It offers advantages over production in aqueous solution and subsequent spray drying. The process is completely anhydrous, avoiding possible drug degradation or instabilities. The production can directly be performed by hot high-pressure homogenization in melted material (38,42). The homogenizers Micron Lab 40, batch and continuous, were equipped with temperature control jackets placed around the sample/product containers. Working temperatures up to 100°C (heated with water) or higher (heated with silicon oil) can be selected depending on the melting temperature of the used matrix material. For batch operation, solidification has to be averted between each homogenizing cycle. For homogenizers working in the continuous mode, the product containers must be also heated. Figure 8 shows the temperature control devices for the continuous and batch versions of the Micron Lab 40.

As the first production step, a presuspension has to be formed consisting of a melted matrix with the addition of the drug powder and surfactant. In the following production

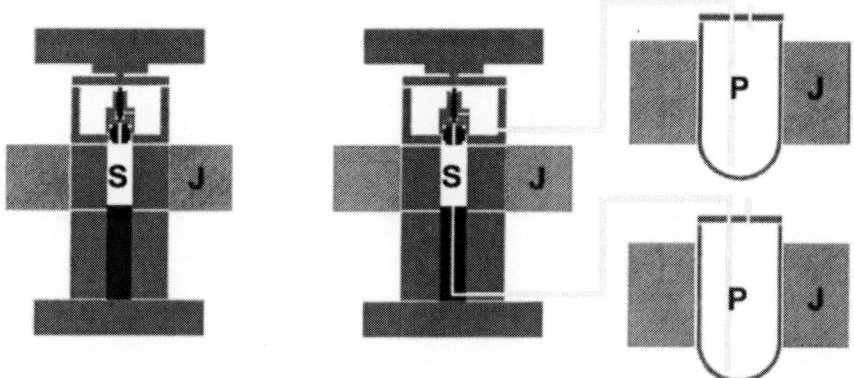

Figure 8 Micron Lab 40 with temperature control jackets (J): temperature control jacket for the discontinuous Micron Lab 40 (*left*) around the sample container (S) containing the suspension to be homogenized, for continuous Micron Lab 40 (*right*) with jackets around the sample container and additional jackets around the two 1000 mL product containers (P).

step, the hot presuspension can be directly homogenized in the temperature-controlled homogenizer. After reaching the envisaged particle size and size distribution, the suspension can be solidified at room temperature by applying controlled cooling. Figure 9 shows the principle production process of drug nanocrystals in hot melts.

Subsequently, the solid nanodispersions obtained can be processed to granulates by milling, for filling capsules or tablet compaction. Alternatively, the hot melt can directly be filled into hard gelatine or hydroxypropyl methylcellulose (HPMC) capsules (Fig. 10).

The absence of water during the whole production process as well as the short processing times and the one-step process to the final product are especially to be noted using the hot melt method. Given these advantages, this technology also has limitations, which have to be compared with the other technologies (e.g., homogenization in water) for the production of solid nanosuspensions. High-pressure homogenization—identical to pearl milling—can only be performed up to a certain drug concentration. Suspensions

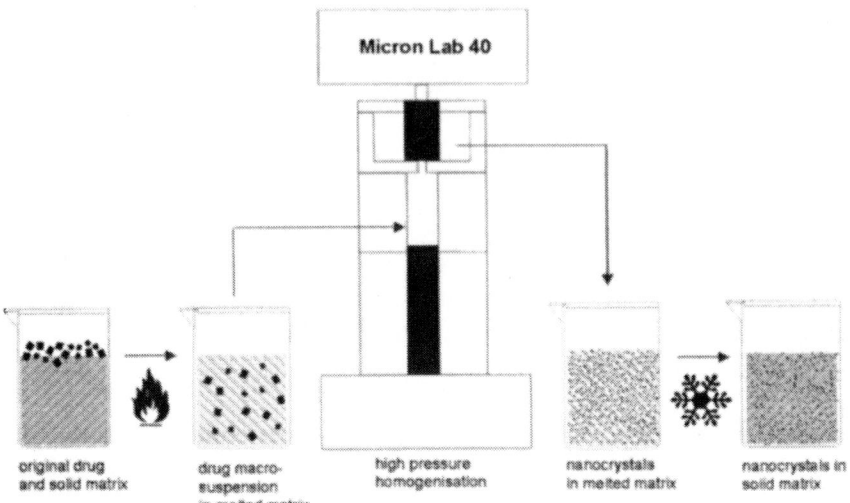

Figure 9 Schematic of the process utilizing melted matrices: the coarse drug material is added to the solid matrix material, which is then melted for dispersing the drug. The nanosuspension is obtained by high-pressure homogenization. Subsequent cooling leads to drug nanocrystals embedded in a solid matrix.

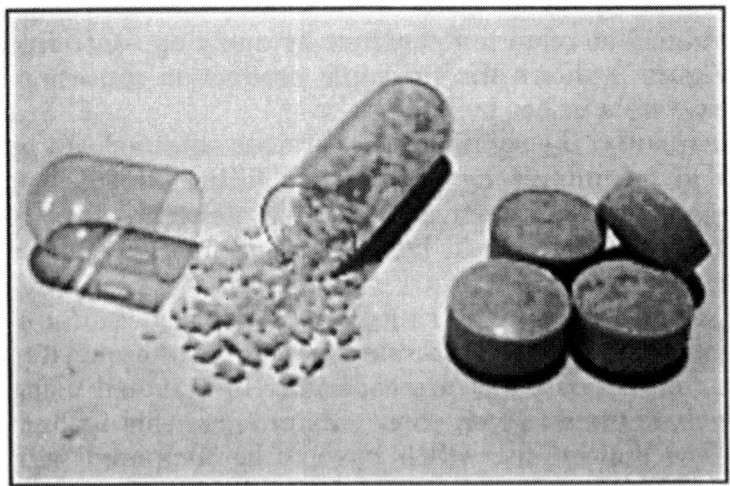

Figure 10 Capsules filled with granulated PEG 2000 containing Amphotericin-B nanocrystals (*left*), tablets produced by compaction of the granulate (*right*). *Abbreviation*: PEG, polyethylene glycol.

can show paste-like properties at high solids content (e.g., >30% or >40%). The resulting rheological properties, especially high viscosity, lead to suboptimal flow conditions within the homogenization gap. Depending on the homogenizer design, some suspensions with higher viscosities can also be processed (e.g., feeding the suspension to the homogenizer by applying air pressure or using a piston, e.g., PANDA range of GEA, Niro-Soavi, Stansted homogenizers) (43). For example, using a Stansted homogenizer a suspension of 40% solid can be processed without any problem (44). However, it should be noted that the viscosity of a suspension does not only depend on the solid content but also on the size, size distribution, and shape of the particles. Depending on these factors particles can form three-dimensional structures in concentrated suspensions with different viscosities. In turn, it is also possible to reduce the viscosity of a highly concentrated suspension by optimizing the size distribution (i.e., making it more polydisperse).

PELLETIZATION TECHNIQUES

Introduction

The nanonization of drugs results in general in a liquid product from most of the techniques described in this chapter. These nanosuspensions have shown excellent long-term stability without Ostwald ripening or chemical alteration (9,45). In some special cases, the nanosuspension can be directly used as a final product, for instance, as pediatric or geriatric dosage forms if the drug absorption rate is limited only by the solubility and dissolution rate of the drug. Apart from this—in case of oral administration—a dry dosage form is clearly preferred for the reasons of convenience (i.e., marketing aspects). There are also other cases in which a more sophisticated dosage form is needed, e.g., to prevent the drug from degradation, to achieve a controlled drug release or to enable better drug targeting. For these reasons, the nanosuspension can be transformed into a solid dosage form by using established techniques, like pelletization, granulation, spray drying, or lyophilization.

Many different pelletization techniques are known, but the most commonly used techniques are the extrusion–spheronization and the drug layering onto sugar spheres. The choice of the pellet type depends on the required drug content, the drug properties, and the available technical equipment. Irrespective of the pelletization technique applied, a multiparticulate dosage form with distinct advantages in comparison to single-unit dosage forms will be obtained. Multiparticulate dosage forms, such as coated pellet systems, show a faster and more predictable gastric emptying and more uniform drug distribution in the GI tract with less inter- and intraindividual variability in bioavailability (46). A broad distribution of the pellets in the gut lumen can enhance the complete redispersion of the nanoparticles from the final solid dosage form. The incorporation of drug nanocrystals in the various pellet systems is schematically shown in Figure 11.

Matrix Pellet Preparation

Aqueous nanosuspensions can be mixed with matrix materials (fillers such as MCC, lactose, or starch); in addition, the nanosuspension works as a binder and wetting fluid for the

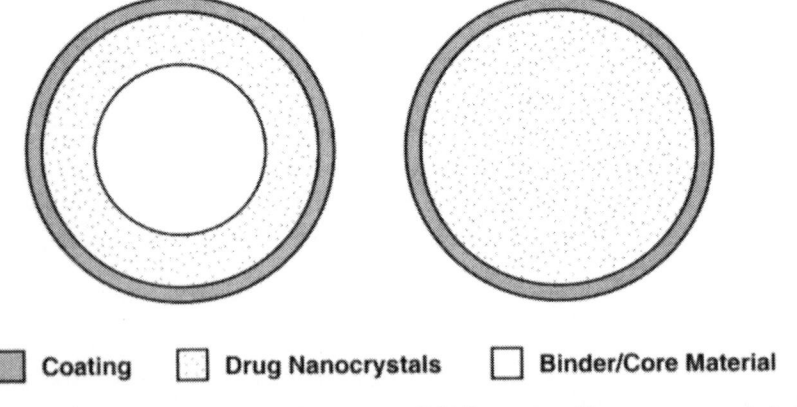

☐ Coating   ☐ Drug Nanocrystals   ☐ Binder/Core Material

Figure 11   Schematic drawing of different pellet types containing drug nanocrystals: coated pellet with a sugar bead as core material and a compact layer consisting of a binder/drug nanocrystal layer (*left*) and a coated matrix pellet with a matrix consisting of a binder/drug nanocrystal mixture as core material (*right*).

extrusion process (27,47–49). Binders like gelatine, HPMC, chitosans, or other polymers can be added to the nanosuspension before the high-pressure homogenization, which simplifies the production process. Alternatively, they can be dispersed in the produced nanosuspension after the high-pressure homogenization. On the one hand these binders are necessary for the extrusion process, but on the other they can also positively influence the properties of the nanosuspension or the nanoparticles. The increased viscosity of the nanosuspension leads to an increased physical stability of the nanosuspension with a decreased tendency of sedimentation, an important factor to obtain reproducible drug content in the final product. Another important point is the possibility to change the zeta potential of the drug nanocrystals by using charged polymers (i.e., chitosan or alginate) to increase the nanosuspension stability under the GI conditions and to achieve better drug targeting (50–53). If the nanosuspension is used as described earlier, one limitation is the maximum achievable drug content of the final product.

In order to overcome this problem an additional drying step, such as spray drying, has to be performed to obtain a fine nanocrystalline powder. This powder can be admixed to the matrix material to obtain a mass highly loaded with drug nanocrystals and ready for the extrusion and subsequent spheronization. Afterward a coating can be applied to the matrix cores to modify their drug release properties. Figure 12 shows the major steps in the production of matrix pellets containing drug nanocrystals. A detailed view on these pellets is given in Figure 13.

Pellet Preparation by Nanosuspension Layering

An alternative way to transfer a prior produced nanosuspension into a pellet formulation is the suspension layering onto sugar spheres (54). The binders that are necessary for this process can also be added before the high-pressure homogenization process resulting in the improved nanosuspension properties mentioned earlier. A schematic production process is shown in Figure 14.

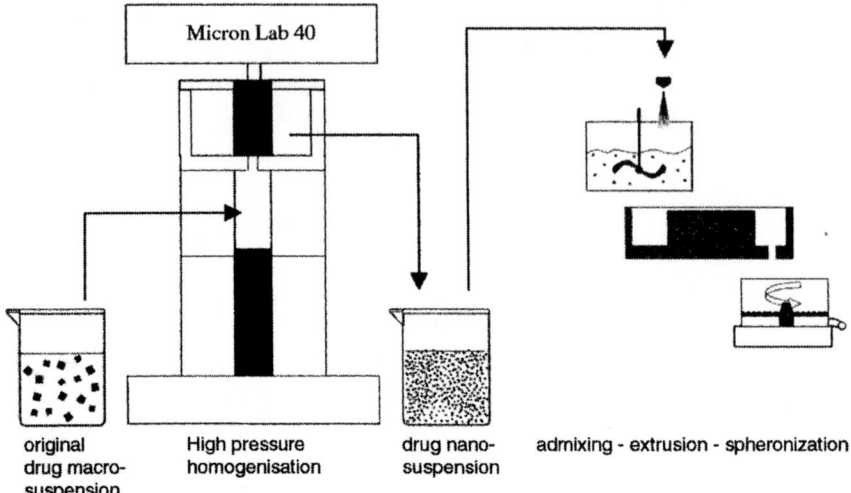

Figure 12 Production of drug nanocrystal-loaded matrix cores: the drug nanosuspension obtained by high-pressure homogenization (Micron Lab 40) is admixed to the matrix material, pellets are prepared by extrusion–spheronization and can be subsequently coated with polymers with the same equipment to modify the drug release properties.

The most important difference between the matrix cores and layered cores is the different drug loading. For the production of matrix cores from an aqueous nanosuspension without an additional drying step, the drug loading is limited to 4.5%, based on Equation (1), whereas the drug loading of layered cores can be raised by increasing the layering level almost without any limitations. (J. Moschwitzer and R. H. Muller, submitted for publication.)

$$\text{Drug content(MC)} = \frac{30\% \text{ drug content}(N) \times 15\,\text{gN}}{100\,\text{g(total pellet mass)}} = 4.5\% \quad (1)$$

Calculated maximal achievable drug content in matrix cores without additional drying steps: MC = matrix core, N = nanosuspension (30% is an achievable drug concentration in the nanosuspension, of course depending on the formulation and equipment).

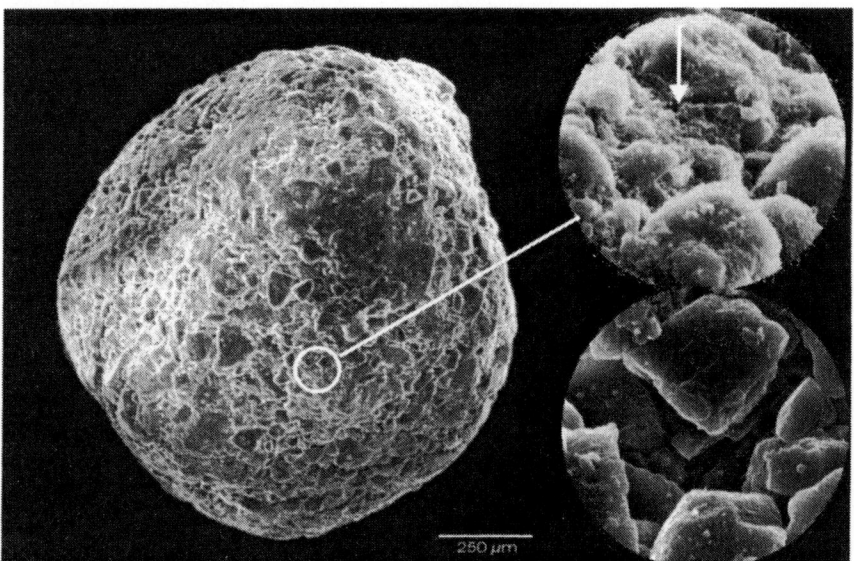

Figure 13 SEM photograph: left—uncoated matrix core containing drug nanocrystals (magnification 60×), right top—detail magnification (1000×) of this matrix core with drug nanocrystals (*arrow*), right bottom—detail magnification (1000×) of matrix core without drug nanocrystals. *Abbreviation*: SEM, scanning electron microscopy.

DIRECT COMPRESS

Spray drying is especially suitable for the transfer of nanosuspensions of drugs that are insensitive to high temperatures. Depending on the spray conditions and formulation, the resulting product possesses a particle size from 1 to100 μm and can easily be filled into a hard gelatine capsule as the final dosage form. In the case of acid labile drugs, the capsule can be coated with enteric polymers to protect the drug from the gastric fluids. Stability tests over a period of several months, even up to one year, have shown a perfect redispersibility for different formulations. An advantage of this method is the resulting drug content in the final dosage form, which can be easily achieved from 20% to 80%.

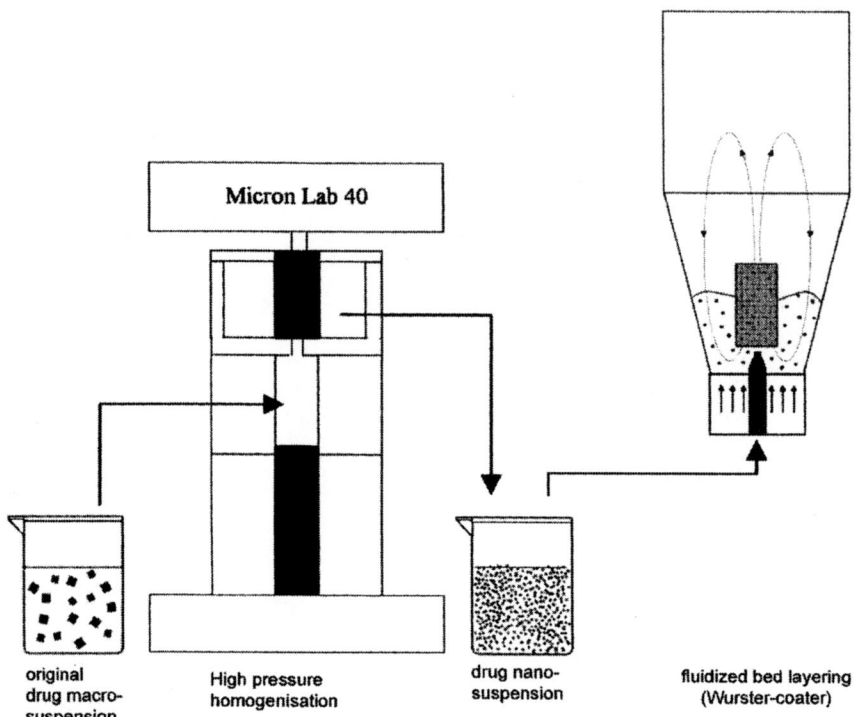

Figure 14 Two-step process for the production of layered cores containing drug nanocrystal: the drug nanosuspension obtained by high-pressure homogenization (Micron Lab 40) is directly layered onto sugar beads and subsequently coated with polymers using the same equipment to modify the drug release properties.

A special form of spray-drying of nanosuspensions is the DirectCompress® technology (55). Lactose and matrix forming materials (like micronized polymer powders or lipids) are admixed to the prior produced nanosuspension. By spray drying, this liquid phase is transferred into a drug–matrix compound. A major advantage of the DirectCompress® technology is the fast transformation of the liquid nanosuspension into a free-flowable powder, ready for direct compression of fast dissolving or prolonged release tablets. Alternatively, the powder obtained from the spray drying can be directly filled into hard gelatine capsules.

REFERENCES

1. Merisko-Liversidge E. Nanocrystals: resolving pharmaceutical formulation issues associated with poorly water-soluble compounds. In: Marty JJ, ed. Particles. Orlando: Marcel Dekker, 2002.

2. Uekama K. Design and evaluation of cyclodextrin-based drug formulation. Chem Pharm Bull 2004; 52(8):900–915.

3. Cydex. Captisol—solubility and so much more. Captisol Brochure.

4. Müller RH, Peters K, Becker R, Kruss B. Nanosuspensions—a novel formulation for the i.v. administration of poorly soluble drugs. In: 1st World Meeting APGI/APV, Budapest, 1995:491–492.

5. Müller RH, Böhm BHL. Nanosuspensions. In: Müller RH, Benita S, Böhm, B, eds. Emulsions and Nanosuspensions for the Formulation of Poorly Soluble Drugs. Stuttgart: Medpharm, 1998:149–174.

6. Grau MJ. Untersuchungen zur Lösungsgeschwindigkeit, Sättigungslöslichkeit und Stabilität von hochdispersen Arzneistoffsuspensionen [PhD Thesis]. Berlin: Freie Universität, 2000.

7. Müller RH, Becker R, Kruss B, Peters K. Pharmaceutical nanosuspensions for medicament administration as systems with increased saturation solubility and rate of solution. United States Patent 5,858,410. USA, 1999.

8. Chen Y, Liu J, et al. Oleanolic acid nanosuspensions: preparation, in-vitro characterization and enhanced hepatoprotective effect. J Pharm Pharmacol 2005; 57(2):259–264.

9. Troester F. Cremophor-free aqueous paclitaxel nanosuspension—production and chemical stability. Controlled Release Society 31st Annual Meeting, Honolulu, HI, 2004.

10. Liversidge E, Wei L. Inventor stabilization of chemical compounds using nanoparticulate formulations. U.S. Patent 2003054042 A1, 2003.

11. Chen X, Young TJ, Sarkari M, Williams RO, 3rd, Johnston KP. Preparation of cyclosporine A nanoparticles by evaporative precipitation into aqueous solution. Int J Pharm 2002; 242(1–2):3–14.

12. Muller RH, Keck CM. Challenges and solutions for the delivery of biotech drugs—a review of drug nanocrystal technology and lipid nanoparticles. J Biotechnol 2004; 113(1–3):151–170.

13. Liversidge GG, Cundy KC, Bishop JF, Czekai DA. Surface modified drug nanoparticles. United States Patent 5,145,684. Sterling Drug Inc., New York: USA, 1992.

14. Wyeth Pharmaceuticals, Inc., Drug Information: Rapamune, Oral Solution and Tablets, internet available, 2004.

15. Merck & Co., Inc., Drug Information: Emend, capsules, internet available, 2004.

16. Keck CMRH, Fichtinger A, Viernstein H. Production and optimisation of oral cyclosporine nanocrystals. 2004 AAPS Annual Meeting and Exposition, Baltimore: MD, 2004.

17. Merisko-Liversidge E, Sarpotdar P, Bruno J, et al. Formulation and antitumor activity evaluation of nanocrystalline suspensions of poorly soluble anticancer drugs. Pharm Res 1996; 13(2):272–278.

18. Merisko-Liversidge E, Liversidge GG, Cooper ER. Nanosizing: a formulation approach for poorly-water-soluble compounds. Eur J Pharm Sci 2003; 18(2):113–120.

19. Buchmann S, Fischli W, Thiel FP, Alex R. Aqueous microsuspension, an alternative intravenous formulation for animal studies. 42nd Annual Congress of the International Association for Pharmaceutical Technology (APV), Mainz, 1996:124.

20. Na GC, Stevens HJ Jr, Yuan BO, Rajagopalan N. Physical stability of ethyl diatrizoate nanocrystalline suspension in steam sterilization. Pharm Res 1999; 16(4):569–574.

21. VMA-Getzmann GmbH, Germany V. Dispermat/Torusmill company brochure, 2003.

22. Müller RH, Heinemann S. Surface modelling of microparticles as parenteral systems with high tissue affinity. In: Gurny RaJ HE, ed. Bioadhesion—Possibilities and Future Trends. Stuttgart: Wissenschaftliche Verlagsgesellschaft, 1989:202–214.

23. Muller RH, Peters K. Nanosuspensions for the formulation of poorly soluble drugs: I. Preparation by a size-reduction technique. Int J Pharm 1998; 160(2):229–237.

24. Muller RH, Jacobs C, Kayser O. Nanosuspensions as particulate drug formulations in therapy: rationale for development and what we can expect for the future. Adv Drug Delivery Rev 2001; 47(1):3–19.

25. Jacobs C, Muller RH. Production and characterization of a budesonide nanosuspension for pulmonary administration. Pharm Res 2002; 19(2):189–194.

26. Jacobs C, Kayser O, Muller RH. Nanosuspensions as a new approach for the formulation for the poorly soluble drug tarazepide. Int J Pharm 2000; 196(2):161–164.

27. Peters K, Müller RH. Nanosuspensions for the oral application of poorly soluble drugs. In: Proceeding European Symposium on Formulation of Poorly-Available Drugs for Oral Administration, APGI, Paris, 1996.

28. Dearn AR, inventor Glaxo Wellcome Inc., assignee. Atovaquone pharmaceutical compositions. U.S. Patent 6,018,080, 2000.

29. Keck C, Bushrab FN, Müller RH. Nanopure nanocrystals for oral delivery of poorly soluble drugs. Particles. Orlando, 2004.

30. Müller RH, Bushrab FN. Drug nanocrystals—production and design of final oral dosage forms. In: 5th European Workshop on Particulate Systems, London, 2004.

31. APV. APV high pressure homogenisers. Company brochure.

32. Muller RH, Dingler A, Schneppe T, Gohla S. Large scale production of solid lipid nanoparticles (SLNTM) and nanosuspensions (DissoCubes^{TM}). In: Wise D, ed. Handbook of Pharmaceutical Controlled Release Technology. 2000:359–376.

33. http.//avestin.com

34. http://www.niro-soavi.com

35. Peters K. Nanosuspension—ein neues Formulierungsprinzip für schwerlösliche Arzneistoffe. Berlin: Freie Universität Berlin, 1999.

36. Freitas C, Muller RH. Spray-drying of solid lipid nanoparticles (SLNTM). Eur J Pharm Biopharm 1998; 46(2):145–151.

37. Bushrab FN, Müller RH. Drug Nanocrystals for Oral Delivery—Compounds by Spray Drying. Philadelphia: AAPS, 2004.

38. Bushrab FN, Müller, RH. Drug nanocrystals: Amphotericin B-containing capsules for oral delivery. Philadelphia: AAPS, 2004.

39. Akkar A, Bushrab FN, Müller RH. Nanosuspensions—ultrafine dispersions of actives and pigments for cosmetics. In: Cosmoderm III, Istanbul, 2003.

40. Jahnke S. Theorie der Hochdruckhomogenisation. Deutschland: GEA, Niro Soavi, 2000.

41. Stieß M. Mechanische Verfahrenstechnik 2. Berlin, Heidelberg: Springer; 1995.

42. Bushrab NF, Müller RH. Nanocrystals of poorly soluble drugs for oral administration. NewDrugs 2003; 5:20–22.

43. Product manual: MODEL NS1001L—PANDA 2k. Deutschland: GEA, Nito Soavi.

44. Krause KP, Muller RH. Production and characterisation of highly concentrated nanosuspensions by high pressure homogenisation. Int J Pharm 2001; 214(1–2):21–24.

45. Moschwitzer J, Achleitner G, Pomper H, Muller RH. Development of an intravenously injectable chemically stable aqueous omeprazole formulation using nanosuspension technology. Eur J Pharm Biopharm 2004; 58(3):615–619.

46. Follonier ND. Biopharmaceutical comparison of oral multiple-unit and single-unit sustained-release dosage forms. STP Pharma Sci 1992; 2(2):141–158.

47. Vergote GJ, Vervaet C, Van Driessche I. In vivo evaluation of matrix pellets containing nanocrystalline ketoprofen. Int J Pharm 2002; 240(1–2):79–84.

48. Vergote GJ, Vervaet C, Van Driessche I. An oral controlled release matrix pellet formulation containing nanocrystalline ketoprofen. Int J Pharm 2001; 219(1–2):81–87.

49. Möschwitzer JM, Muller RH. Final formulations for drug nanocrystals: pellets. In: AAPS Pharmaceutics and Drug Delivery Conference, Philadelphia, 2004.

50. Möschwitzer J, Müller RH. From the drug nanocrystal to the final mucoadhesive oral dosage form. International Meeting

on Pharmaceutics, Biopharmaceutics and Pharmaceutical Technology, Nuremberg, 2004.

51. Muller RH, Jacobs C. Buparvaquone mucoadhesive nanosuspension: preparation, optimisation and long-term stability. Int J Pharm 2002; 237(1-2):151–161.

52. Jacobs C, Kayser O, Muller RH. Production and characterisation of mucoadhesive nanosuspensions for the formulation of bupravaquone. Int J Pharm 2001; 214(1-2):3–7.

53. Muller RH, Keck CM. Drug delivery to the brain—realization by novel drug carriers. J Nanosci Nanotechnol 2004; 4(5):471–483.

54. Möschwitzer J, Müller RH. Controlled drug delivery system for oral application of drug nanocrystals. In: 2004 AAPS Annual Meeting and Exposition, Baltimore, MD, 2004.

55. Muller RH, inventor. Preparation of a matrix material-excipient compound containing a drug. Patent WO 9825590, 1998.

3

Supercritical Fluid Technology for Particle Engineering

RAM B. GUPTA

Department of Chemical Engineering,
Auburn University, Auburn, Alabama, U.S.A.

INTRODUCTION

Design and fabrication of pharmaceutical particulate systems is still largely an art as opposed to a fundamental science. However, a more systematic design and manufacture of particulate systems including nanoparticles is being enabled by the application of novel technologies, such as supercritical fluid (SCF) technology, which is the focus of this chapter (1). A fluid is supercritical when it is compressed beyond its critical pressure (P_c) and heated beyond its critical temperature (T_c). SCF technology has emerged as an important technique for particle manufacturing. In many industrial applications, it is poised to replace the conventional recrystallization and

Table 1 Critical Constants and Safety Data for Various Supercritical Solvents

SCF	T_c (°C)	P_c (bar)	Safety hazard
Ethylene	9.3	50.3	Flammable gas
Trifluoromethane (fluoroform)	25.9	47.5	
Chlorotrifluoromethane	28.9	39.2	
Ethane	32.3	48.8	Flammable gas
Carbon dioxide	31.1	73.7	
Dinitrogen monoxide (laughing gas)	36.5	72.6	Not combustible but enhances combustion of other substances
Sulfur hexafluoride	45.5	37.6	
Chlorodifluoromethane (HCFC 22; R 22)	96.4	49.1	Combustible under specific conditions
Propane	96.8	43.0	Extremely flammable
Ammonia	132.4	112.7	Flammable and toxic
Dimethyl ether (wood ether)	126.8	52.4	Extremely flammable
Trichlorofluoromethane (CFC 11, R 11)	198.0	44.1	
Isopropanol	235.2	47.6	Highly flammable
Cyclohexane	280.3	40.7	Highly flammable
Toluene	318.6	41.1	Highly flammable
Water	374.0	220.5	

Abbreviation: SCF, supercritical fluid.

milling processes, mainly because of the quality and the purity of the final particles and environmental benefits. There are a variety of SCFs available as listed in Table 1.

SUPERCRITICAL CO$_2$

Out of the fluids listed in Table 1, carbon dioxide is the SCF of choice because it is nonflammable, nontoxic, inexpensive, and has mild critical temperature. Hence, much of the attention has been given to supercritical carbon dioxide for pharmaceutical particle formation.

No amount of compression can liquefy the SCF. In fact, pressure can be used to continuously change the density from

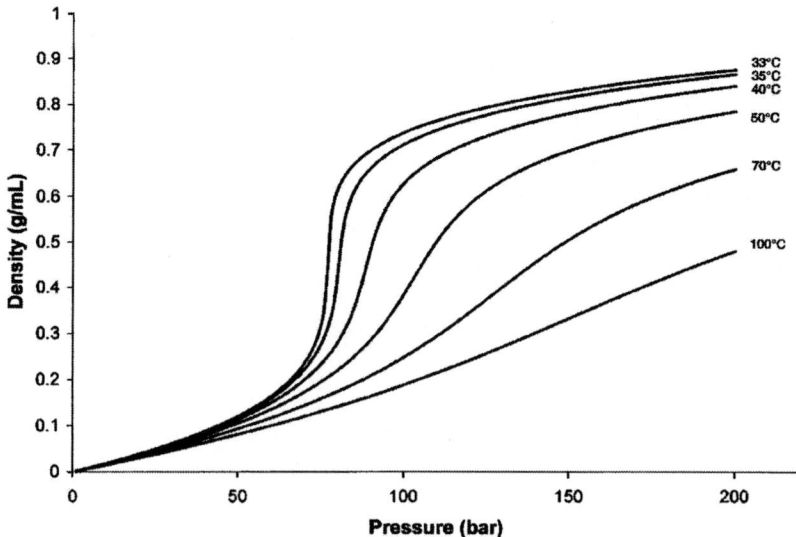

Figure 1 Density dependence of CO_2 at various temperatures. *Abbreviation*: CO_2, carbon dioxide.

gas-like conditions to liquid-like conditions. Near the critical region, small changes in the pressure can give rise to large changes in the density. Figure 1 shows how the density of carbon dioxide is varied by pressure at different temperatures.

In addition to density, diffusivity of the SCFs is higher than that of liquid solvents, and can be easily varied. For typical conditions, diffusivity in SCFs is of the order of 10^{-3} cm^2/sec as compared to 10^{-1} for gases and 10^{-5} for liquids. Typical viscosity of SCFs is of the order of 10^{-4} g/cm/sec, similar to that of gases, and about 100-fold lower than that of liquids. High diffusivity and low viscosity provide rapid equilibration of the fluid.

SOLUBILITY IN SUPERCRITICAL CO$_2$

Carbon dioxide (O=C=O) is a nonpolar molecule with a small polarity due to the quadrupole moment. Hence, nonpolar or light molecules (e.g., menthol, methanol, acetone, toluene, and hexanes) easily dissolve in CO_2, whereas the polar or heavy molecules (e.g., griseofulvin, paclitaxel, tetracycline, and dexamethasone phosphate) have a very poor solubility. For example, solubility of menthol in CO_2 is as high as 5 mol% (Fig. 2), whereas the solubility of griseofulvin in CO_2 is only about 18 ppm (Fig. 3). Solubilities of some other pharmaceutical compounds are shown in Figures 4–7. A comprehensive compilation of solubility data in supercritical CO_2 is given in a recent book by Gupta and Shim (6).

Three important factors that govern drug solubility in supercritical CO_2 are the vapor pressure of drug, drug–CO_2 interaction, and density of CO_2. Drug vapor pressure is a function of temperature (T), and CO_2 density is a function of pressure (P) and T. Mendez–Santiago and Teja (8) observed that the solubility (y_2 μmol/mol) can be correlated using the following equation:

$$y_2 = \frac{10^6}{P} \exp\left(\frac{A}{T} + \frac{B\rho_1}{T} + C\right) \tag{1}$$

where P is in bars, T is in Kelvin, ρ_1 is CO_2 density in moles per milliliter. Constants A, B, and C are listed in Table 2

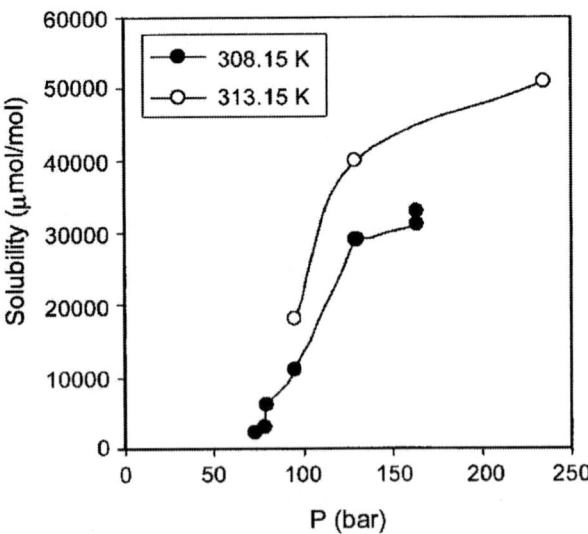

Figure 2 Solubility of menthol in CO_2. *Abbreviation*: CO_2, carbon dioxide. *Source*: From Ref. 2.

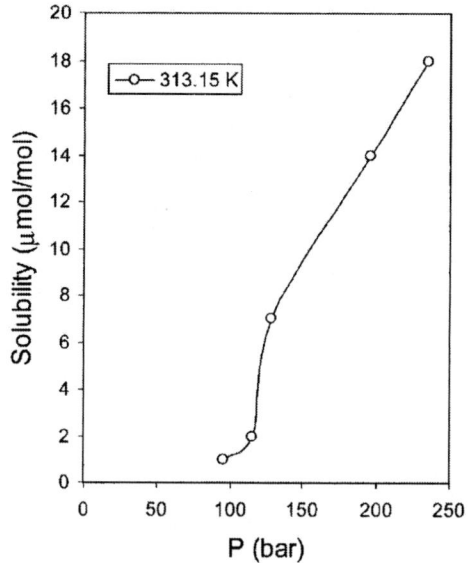

Figure 3 Solubility of griseofulvin in CO_2. *Abbreviation*: CO_2, carbon dioxide. *Source*: From Ref. 2.

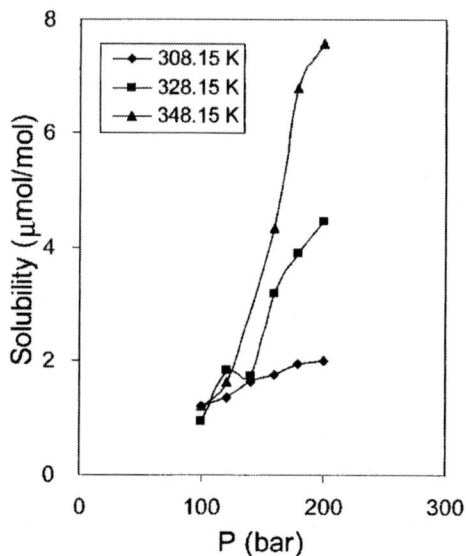

Figure 4 Solubility of nicotinic acid in CO_2. *Abbreviation*: CO_2, carbon dioxide. *Source*: From Ref. 4.

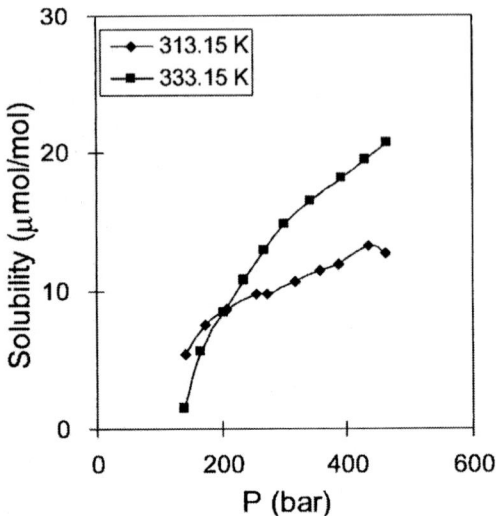

Figure 5 Solubility of chloramphenicol in CO_2. *Abbreviation*: CO_2, carbon dioxide. *Source*: From Ref. 5.

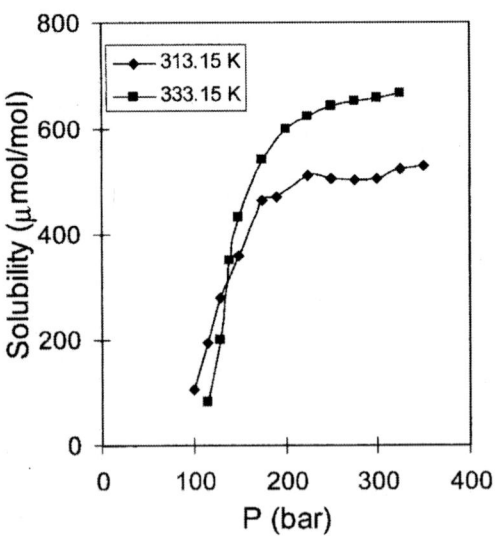

Figure 6 Solubility of salicylic acid in CO_2. *Abbreviation*: CO_2, carbon dioxide. *Source*: From Ref. 3.

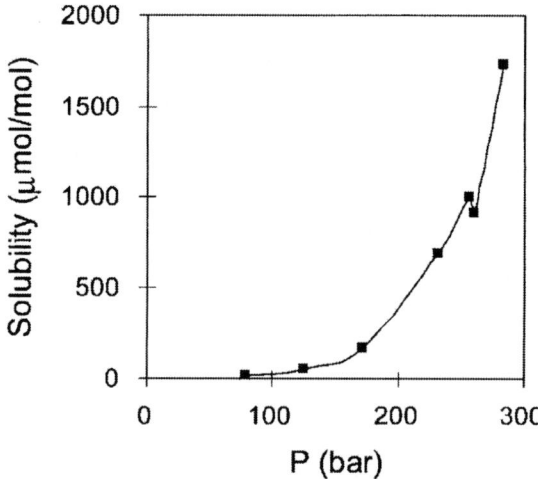

Figure 7 Solubility of α-tocopherol in CO_2 at 333 K. *Abbreviation*: CO_2, carbon dioxide. *Source*: From Ref. 7.

for selected drugs. Density of pure CO_2 can be obtained from NIST Standard Reference Database (http://webbook.nist.gov/chemistry/) at the desired T and P. Alternatively, the following empirical expression can be used (9):

$$\rho_1 = \frac{1}{44}\exp\left(-27.091 + 0.609\sqrt{T} + \frac{3966.170}{T}\right.$$
$$\left.- \frac{3.445P}{T} + 0.401\sqrt{P}\right) \qquad (2)$$

RAPID EXPANSION OF SUPERCRITICAL SOLUTION FOR PARTICLE FORMATION

From the previous section it is evident that the solubility of pharmaceutical compounds is highly dependent on CO_2 pressure. As the pressure is reduced, solubility decreases because of a reduction in the CO_2 density, which is closely related to its solubility power (8–11). At a high pressure, the drug can be dissolved in CO_2 and if the pressure is reduced to ambient, the drug precipitates out as fine particles. The depressurization

Table 2 Values of the Constants for Equation (1)

Drug	A	B	C
7-Azaindole	−8,412	87,110	20.66
Behenic acid	−4,473	61,240	6.80
Biphenyl	−10,200	132,800	25.75
Brassylic acid	−10,860	146,100	21.01
Capsaisin	−7,172	70,830	19.54
Cholecalciferol	−9,784	172,500	18.42
Diphenylamine	−18,720	397,100	33.40
Eicosanoic acid	−15,990	161,600	36.97
1-Eicosanol	−14,530	122,500	36.15
Endrin	−9,912	167,800	20.29
Ergocalciferol	−1,092	173,500	21.51
Flavone	−11,430	110,100	27.38
D(−)-Fructose	−871.2	10,740	−4.29
D(+)-Glucose	847.1	2,471	−9.12
3-Hydroxyflavone	−9,746	81,530	21.31
Ketoprofen	−12,090	157,500	24.72
Medroxyprogesterone acetate	−10,270	186,100	17.77
Methoxychlor	−12,670	184,100	27.38
Monocrotaline	−10,440	8,057	20.28
Mystiric acid	−17,250	173,100	44.84
Naproxen	−9,723	122,900	18.11
Narasin	−8,529	124,900	13.86
Nifedipine	−10,020	168,500	15.92
Nimesulide	−13,820	186,900	28.14
Nitrendipine	−9,546	151,400	15.91
Octacosane	−19,860	123,000	52.555
1-Octadecanol	−17,290	141,000	45.32
Palmityl behenate	−8,378	59,180	18.44
Penicillin V	−6,459	73,730	13.29
Phenylacetic acid	−13,730	14,450	35.78
Piroxicam	−10,560	18,130	17.57
Progesterone	−12,090	21,040	23.43
t-Retinol	−8,717	168,900	16.60
Salinomycin	−18,990	185,500	42.05
Stigmasterol	−13,010	169,000	25.23
Testosterone	−14,330	238,300	26.42
Theobromine	−7,443	114,000	8.31
Theophyline	−6,957	94	760
Triacontane	−22,965	199,800	57.22
Trioctylphosphine oxide	−9,378	211,900	17.65
Vanillin	−7,334	136,500	14.53

Source: From Ref. 8.

can be done very fast; so fast that CO_2 comes out of the nozzle at the speed of sound. The fast depressurization results in a very fast rate of precipitation providing small drug particles. This process is termed as rapid expansion of supercritical solution (RESS) and has been tested for a wide variety of drugs. A schematic of the RESS process is shown in Figure 8.

The bulk drug is solubilized in CO_2 in a high-pressure chamber. The solution is then passed through a nozzle to rapidly reduce the pressure. In some applications, the nozzle is also heated to avoid clogging due to freezing of CO_2 by sudden expansion. The precipitated drug particles are collected in an ambient pressure bag filter. The morphology of the resulting particles (crystalline or amorphous) depends on the molecular structure of the drug and RESS process conditions (solubilization temperature, expansion temperature, pressure drop across nozzle, nozzle geometry, impact distance of the jet against collection surface, etc.).

Most of the drug particles produced by RESS, have been in the 1–5 µm-size range. The rapid expansion of supercritical CO_2 does produce nuclei 5–10 nm in diameter, but these nuclei grow because of coagulation and condensation to produce the final micrometer-size particle. The micronized drugs include 2–5 µm aspirin, 3–5 µm caffeine, 2–3 µm cholesterol, 2 µm ibuprofen, 1–3 µm nifedipine, 2–5 µm progesterone, 1–5 µm salicylic acid, 2–5 µm testosterone, 4–12 µm theophyline, and 1–2 µm α-tocopherol (3,12–19).

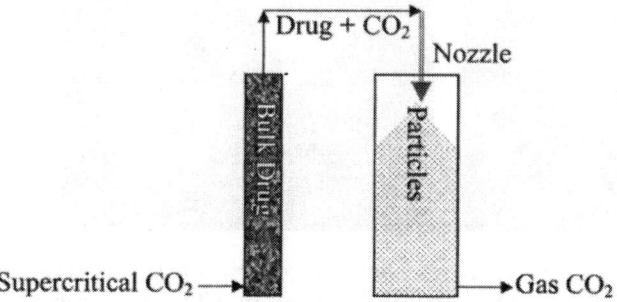

Figure 8 Schematic of RESS process. *Abbreviations*: RESS, rapid expansion of supercritical solution; CO_2, carbon dioxide.

For a few drugs, nanoparticles have also been obtained using RESS. These nanonized drugs include 100 nm lidocaine, 200 nm griseofulvin, 200 nm β-sitosterol (20,21). Recently, by expanding the drug CO_2 mixture in a liquid medium containing stabilizers, Pathak et al. (22) have obtained small nanoparticles of ibuprofen and naproxen.

As the obtained particles are free of organic solvents and the high-pressure part of the equipment is not too expensive, theoretically RESS process is very useful. Unfortunately, for most drugs, nanoparticles are not obtained. Instead, oriented-fused particles are obtained (Fig. 9).

Another major drawback of the RESS process is the low solubility of most drugs in supercritical carbon dioxide. For example, solubility of griseofulin is only 18 ppm. Hence, to obtain 18 mol of griseofulvin, one needs to use one million mol of CO_2 (i.e., 1 g griseofulvin particles from about 7 kg CO_2). The worst part is the collection problem. For the earlier example, 1 g of powder would be dispersed in 3573 L of gaseous CO_2 requiring efficient filtration.

Addition of cosolvents, such as methanol, acetone, or ethanol, can enhance the drug solubility to some extent.

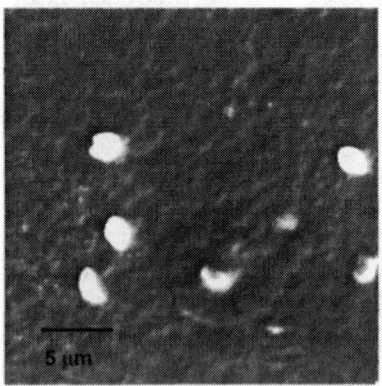 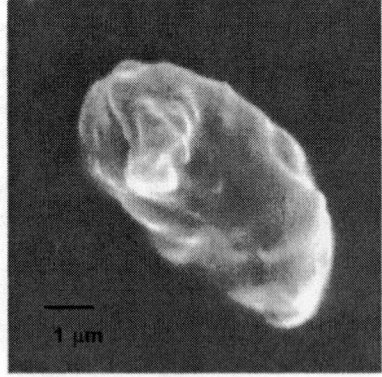

Figure 9 Scanning electron micrograph of griseofulvin particles obtained from RESS process (solubilization in CO_2 was done at 196 bar, 40°C). *Abbreviations*: RESS, rapid expansion of supercritical solution; CO_2, carbon dioxide.

But, the presence of such a cosolvent in the expansion chamber is not desired, as it will lead to solubilization of the particles in the cosolvent.

RESS WITH SOLID COSOLVENT FOR NANOPARTICLE FORMATION

Recently, Thakur and Gupta (2) have addressed both the challenges of RESS (low solubility and growth by coagulation) by utilizing a cosolvent that is solid at the nozzle exit conditions. The solid cosolvent (SC) enhances the solubility in supercritical carbon dioxide and provides a barrier for coagulation in the expansion chamber. The SC is later removed from the solute particles by lyophilization (sublimation). The new process is termed as RESS–SC.

In RESS, all the nuclei or small particles of solute are surrounded by the same kind of particles as in Figure 10(A). But in the RESS–SC process, nuclei or small particles of the solute are surrounded by excess SC particles. This reduces the probability of solute particle growth by coagulation. The

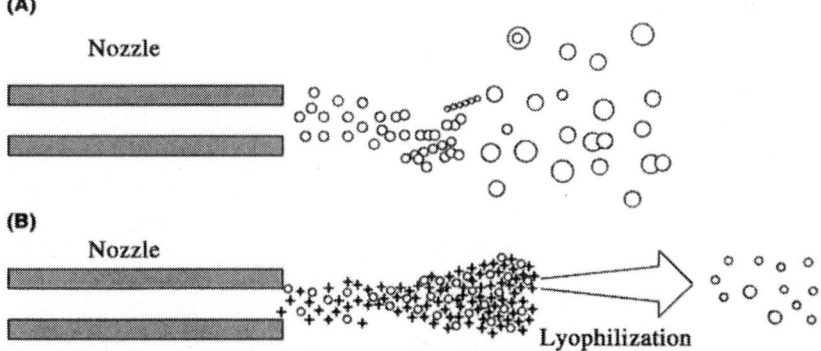

Figure 10 (A) Magnified view of the RESS nozzle. (B) Schematic of RESS–SC process. Circles represent drug particles, whereas stars represent solid–cosolvent particles. *Abbreviations*: RESS, rapid expansion of supercritical solution; RESS–SC, rapid expansion of supercritical solution solid cosolvent.

RESS–SC concept is depicted in Figure 10(B). The lyophilization step shown in the figure is carried out separately after
the expansion.

The choice of a proper SC is the key for successful RESS-
SC. Various requirements for the selection of the SC are

- good solubility in supercritical CO_2,
- solid at nozzle exit condition (5–30 °C),
- good vapor pressure for easy removal by sublimation,
- should be nonreactive with drugs or CO_2, and
- inexpensive.

Menthol is a solid compound (melting point, 42°C) that
satisfies the requirements mentioned earlier. It has appreciable solubility in CO_2 (Fig. 2) and can easily sublime under
vacuum. Menthol naturally occurs in mint-flavored plants,
and is widely used in antipruritic agents, mouthwashes, nasal
sprays, food, etc. Because of its wide use in food and pharmaceutics, menthol does not seem to possess harmful effects
and its use as a cosolvent with supercritical carbon dioxide
still carries the benign benefit of the technology. The following are two examples of the RESS-SC process using menthol
solid cosolvent.

Griseofulvin Nanoparticles

Using menthol cosolvent, griseofulvin solubility can be
enhanced by up to 28-fold, as shown in Table 3.

The nanoparticles obtained from the RESS–SC process
are in the size range of 50–250 nm (Fig. 11), which is about
10-fold smaller than in RESS. In addition, due to the solubility
enhancement, the CO_2 requirement is about 28-fold lower.

Aminobenzoic Acid Nanoparticles

By using menthol cosolvent, the solubility of 2-aminobenzoic
acid can be enhanced by up to 100-fold as shown in Figure 12
(23).

The RESS–SC process produced ~80 nm size nanoparticles, which is significantly smaller than the ~610 nm size
nanoparticles obtained from the RESS process. Menthol is

Table 3 Solubility of Griseofulvin in Supercritical CO_2 with Menthol Cosolvent

P (bar)	T (°C)	Menthol amount (μmol/mol)	Griseofulvin solubility (μmol/mol)	Enhancement factor[a]
96	40	21,000	27	28
117	40	25,000	71	–
130	40	37,000	133	20
198	40	42,000	217	15
239	40	60,000	266	15
96	50	5,000	2	15
130	50	24,000	43	12
164	50	34,000	110	15

[a]Ratio of griseofulvin solubility in menthol/CO_2 to that in pure CO_2.
Abbreviation: CO_2, carbon dioxide.

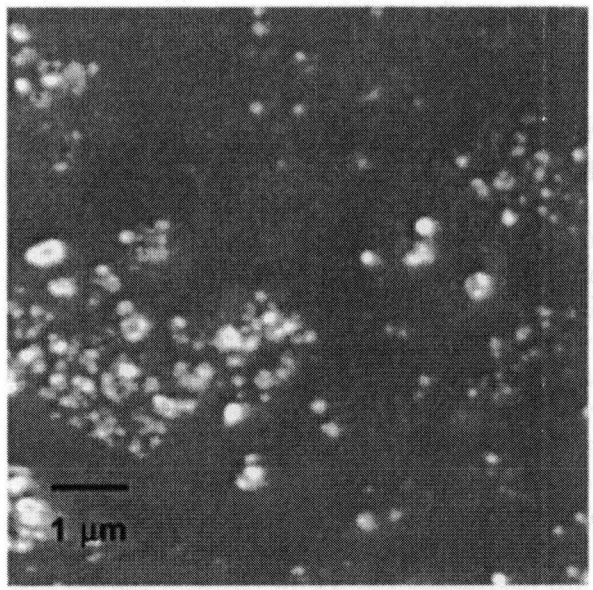

Figure 11 Griseofulvin nanoparticles from RESS–SC process. *Abbreviation*: RESS–SC, rapid expansion of solid supercritical solution solid cosolvent.

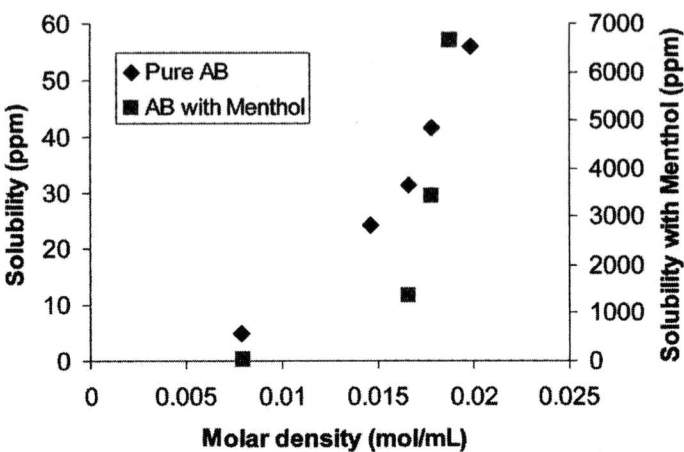

Figure 12 Solubility of 2-aminobenzoic acid in pure CO_2 and with menthol cosolvent versus fluid density. *Abbreviation*: CO_2, carbon dioxide.

easily removed from 2-aminobenzoic acid nanoparticles by sublimation (lyophilization) (Fig. 13).

SUPERCRITICAL ANTISOLVENT PROCESS FOR PARTICLE FORMATION

Before the invention of the RESS–SC process, the low-solubility aspect of supercritical CO_2 was utilized to produce particles by its antisolvent action. The drug is dissolved in an organic solvent, and then the solution is injected into supercritical carbon dioxide. The SCF, due to its high diffusivity, rapidly extracts the solvent precipitating the drug particles. A schematic of the supercritical antisolvent (SAS) concept is shown in Figure 14.

The SAS process has been proposed with numerous acronyms (SAA, SEDS, GAS, ASES, etc.) in the literature, but the basic concepts remain the same. Typically, 50–200 µm nozzles have been utilized in SAS. When the injection of the drug solution is complete, a washing step is carried out to remove the organic solvent so as to prevent it from condensing during

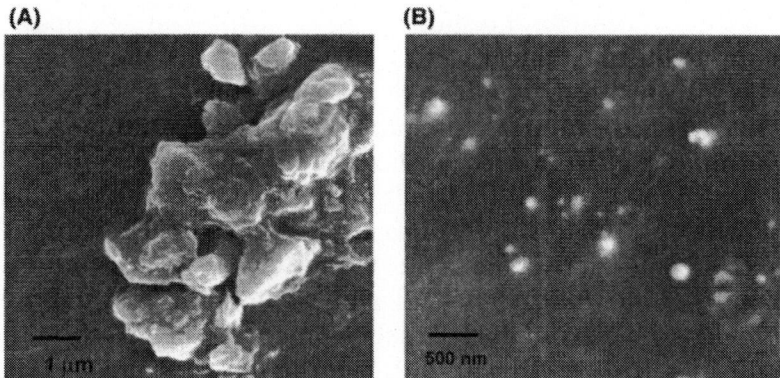

Figure 13 2-Aminobenzoic acid particles from (**A**) RESS and (**B**) RESS–SC processes. *Abbreviations*: RESS, rapid expansion of supercritical solution; RESS–SC, rapid expansion of supercritical solution solid cosolvent.

the depressurizing step. For this purpose, the feed of supercritical CO_2 is maintained to carry out the residual solvent. Once all the residual solvent is removed, the vessel pressure is reduced to atmospheric pressure, and the solid particles are collected on a filter at the bottom of the vessel. A review of SAS-based processes is provided by Jung and Perrut and by Charbit et al. (25). A polymer can be coprecipitated along with the drug to obtain controlled release formulation (26,27).

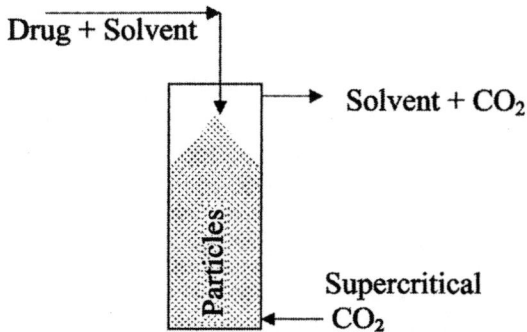

Figure 14 Schematic of SAS process. *Abbreviations*: SAS, supercritical antisolvent; CO_2, carbon dioxide.

The particle size and morphology depends on the nozzle geometry, solution velocity, CO_2 pressure, and the type of organic solvent used. The SAS process provides mostly 15 μm drug particles. Examples include 10–40 μm acetaminophen from ethanol, 1–10 μm ascorbic acid and aspirin from ethanol, 1.2–2 μm budesonide from methylene chloride, 0.5–20 μm camptothecin from dimthyl sulfoxide, 1–5 μm chlorpeniramine maleate from methylene chloride, 1.7 μm fluticasone-17-propionate from methylene chloride, 14 μm ibuprofen from methanol, 1–5 μm indomethacine from methylene chloride, 1–10 μm insulin from hexafluoro isopropanol, 1–5 μm insulin from dimethyl sulfoxide, 0.5–5 μm insulin from ethanol, 1–5 μm lysozyme from dimethyl sulfoxide, 1–10 μm para-cetamol and saccharose from ethanol, 2–20 μm sulfathiazole from acetone and methanol, and 1.5 μm trypsin from ethanol (27–38,63).

A few SAS studies have produced nanoparticles. These are listed in Table 4, along with the process conditions used.

In SAS, the inability to form small nanoparticles and to have a narrow size distribution can be attributed to particle growth after nuclei formation. The main phenomenon in

Table 4 Drug Nanoparticles from SAS-Based Precipitation Processes

Drug	Solvent	P (bar)	T (K)	Particle size (nm)	References
Albumin	Water/ethanol			50–500	39
Amoxicillin	N-Methylpyrrolidone	150	313	300–1200	40
Gentamicin/PLA	Methylene chloride	85	308	200–1000	41
Hydrocortisone	Dimethyl sulfoxide	100	308	600	29
Ibuprofen	Dimethyl sulfoxide	100	308	500–1000	29
Naloxone/l-PLA	Methylene chloride	85	308	200–1000	41
Insulin	Water/ethanol			50–500	39
Naltrexen/l-PLA	Methylene chloride	85	308	200–1000	41
Nicotinic acid	Ethanol			400–750	42
RhDNase	Ethanol			50–500	39
Salbutamol	Methanol/acetone	100	333	500	42

Abbreviation: SAS, supercritical antisolvent.

RESS is the high rate of pressure reduction, where in SAS, it is the high diffusivity of supercritical CO_2. The antisolvent action (mixing or mass transfer of solvent and antisolvent) needs to be even faster than SAS, in order to produce smaller particles of < 300 nm in size.

SAS WITH ENHANCED MASS (EM) TRANSFER (SAS-EM) PROCESS FOR NANOPARTICLE FORMATION

A significant improvement in the SAS process is introduced by Gupta and Chattopadhyay leading to nanoparticles of controllable size that are up to an order of magnitude smaller than those resulting from the conventional SAS process, and have a narrower size distribution (43). Like the SAS, this process, SAS–EM, utilizes supercritical carbon dioxide as the antisolvent, but in this case the solution jet is deflected by a surface vibrating at an ultrasonic frequency that atomizes the jet into much smaller droplets. Furthermore, the ultrasound field generated by the vibrating surface enhances mass transfer and prevents agglomeration through increased mixing. The particle size is controlled by varying the vibration intensity of the deflecting surface, which in turn is easily adjusted by changing the power supplied to the attached ultrasound transducer. The SAS–EM process is shown in Figure 15.

The SAS–EM process has been demonstrated by the formation of tetracycline, griseofulvin, lysozyme, and dexamethasone phosphate nanoparticles (44–46). The size is easily varied from 100 to 1000 nm by the power supply knob on the ultrasonic processor. These results are summarized in Table 5.

SAS–EM has been scaled up by Thar Technologies (www.thartech.com) for production at pilot scale (Fig. 16). This unit can produce up to 1 kg nanoparticle/day. It has one precipitation vessel and two separate collection vessels. One collection vessel can be used to collect the nanoparticles, while the other can be used to remove the nanoparticles for final use. The system is fully automated and can provide nanoparticles

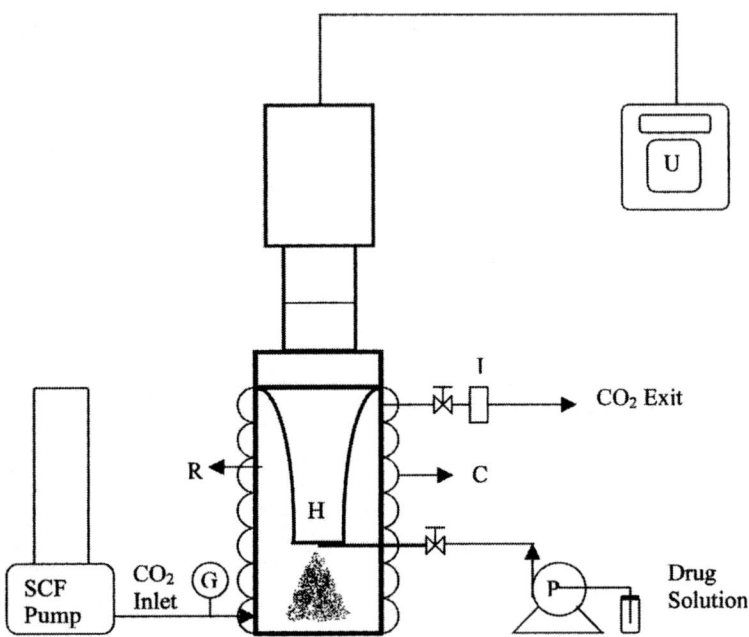

Figure 15 SAS-EM process. R, precipitation chamber; SCF pump, supply of supercritical CO_2; I, inline filter; U, ultrasonic processor; P, pump for drug solution; G, pressure gauge; C, heating coil with temperature controller; and H, ultrasonic horn. *Abbreviations*: SAS–EM, supercritical antisolvent with enhanced mass transfer; SCF, supercritical fluid; CO_2, carbon dioxide.

continuously. The ultrasound power supply is controlled by a computer, which in turn controls the nanoparticle size.

FUNDAMENTALS GOVERNING PARTICLE FORMATION WITH RESS AND SAS

Both SAS and RESS are complex processes involving the interaction of jet hydrodynamics, phase equilibrium, nucleation and growth (48,49). In SAS, additional complexity arises because of droplet formation, and mass transfer into and out of the droplets. In both cases, a high supersaturation is achieved, which results in rapid precipitation of the dissolved drug. In RESS, a sudden change in the fluid pressure causes

Table 5 Drug Nanoparticles from SAS–EM Process

Drug	Solvent	P (bar)	T (°C)	Ultra-sound power (W)	Par-ticle size (nm)	References
Dexametha-sone phosphate	Methanol	102	40	90	175	46
Griseofulvin	Dichloromethane	96.5	35	90	510	47
Griseofulvin	Dichloromethane	96.5	35	150	520	47
Griseofulvin	Dichloromethane	96.5	35	180	310	47
Griseofulvin	Tetrahydrofuran	96.5	35	120	200	47
Griseofulvin	Tetrahydrofuran	96.5	35	150	280	47
Griseofulvin	Tetrahydrofuran	96.5	35	180	210	47
Lysozyme	Dimethylsulfoxide	96.5	37	12	730	45
Lysozyme	Dimethylsulfoxide	96.5	37	30	650	45
Lysozyme	Dimethylsulfoxide	96.5	37	60	240	45
Lysozyme	Dimethylsulfoxide	96.5	37	90	190	45
Tetracycline	Tetrahydrofuran	96.5	37	30	270	44
Tetracycline	Tetrahydrofuran	96.5	37	60	200	44
Tetracycline	Tetrahydrofuran	96.5	37	90	184	44
Tetracycline	Tetrahydrofuran	96.5	37	120	110	44

Abbreviation: SAS–EM, supercritical antisolvent with enhanced mass transfer.

rapid precipitation, whereas in SAS the sudden diffusion of CO_2 into a drug solution causes drug precipitation. For RESS, the nanoparticle population balance equation accounting for particle nucleation and growth dynamics is as follows (50).

$$
\frac{\partial n}{\partial t} = J(\nu^*)\delta(\nu - \nu^*) - \frac{\partial(G_g n)}{\delta \nu}
$$
$$
+ \frac{1}{2}\int_0^\nu \beta(\nu - \bar{\nu}, \bar{\nu})n(\nu - \bar{\nu}, t)n(\bar{\nu}, t)d\bar{\nu} - n(\nu, t)
$$
$$
\times \int_0^\infty \beta(\nu, \bar{\nu})n(\bar{\nu}, t)d\bar{\nu} \tag{3}
$$

to obtain the number concentration of the particles from nuclea-tion, condensation, coagulation, and decoagulation. Where n

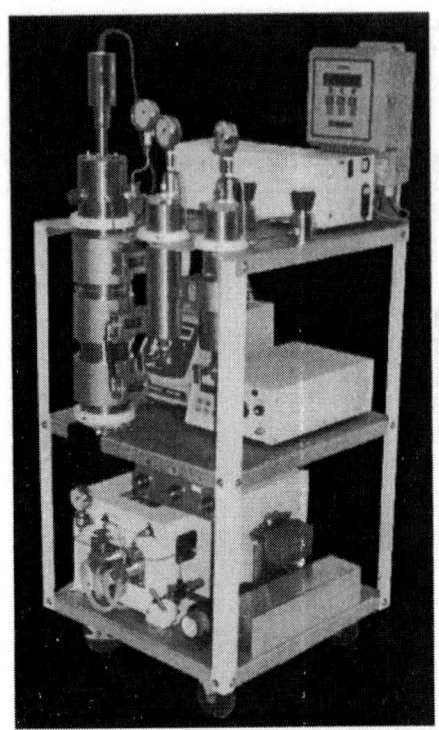
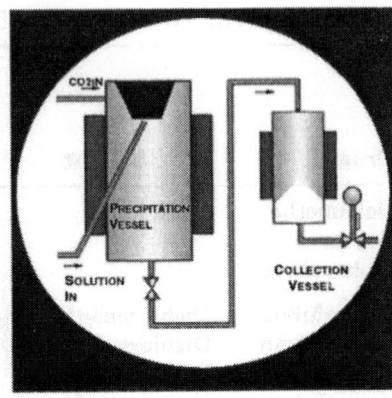

Figure 16 SAS–EM commercial unit by Thar Technologies, Inc. *Abbreviation*: SAS–EM, supercritical antisolvent with enhanced mass transfer.

is the number concentration, t is the time, J is the nucleation rate, δ is the delta function, v is the nanoparticle volume, G_g is the condensation rate, and β is the coagulation function.

Nucleation rate, J, is obtained from supersaturation (51)

$$J = 2N_2 \frac{P y_2}{\sqrt{2\pi m_2 kTL^{-1}}} \sqrt{\frac{\sigma \left(v_2^s\right)^2}{kT}} \exp\left\{ -\frac{16\pi}{3} \left(\frac{\sigma \left(v_2^s\right)^{2/3}}{kT}\right)^3 \right.$$
$$\left. \times \left[\frac{1}{\ln S - K y_2^{\text{eq}}(S-1)}\right]^2 \right\} \tag{4}$$

where y_2 is the actual drug mole fraction in CO_2 phase; y_2^{eq} is the equilibrium drug mole fraction over a flat surface (i.e., solubility); S is the supersaturation ratio, y_2/y_2^{eq}; k is the Boltzmann constant; N_2 is the number concentration of the solute in the fluid phase; and P is the pressure. The equilibrium solubility can be obtained from Equation (1) as discussed earlier. It will be a function of pressure, temperature, and cosolvent if present.

Particles grow by the condensation of solute from the fluid phase onto the particle surface. The net rate of a single molecule condensation onto a spherical particle is given by (52),

$$G_g = 2\pi d_p D \left[N_2 - N_2^{eq}(g) \right] \tag{5}$$

where d_p is the diameter of spherical particles containing g molecules and D is the diffusion coefficient for the solute molecule in the fluid phase.

The particle size and concentration can also change by coagulation and decoagulation. For coagulation of two particles (1 and 2), rate of coagulation (J') can be expressed as (53)

$$J' = K_{12} N_1 N_2 \tag{6}$$

where N_1 and N_2 are the number concentrations of the coagulating particles and K_{12} is the effective coagulation coefficient given as

$$K_{12} = \left[\frac{2kT}{3\mu} \frac{(D_{p1} + D_{p2})^2}{D_{p1} D_{p2}} \right] + \left[\frac{du}{dy} \frac{(D_{p1} + D_{p2})^3}{6} \right]$$
$$+ \left[\left(\frac{\pi \varepsilon_k}{120 v} \right)^{1/2} (D_{p1} + D_{p2})^3 \right] \tag{7}$$

which is the sum of Brownian, laminar shear, and turbulent coefficients. And

$$N_i(r,t) = N_i(0) \left[1 - \frac{D_{p1} + D_{p2}}{2r} erfc \left(\frac{2r - (D_{p1} + D_{p2})}{4\sqrt{D_{12}t}} \right) \right] \tag{8}$$

where du/dy is the velocity gradient in the case of laminar flow; ε_k is the rate of dissipation of kinetic energy per unit mass; v is the kinematic viscosity of the fluid; r is the distance of the particle from the center of the fixed particle; and D_{12} is the effective diffusion coefficient.

OTHER APPLICATIONS OF SCFs FOR PARTICLE ENGINEERING

SCFs can be applied to a variety of other applications where nano- and microdimensions of the drug material in excipient are important for drug release (54). These include the following.

Porous Particles and Polymer Foams

Since a fast removal of dissolved CO_2 can be achieved by rapid depressurization, this behavior can be used to create foams, especially that of poly(lactide–co–glycolide) (PLGA) polymer, because CO_2 has a good solubility in this approved polymer. Hile et al. (55) prepared PLGA foam capable of sustained release of basic fibroblast growth factor for tissue engineering applications. To prepare the foam, a water-in-oil microemulsion consisting of an aqueous protein phase (typical reverse micelle domain size of 5–10 nm) and an organic polymer solution was prepared. The microemulsion was filled in molds and then placed in a pressure vessel. Now, the pressure vessel was pressurized with supercritical CO_2, to extract the organic phase, causing the polymer to precipitate onto the protein droplets. Now the vessel is purged with more CO_2 to remove the solvent from the system. Finally, the vessel is depressurized in 10–12 sec causing rapid removal of the CO_2 that was dissolved in the polymer, making a porous foamy structure.

Koushik and Kompella (56) employed an SCF pressure-quench technique to create porous peptide (deslorelin) encapsulating PLGA particles (Fig. 17). On SC CO_2 treatment (1200 psi, 33°C for 30 min) of deslorelin, PLGA particles prepared using emulsion–solvent evaporation, the mean particle size of the deslorelin PLGA microparticles increased from 2.2 to 13.8 µm, the mean porosity increased from 39% to 92%, the

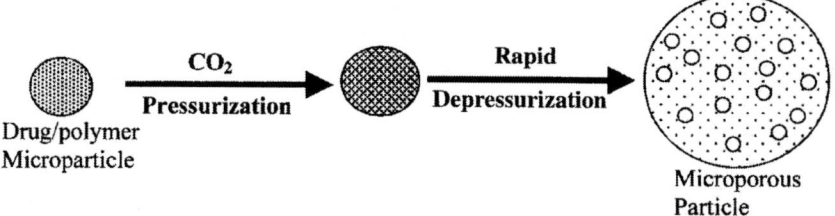

Figure 17 Supercritical-fluid pressure-quench technique to create porous microparticles. *Abbreviation*: CO_2, carbon dioxide. *Source*: From Ref. 56.

mean bulk density reduced from 0.7 to $0.08 \, g/cm^3$, mass spectrometry indicated structural integrity of released deslorelin, the circular dichroism spectrum indicated stabilization of β-turn conformation of the peptide, and the scanning electron microscopy confirmed increased particle size and pore formation. Further, the deslorelin release was sustained during the seven-day study period and the residual solvent content was reduced from 4500 ppm to below the detection limit (< 25 ppm).

Liposomes

Liposomes, in which nanodomains of drug are stabilized using lipids, are useful drug carriers for both small and macromolecular drugs. Unfortunately, the conventional methods of making liposomes require large amounts of organic solvents and have difficulty with scale-up for hydrophilic drugs. Lipids actually have some solubility in supercritical CO_2, and this behavior has been used to form liposomes without using organic solvents. For example, Fredereksen et al. (52) dissolved a phospholipid (1-palmitoyl-2-oleoylphosphatidylcholine) and cholesterol in supercritical CO_2 using 7% ethanol cosolvent. The mixture is expanded into an aqueous state containing fluorescein isothiocyanate (FITC)–dextran at low pressure. Because of the sudden reduction in the solubility of the phospholipid and the cholesterol at the nozzle tip, liposome-encapsulating FITC–dextran was formed. The process yielded 200-nm-size liposomes (termed as critical fluid liposomes) with 20% encapsulation efficiency. The main benefit of this process is the significantly reduced use of organic solvent. Later, Castor and Chu (57) prepared liposomes containing hydrophobic drugs, such as paclitaxel, camptothecins, doxorubicin, vincristine, and cisplatin. These formulations including 150–250-nm paclitaxel liposomes are claimed to be more effective against tumors in animals compared to commercial formulations.

Inclusion Complexes

Inclusion compounds, such as inclusion of poorly water-soluble drugs in cyclodextrin, are useful in enhancing bioavailability. Basically, the lipophilic drug is included in the lipophilic interior of the cyclodextrin molecule. The exterior of the cyclodextrin molecule is hydrophilic, and hence the whole complex can be dissolved in water. Inclusion can be achieved when both the drug and the cyclodextrin molecules are in a dissolved state, i.e., have a higher molecular mobility as compared to the solid forms. In conventional technique, both are dissolved in an organic solvent and then the solvent is removed. Unfortunately, the concentration of the residual solvent is high in the final product (58).

Supercritical CO_2 processes allow preparation of drug–cyclodextrin inclusion complexes without the use of organic solvents. This is because the interaction of supercritical CO_2 with solid cyclodextrin makes the cyclodextrin molecules more fluid. This interesting plasticizing effect of supercritical CO_2 has been well known for organic polymers, for which the glass transition or melting can be achieved at a lower temperature with SC CO_2. To make inclusion compounds, the physical solid mixture of the drug and cyclodextrin is exposed to supercritical CO_2, and then rapidly CO_2 is removed by depressurization.

Bandi et al. (59) prepared budesonide and indomethacin hydroxypropyl–cyclodextrin (HPBCD) complexes using an organic solvent-free SCF process (59,60). The process involved the exposure of drug–HPBCD mixtures to supercritical carbon dioxide. The ability of the SCF process to form complexes was assessed by determining drug dissolution using a high-performance liquid chromatography assay, crystallinity using powder x-ray diffraction (PXRD) and differential scanning calorimetry, and drug–excipient interactions using Fourier transform infrared spectroscopy (FTIR). The SC CO_2 process did not alter the dissolution rate of pure drugs but resulted in two- and threefold higher dissolution rates for budesonide and indomethacin–HPBCD mixtures, respectively. SCF-processed mixtures exhibited a disappearance of the crystalline peaks of the drugs (PXRD), a partial or a complete absence of the melting endotherm of the drugs (DSC), and a shift in the C=O stretching of the carboxyl groups of the drugs (FTIR), consistent with the loss of drug crystallinity and the formation of intermolecular bonds with HPBCD. Thus, budesonide and indomethacin–HPBCD complexes with an enhanced dissolution rate can be formed using a single-step, organic solvent-free SC CO_2 process. Similar inclusion complexes were also reported for piroxicam using a supercritical CO_2 process (61).

Solid Dispersions

In many delivery applications, molecularly intimate mixtures (i.e., solid dispersion) of drug with excipients, such as polymers are needed. An organic solvent, which can dissolve both, does bring the two in intimate contact while in solution. Unfortunately, when the solvent is removed by evaporation or by addition of a liquid antisolvent, the drug and the polymer phases precipitate out or separate. Hence, the dispersion of the two is poor in the solid state. Supercritical CO_2 antisolvent induces the precipitation about 100-fold faster than the liquid antisolvent, not allowing enough time for the drug and the polymer domains to separate out. Thus, supercritical CO_2 precipitation can provide a more dispersed solid mixture. Supercritical CO_2-based precipitation is superior to the liquid-based precipitation or the milling process. For example, a solid dispersion of carbamazepine in polyethyleneglycol

(PEG)-4000, produced by CO_2 method, increased the rate and the extent of dissolution of carbamazepine (62). In this method, a solution of carbamazepine and PEG4000 in acetone was loaded in a pressure vessel, in which supercritical CO_2 was added from the bottom to obtain solvent-free particles.

SAFETY AND HEALTH ISSUES

When dealing with supercritical carbon dioxide, there are two safety and health issues that are to be kept in mind when designing and operating the equipment: (i) the high pressure involved requires that personnel is protected from the plant by proper isolating walls and (ii) if carbon dioxide is released in the closed atmosphere it can lead to asphyxiation, as it can replace the oxygen in the surroundings.

CONCLUSIONS

For particle formation, SCF technology offers two processes: (i) RESS for drugs that are soluble in supercritical CO_2 and (ii) SAS for drugs that are poorly soluble in supercritical CO_2. In RESS, a sudden change in the fluid pressure causes rapid precipitation, whereas in SAS the sudden diffusion of CO_2 into a drug solution causes drug precipitation. Conventionally, both the technologies have produced microparticles in the 1–5-µm-size range. With enhancement in mixing, SAS-EM process produces nanoparticles of controllable size. With the reduction in particle coagulation, the RESS–SC process produces nanoparticles with a high yield. The RESS–SC equipment is expected to be cheaper than SAS–EM, because the residence time of the drug in the high-pressure chamber is lower in the former. The particle formation techniques can also be employed for the preparation of liposomes and solid dispersions of drugs and solubility enhancing carriers. In addition, SCF exposure or pressure-quench techniques can be employed to form porous structures or inclusion complexes and to remove residual solvents in pharmaceutical particulate systems.

REFERENCES

1. York P, Kompella UB, Shekunov, B. Supercritical Fluid Technology for Drug Product Development. New York: Marcel Dekker, 2004.

2. Thakur R, Gupta RB. Rapid expansion of supercritical solution with solid cosolvent (RESS-SC) process: formation of griseoful-vin nanoparticles. Ind Eng Chem Res 2005; 44:7380–7387.

3. Reverchon E, Donsi G, Gorgoglione D. Salicylic acid solubiliza-tion in supercritical CO_2 and its micronization by RESS. J Supercrit Fluids 1993; 6(suppl 4):241–248.

4. Jouyban A, Chan H-K, Foster NR. Mathematical representation of solute solubility in supercritical carbon dioxide using empiri-cal expressions. J Supercrit Fluids 2002; 24(1):19–35.

5. Li S, Maxwell RJ, Shadwell RJ. Solubility of amphenicol bacteriostats in CO_2. Fluid Phase Equilib 2002; 198(1): 67–80.

6. Gupta RB, Shim J-J. Solubility in Supercritical CO_2. Boca Raton: CRC Press, 2006.

7. Kerget M, Kotnik P, Knez Z. Phase equilibria in systems con-taining a-tocopherol and dense gas. J Supercrit Fluids 2003; 26(3):181–191.

8. Mendez-Santiago J, Teja AS. The solubility of solids in super-critical fluids. Fluid Phase Equilib 1999; 158–160:501–510.

9. Jouyban A, Rehman M, Shekunov BY, Chan H-K, Clark BJ, York P. Solubility prediction in supercritical CO_2 using minimum number of experiments. J Pharm Sci 2002; 91(5): 1287–1295.

10. Dixon DJ, Johnston KP. Supercritical fluids. In: Ruthven DM, ed. Encyclopedia of Separation Technology. John Wiley, 1997:1544–1569.

11. McHugh MA, Krukonis VJ. Supercritical Fluid Extraction. 2nd ed. Elsevier, 1994.

12. Domingo C, Berends E, van Rosmalen GM. Precipitation of ultrafine organic crystals from the rapid expansion of supercri-tical solutions over a capillary and a frit nozzle. J Supercrit Fluids 1997; 10:39–55.

13. Subra P, Boissinot P, Benzaghou S. Precipitation of pure and mixed caffeine and anthracene by rapid expansion of supercritical solutions. In Perrut M, Subra P, eds. Proceedings of the 5th Meeting on Supercritical Fluids. Vol. 1. Nice, France, 1998:307–312.

14. Sievers RE, Hybertson B, Hansen B. European Patent EP 0,627,910 B1, 1993.

15. Charoenchaitrakool M, Dehghani F, Foster NR. Micronisation of ibuprofen using the rapid expansion of supercritical solutions (RESS) process. CISF 99, 5th Conference on Supercritical Fluids and Their Applications, Garda, Italy, Jun 13–16, 1999:485–492.

16. Stahl E, Quirin KW, Gerard D. IV. High Pressure Micronising in Dense Gases for Extraction and Refining. Berlin Heidelberg: Springer, 1988, Chapter V.

17. Coffey MP, Krukonis VJ. Supercritical Fluid Nucleation. An Improved Ultrafine Particle Formation Process. Phasex Corp. Final Report to NSF, 1988, Contr. ISI 8660823.

18. Subra P, Debenedetti P. Application of RESS to several low molecular weight compounds. In: Rudolf P, von Rohr Trepp C, eds. High Pressure Chemical Engineering. Amsterdam: Elsevier Science B.V., 1996:49–54.

19. Hybertson BM, Repine JE, Beehler CJ, Rutledge KS, Lagalante AF, Sievers RE. Pulmonary drug delivery of fine aerosol particles from supercritical fluids. J Aerosol Med 1993; 8(4):275–286.

20. Mohamed RS, Halverson DS, Debenedetti PG, Prud'homme RK. Solids formation after the expansion of supercritical mixtures. In: Johnston KP, Penninger JML, eds. Supercritical Fluid Science and Technology. Washington, DC: ACS Symposium Series 406, 1989:355–378.

21. Turk M, Hils P, Helfgen B, Schaber K, Martin H-J, Wahl MA. Micronization of pharmaceutical substances by the rapid expansion of supercritical solutions (RESS): a promising method to improve the bioavailability of poorly soluble pharmaceutical agent. J Supercrit Fluids 2002; 22:75.

22. Pathak P, Meziani MJ, Desai T, Sun Y-P. Nanosizing drug particles in supercritical fluid processing. J Am Chem Soc 2004; 126:10,842.

23. Thakur R. Nanoparticles and nanofibers production using supercritical CO_2. Ph.D. dissertation, Auburn University, 2005.

24. Jung J, Perrut M. Particle design using supercritical fluids: literature and patent survey. J Supercrit Fluids 2001; 20:179–219.

25. Charbit G, Badens E, Boutin O. Methods of particle production. In: Peter Y, Uday BK, Boris S, eds. Supercritical Fluid Technology for Drug Product Development. New York: Marcel Dekker, 2004:367–410.

26. Bandi N, Gupta RB, Roberts CB, Kompella UB. Formulation of controlled-release drug delivery systems. In: Peter Y, Uday BK, Boris S, eds. Supercritical Fluid Technology for Drug Product Development. New York: Marcel Dekker, 2004: 367–410.

27. Martin TM, Bandi N, Schulz R, Roberts C, Kompella UB. Preparation of budesonide and budesonide-PLA microparticles using supercritical fluid precipitation technology. AAPS Pharm Sci Technol 2002; 3(3), Article 18.

28. Shekunov BY, Baldyga J, York P. Particle formation by mixing with supercritical antisolvent at high Reynolds numbers. Chem Eng Sci 2001; 56:2421–2433.

29. Weber A, Weiss C, Tschernjaew J, Kummel R. Gas antisolvent crystallization-from fundamentals to industrial applications. Proceedings of High Pressure Chemical Engineering, Karlsruhe, Germany, 1999: 235–238.

30. Said S, Rajewski RA, Stella V, Subramanian B. World Patent WO 9,731,691, 1997.

31. Bodmeier R, Wang H, Dixon DJ, Mawson S, Johnston KP. Polymeric microspheres prepared by spraying into compressed carbon dioxide. Pharm Res 1995; 12(8):1211–1217.

32. Steckel H, Muller BW. Metered-dose inhaler formulations of fluticasone-17-propionate micronized with supercritical carbon dioxide using the alternative propellant HFA-227. Int J Pharm 1998; 46(1):77–83.

33. Hanna M, York P. Method and apparatus for the formation of particles. World Patent WO 99/59710, 1999.

34. Niu F, Rejewski R, Snaveley WK, Subramanian B. World Patent WO 0,235,941, 2002.

35. Moshashaee S, Bisrat M, Forbes RT, Nyqvist H, York P. Supercritical fluid processing of proteins. I Lysosyme precipitation from organic solution. Eur J Pharm Sci 2000; 11:239–245.

36. Gilbert R, Palakodaty R, Sloan R, York P. Particle engineering for pharmaceutical applications-process scale-up. Proceedings of the 5th International Symposium on Supercritical Fluids, Atlanta, 2000.

37. Kordikowski A, Shkunov T, York P. Polymorph control of sulfathiazole in supercritical CO_2. Pharm Res 2001; 18:685–688.

38. Sloan R, Hollowood ME, Humphreys GO, Ashraf W, York P. Supercritical fluid processing: preparation of stable protein particles. In: Perrut M, Subra P, eds. Proceedings of the 5th Meeting of Supercritical Fluids. Vol 1. Nice, France, 1998: 301–306.

39. Bustani RT, Chan HK, Dehghani F, Foster NR. Generation of protein microparticles using high pressure modified carbon dioxide. Proceedings of the 5th International Symposium on Supercritical Fluids, Atlanta, 2000.

40. Reverchon E, De Marco I, Caputo G, Della Porta G. Pilot scale micronization of amoxicillin by supercritical antisolvent precipitation. J Supercrit Fluids 2003; 26(1):1–7.

41. Falk R, Randolph TW, Meyer JD, Kelly RM, Manning MC. Controlled release of ionic compounds from poly(l-lactide) microspheres produced by precipitation with a compressed antisolvent. J Contolled Release 1997; 44:77–85.

42. Hanna M, York P. Method and apparatus for the formation of particles. European Patent WO 98/36825, 1998.

43. Gupta RB, Chattopadhyay P. Method of forming nanoparticles and microparticles of controllable size using supercritical fluids with enhanced mass transfer. U.S. Patent 6,620,351, Sep 16, 2003.

44. Chattopadhyay P, Gupta RB. Production of antibiotic nanoparticles using supercritical CO_2 as antisolvent with enhanced mass transfer. Ind Eng Chem Res 2001; 40(16):3530–3539.

45. Chattopadhyay P, Gupta RB. Protein nanoparticles formation by supercritical antisolvent with enhanced mass transfer. AIChE J 2002; 48:235–244.

46. Thote, A. Gupta, RB. Formation of nanoparticles of a hydrophilic drug using supercritical CO_2 and microencapsulation for sustained release. Nanomed: Nanotechnol Biol Med 2005; 1:85–90.

47. Chattopadhyay P, Gupta RB. Production of griseofulvin nanoparticles using supercritical CO_2 antisolvent with enhanced mass transfer. Int J Pharm 2001; 228(1–2):19–31.

48. Chavez F, Debenedetti PG, Luo JJ, Dave RN, Pfeffer R. Estimation of the characteristic time scales in the supercritical antisolvent process. Ind Eng Chem Res 2003; 42: 3156–3162.

49. Werling JO, Debenedetti PG. Numerical modeling of mass transfer in the supercritical antisolvent process: miscible conditions. J Supercrit Fluid 2000; 18:11–24.

50. Helfgen B, Hils P, Holzknecht CH, Turk M, Schaber K. Simulation of particle formation during expansion of supercritical solutions. Aerosol Sci 2001; 32:295–319.

51. Debenedetti PG. Homogenous nucleation in supercritical fluids. AIChE J 1990; 36:1289.

52. Frederiksen L, Anton K, Hoogevest PV, Keller HR, Leuenberger H. Preparation of liposomes encapsulating water-soluble compounds using supercritical carbon dioxide. J Pharm Sci 1997; 86(8):921–928.

53. Friedlander SK. Smoke, Dust, and Haze: Fundamentals of Aerosol Behavior. New York: Wiley-Interscience, 1977: Chapter 9.

54. Sunkara G, Kompella UB. Drug Delivery Applications of Supercritical Fluid Technology, Drug Delivery Technology, www.drugdeliverytech.com, 2005.

55. Hile DD, Amirpour ML, Akgerman A, Pishko MV. Active growth factory delivery from poly(D,L-lactide-co-glycolide) foams prepared in supercritical CO2. J Control Rel 2000; 66(2–3):177–185.

56. Koushik K, Kompella UB. Preparation of large porous deslor-
    elin-PLGA microparticles with reduced residual solvent and
    cellular uptake using a supercritical CO_2 process. Pharm Res
    2004; 21:524–535.

57. Castor TP, Chu L. Methods and apparatus for making lipo-
    somes containing hydrophobic drugs. U.S. Patent 5,776,486,
    1998.

58. Lin SY, Kao YH. Solid particulates of drug-b-cyclodextrin
    inclusion complexes directly prepared by a spray-drying tech-
    nique. Int J Pharm 1989; 56:249–259.

59. Bandi N, Wei W, Roberts CB, Kotra LP, Kompella, UB.
    Preparation of budesonide- and indomethacin-hydroxypropyl-
    β-cyclodextrin (HPβCD) complexes using an organic-solvent-
    free, single-step supercritical fluid process. Eur J Pharm Sci
    2004; 23(2):159–168.

60. Mayo A, Kompella UB. Supercritical fluid technology in
    pharmaceutical research. In: James S, ed. Encyclopedia of
    Pharmaceutical Technology. New York: Marcek Dekker Inc.,
    2004:1–17.

61. Van Hees T, Piel G, Evrard B, Otte X, Thunus T, Delattre L.
    Application of supercritical carbon dioxide for the preparation
    of a piroxicam-beta-cyclodextrin inclusion compound. Pharm
    Res 1999; 16(12):1864–1870.

62. Moneghini M, Kikic I, Voinovich D, Perissutti B, Filipovic-Grcic
    J. Processing of carbamazepine-PEG 4000 solid dispersions
    with supercritical carbon dioxide: preparation, characteriza-
    tion, and in vitro dissolution. Int J Pharm 2001; 222(1):
    129–138.

63. Winters MA, Knutson BL, Debenedetti PG, et al. Precipitation
    of proteins in supercritical carbon dioxide. J Pharm Sci 1996;
    85(6):586–594.

4

Polymer or Protein Stabilized Nanoparticles from Emulsions

RAM B. GUPTA

Department of Chemical Engineering,
Auburn University, Auburn, Alabama, U.S.A.

INTRODUCTION

Poorly water-soluble drugs pose a significant challenge in their delivery. A large number of drugs are discarded from consideration in their early stages of development owing to poor bioavailability. Such drugs are an excellent candidate for nanoparticle delivery, which can avoid the allergic side effects due to the use of cremaphors (e.g., polyethyoxylated castor oil) in conventional formulations. However, for drugs with crystal forming habits, there is always the hazard of the formation of large microparticles (>10–$15\,\mu m$) from aggregation/bonding of nanoparticles; this can lead to infarction or blockage of the capillaries, resulting in ischemia or oxygen deprivation and possible tissue death. Hence, the nanoparticles need to be stabilized using biocompatible proteins (e.g., human serum albumin) or polymers (e.g., polylactide, polycaprolactone). An example is the recently approved drug Abraxane^{TM} for cancer therapy, which is composed of 130-nm albumin-stabilized paclitaxel nanoparticles. This chapter discusses the technology aspect of the protein and polymer-stabilized nanoparticle formation. Though proteins and polymers can be added to the drug nanoparticles in supercritical fluid or milling based technologies, this chapter focuses on the use of emulsions for making stabilized nanoparticles.

EMULSIFICATION SOLVENT EVAPORATION PROCESS

In a typical emulsification solvent evaporation process to produce nanoparticles (Fig. 1), drug and polymer [e.g., poly(d,l-lactide-co-glycolide) (PLGA), poly(lactic acid) (PLA), poly-methacrylate] dissolved in a water-immiscible solvent (e.g, methylene chloride, chloroform, ethyl acetate) are added dropwise

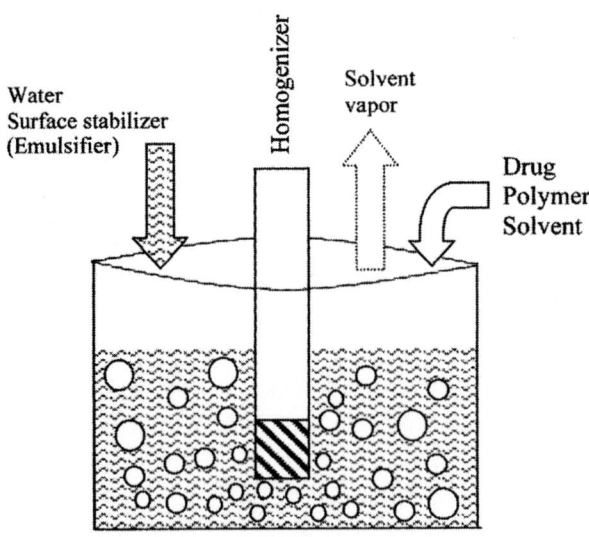

Figure 1   Emulsification solvent evaporation process for particle formation.

to aqueous phase containing a surface stabilizer (e.g., polysorbate, polyvinyl alcohol, methyl cellulose, genatin, albumin, poloxomar) (1,2). A high shear is provided using a homogenizer, which reduces the droplet size of the organic dispersed phase. The evaporation of solvent hardens the nanoparticles. Formed nanoparticles are harvested from the aqueous slurry by lyophilization.

In a variation of the above process, the solvent removal is done by adding a large quantity of aqueous phase, which induces the rapid diffusion of the solvent from the internal into the external phase. In yet another variation, a water-miscible solvent such as acetone is added to the organic phase, which influences the droplet hardening process.

For the water-soluble drugs, a double-emulsion (water/oil/water) variation of the process is utilized. First, the drug is dissolved in water and then emulsified in water to obtain drug/water as the dispersed phase and organic solvent as the continuous phase. Then, this emulsion is added to the large aqueous phase with emulsifier to create double emulsion. The emulsifier amount is much higher in the first emulsion than in the second emulsion, because the droplet size of the first emulsion needs to be much smaller than in the second outer emulsion.

Various parameters in the emulsification solvent evaporation process that affect particle size, zeta potential, hydrophilicity, and drug loading include

1. homogenization intensity and duration,
2. type and amounts of emulsifier, polymer, and drug, and
3. particle hardening (solvent removal) profile.

EMULSIFICATION

Emulsions are metastable colloids composed of two immiscible liquids, one dispersed in the other, in the presence of surface-active agents. Emulsions are different from microemulsions, which are thermodynamically stable and are formed by using surfactant concentrations above the critical micelle concentration. Emulsion droplets exhibit all the classical behaviors of metastable colloids, including Brownian motion, reversible phase transitions as a result of droplet interactions that may be strongly modified, and irreversible transitions that generally involve their destruction (3). Emulsions are obtained by shearing two immiscible liquids, leading to the fragmentation of one phase into the other. The lifetime of emulsions is limited, hence a small change in the process conditions may yield varying emulsion droplets.

In emulsification, shear forces help create more surface and hence smaller droplet size emulsion, whereas the surface tension opposes the formation of more surface. Surface energy of an emulsion (E_S) with droplet diameter d is given as

$$E_S = N\sigma\pi d^2 \qquad (1)$$

where N is the number of droplets, and σ is the interfacial tension between the two phases. If the total volume of the dispersed phase is V, then $N = 6V/(\pi d^3)$, and

$$E_S = \frac{6V\sigma}{d} \qquad (2)$$

A part of the shearing energy is utilized to provide the surface energy and the remaining energy goes toward creating turbulence, which ultimately is dissipated as heat. Hence, it is clear that if a smaller droplet size is desired, then a high shear energy is needed. This energy requirement can be reduced if the surface tension is reduced, which is a function of temperature and composition of both the phases. Typically, the reduction is achieved by adding a surfactant or surface stabilizing agent such as albumin, poly(vinyl alcohol) (PVA), poly(acrylic acid) (Carbopol®), poly(oxyethylene-*b*-oxypropylene-*b*-oxyethylene) (Poloxamer or Pluronic®). Both Carbopol and Poloxamer show mucoadhesive properties in addition to surface stabilization, which may be of significance in oral drug delivery applications. Once the droplets are created, it is then important to solidify them to avoid coalescence. Typically, two droplets will coalesce, if they are less than 1 nm apart; the liquid bridge formation occurs in 10 ps and the coalescence is complete in 200 ps (4). Hence, a fast solidification process is needed to keep particles in a small size.

Shear or Homogenization

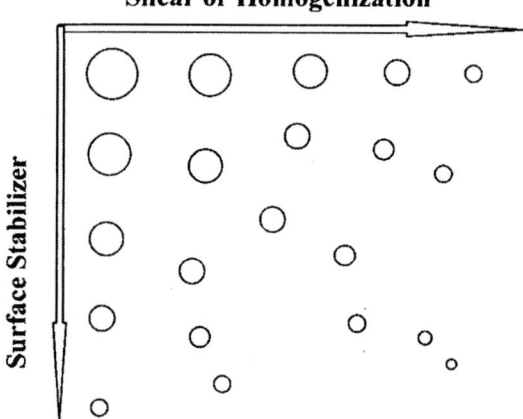

Figure 2 Effects of surface stabilizer and shear on the emulsion droplet size.

The final particle size is directly proportional to emulsion droplet size and the coalescence during hardening. The emulsion droplet size is mostly determined by the amounts of shear and surface stabilizer used. Figure 2 shows how the two affect the emulsion droplet size.

An increase in the amount of emulsifier used reduces the droplet size, which in turn reduces the final particle size. For example, the effect of emulsifier concentration on the PLA nanoparticles from propylene carbonate solvent was studied by Quintanar-Guerrero et al. (5) as shown in Figure 3.

For creating fine emulsion for obtaining nanoparticles, the use of a high amount of surface stabilizer is avoided to reduce the high load of the polymer exipients, as some of these exipients have shown toxicity. This leaves us to the use of high shear to generate fine emulsions for which sonication and homogenization techniques are available (6).

Sonication

Sonication generates emulsions through ultrasound-driven mechanical vibrations, which causes cavitation. Rarefaction and compression cycles of sonication create vapor bubbles, which grow with time. Once a critical size is achieved, the

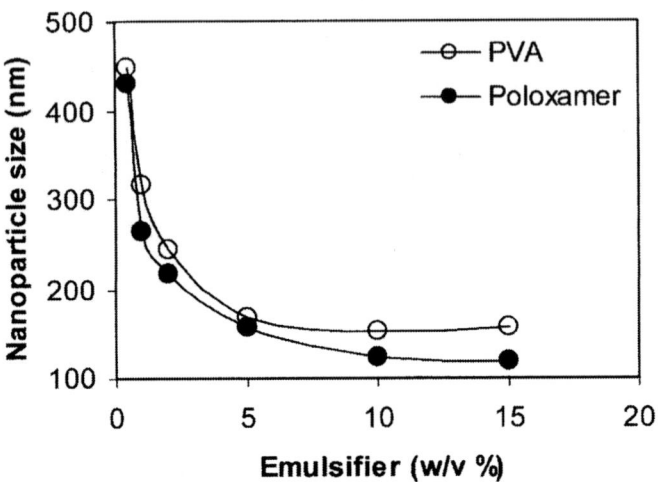

Figure 3 Influence of the emulsifier (PVA 26,000 MW and poloxamer 188) concentration on the PLA particle size from propylene
carbonate solvent. *Abbreviations*: PVA, polyvinyl alcohol; PLA,
poly-lactic acid. *Source*: From Ref. 5.

bubble collapses violently, releasing the energy creating hot
spots and hydroxyl free radicals. In addition, jets of fluid ele-
ment propel out. The turbulence and the high-speed jets cause
the oil phase to finely divide and disperse in the water phase. An
increase in viscosity of the oil phase improves the sonicator's
emulsification capability, but an increase in the viscosity of
the water phase decreases the sonicator's emulsification capability. The duration and intensity of sonication can be used
to create varying emulsion droplet sizes. For example, Mainardes and Evangelista utilized sonication to form praziquan-
tel-loaded PLGA nanoparticles from methylene chloride
solvent and PVA emulsifier (7). For a fixed sonication intensity
of 5 W/mL, 380-, 335-, 298-, and 255-nm particles were obtained
for a sonication time of 1-, 5-, 10-, and 20-minutes, respectively.

Homogenization

Although sonication is comparable to homogenization in
terms of emulsification efficiency, homogenization is relatively
more effective in emulsifying viscous solutions. Ambient

pressure homogenizers use rotor–stator types of mixers, which can go to very high rotational speeds. High-pressure homogenization uses high pressure to force the fluid into microchannels of a special configuration and initiates emulsification via a combined mechanism of cavitation, shear, and impact, exhibiting excellent emulsification efficiency. Sonication usually generates more heat, and hence is less suitable for heat-sensitive materials. Homogenization is generally more effective in making fine emulsions. Usually, multiple passes are needed to achieve the desired emulsion droplet size.

The influence of process parameters on the emulsion droplet size was studied by Maa and Hsu (8). The change in the emulsion droplet size was found to reduce initially with homogenization and then reach a steady value. The emulsion droplet size decreases with increasing homogenization intensity. Using a rotor–stator homogenizer, the emulsion droplet size was found to be viscosity (μ) dependent and proportional to $\mu^{0.11}$ of the dispersed phase and $\mu^{-0.43}$ of the continuous phase.

The effect of a high-speed homogenizer for producing cystatin-load PLGA nanoparticles was studied by Cegnar et al. (9). When the stirring speed was increased from 5000 to 15,000 rpm, the particle reduced from micro to nano size (Fig. 4). When the stirring was combined with bath

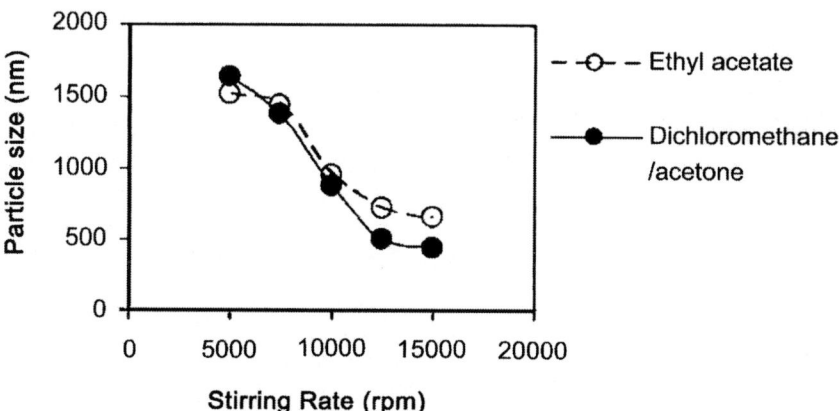

Figure 4 Cystatin-load PLGA nanoparticles from high stirring. *Abbreviation*: PLGA, poly-lactic/glycolic acid. *Source*: From Ref. 9.

sonication, the particle size went further down to about 250 nm. For similar stirring speeds, much smaller PLA particles were obtained by Quintanar-Guerrero et al. (5) because the PLA is a much less sticky polymer than PLGA. During the hardening process, as explained later, many more PLGA droplets will come to form a particle than the PLA droplets.

High-pressure homogenization is described in detail in chapter 2. The intensity and duration of homogenization can have a profound effect on the particle size. The general trend in this effect is independent of the emulsifier used. However, emulsifier type has its own effect on the nanoparticle formation. A good example is the study of Yoncheva et al. (10) for encapsulation of pilocarpin hydrochloride in PLGA by using a combination of a double emulsification and homogenization procedure. First, the aqueous solution of drug was emulsified in PLGA/methylene chloride using sonication, to form dispersed aqueous phase and continuous organic phase. This emulsion was then further emulsified with an aqueous stabilizer solution and subjected to high-pressure homogenization using a microfluidizer. The particle size decreases with the homogenization pressure and/or the number of homogenization cycles (Table 1).

As shown above, the homogenization conditions and the choice of stabilizer can be used to vary the nanoparticle properties. In addition to the size, zeta potential, drug loading, and drug release also depend on the process conditions. For example, higher loading with smaller-size pilocarpine HCL-load PLGA nanoparticles can be obtained by using Carbopol® stabilizer as compared to PVA or Poloxamer (11).

Dillen et al. (12) carried out a 2^4 full factorial design for the production of ciprofloxacin HCL-load PLGA nanoparticles. The effect of process parameters (homogenization cycles, addition of boric acid to the inner water phase, drug concentration, and oil:outer water phase ratio) on particle size, zeta potential, drug loading efficiency, and drug release kinetics was studied. Gamma radiation, used for terminal sterilization, results in a small increase in the particle size.

Table 1 The Effect of Shear Intensity and Duration on the Particle Size for Pilocarpin Hydrochloride Encapsulation in PLGA Using Different Emulsifiers

Emulsifier	Pressure, gauge (bar)	Cycles	Particle diameter (nm)	Polydispersity index
Polyvinyl alcohol	0		332	0.08
	100	1	283	0.12
	100	3	232	0.10
	500	1	231	0.08
	500	3	204	0.31
Carbopol	0		1125	0.78
	100	1	631	0.64
	100	3	366	0.54
	500	1	467	0.66
	500	3	309	0.05
Poloxamer	0		572	0.80
	100	1	692	0.76
	100	3	424	0.53
	500	1	467	0.81
	500	3	304	0.31

Source: From Ref. 10.

NANOPARTICLE HARDENING

Particle hardening due to solvent evaporation plays an important role in the growth of the particle during coalescence. The particle stickiness comes from the solvent associated with the polymer and drug. In the beginning of the process, the droplets are liquid and coalesce if they come any closer than about 1 nm. When part of the solvent is removed, the droplets are still sticky, but the particle bridging is slowed down owing to the increased viscosity of the drop interior. Once most of the solvent is removed, the particles become hard and now they can start to bounce off from other colliding particles. Wang and Schwendeman (13) measured the removal rate of the solvent from particles with respect to time as shown in Figure 5. Initially, the solvent removal is fast, owing to the high diffusivity of solvent and the dissolution of the solvent in the aqueous media. With time the droplets become hard on the surface

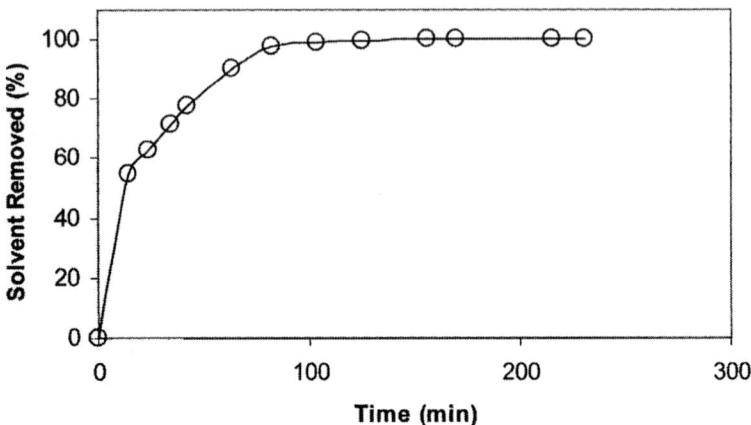

Figure 5 Methylene chloride removal profile from encapsulation of triamcinolone acetonide in PLGA particles. Values are normalized with the final amount of solvent removed. *Abbreviation*: PLGA, poly-lactic glycolic acid. *Source*: From Ref. 13.

due to polymer precipitation, which slows down the solvent diffusion.

Particle growth continues as a result of coalescing for the duration in which the solvent is not completely removed to the point when particles are not sticky. For example, Desgouilles et al. (14) have studied the formation of PLA and ethyl cellulose nanoparticles from ethyl acetate solvent. The change in the particle size with respect to time, as the solvent is removed, is shown in Figure 6. For PLA, the particle size decreases as the solvent leaves the droplet, finally yielding to a constant size when all the solvent is removed. For ethyl cellulose, the particle size first decreases and then increases. The difference is attributed to the softer/stickier nature of ethyl cellulose as compared to PLA. More ethyl cellulose droplets come together to make one particle than the PLA droplets (Fig. 7). This number, aggregation ratio A, can be calculated as

$$A = \frac{c_0}{c} \left(\frac{d_{\text{droplet}}}{d_{\text{nanoparticle}}} \right)^3 \tag{3}$$

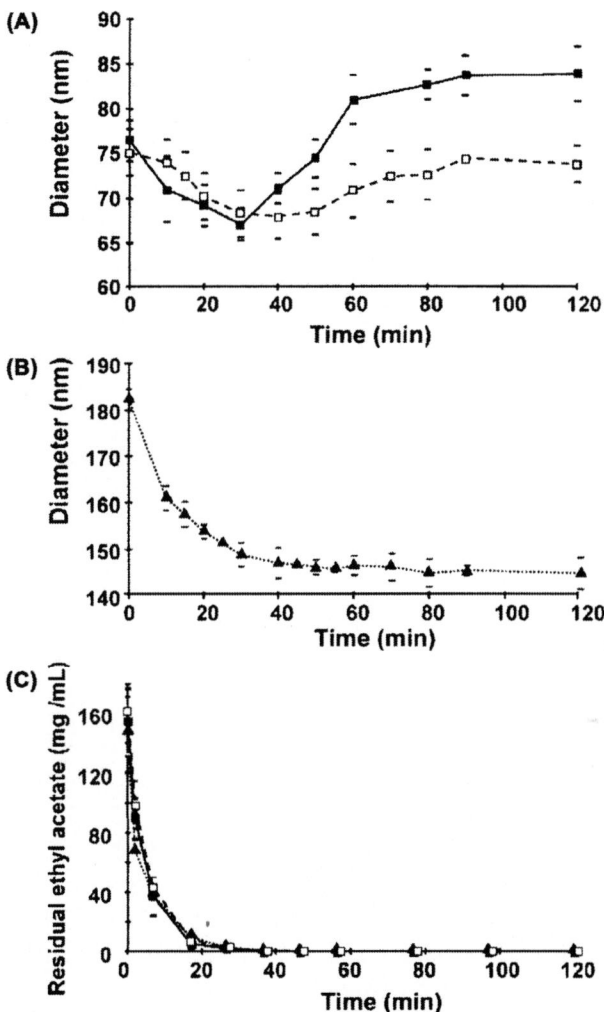

Figure 6 Variation of the hydrodynamic diameter and of the ethyl acetate content of the emulsion/nanoparticle suspension during the course of the evaporation of ethyl acetate. Hydrodynamic diameter of the systems prepared with (**A**) EC7 (□) and EC22 (■) and (**B**) PLA ( ) is shown. (**C**) shows the residual ethyl acetate remaining in the emulsion/nanoparticle suspension during the course of the preparation of the nanoparticles by the emulsion solvent evaporation method. Viscosity of polymer solutions in ethyl acetate was 0.08 Pa s. *Abbreviation*: PLA, poly-lactic acid. *Source*: From Ref. 14.

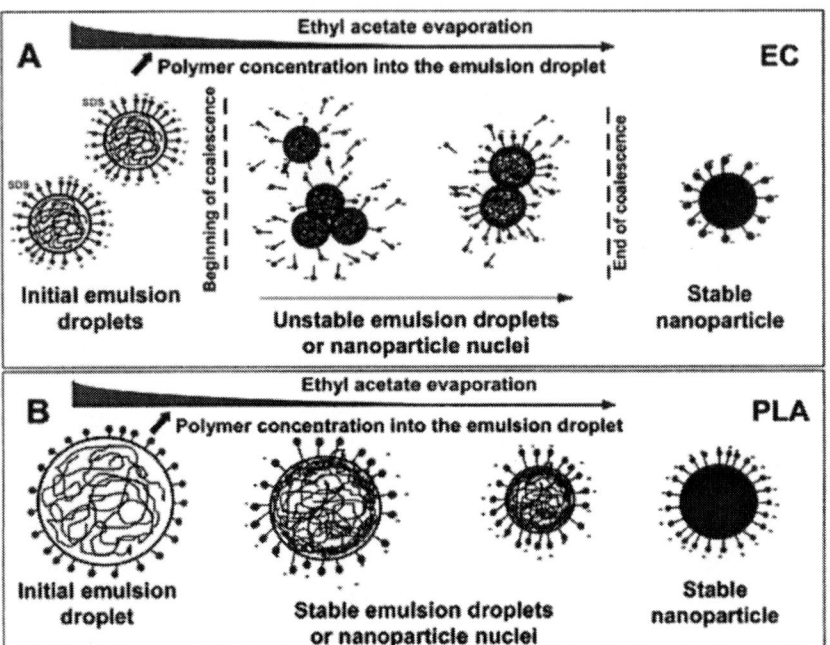

Figure 7 Hypothesis about the mechanisms of formation of the nanoparticles by emulsion solvent evaporation using solutions (**A**) of EC in ethyl acetate and (**B**) of PLA in ethyl acetate. *Abbreviation*: PLA, poly-lactic acid. *Source*: From Ref. 14.

where c_0 is the concentration of polymer in solid state (i.e., density of polymer), c is the concentration of polymer in the organic phase, d_{droplet} is the droplet diameter, and $d_{\text{nanoparticle}}$ is the final nanoparticle diameter. Desgouilles et al. (14) observed that the value of A is about 4 for PLA and 9–11 for ethyl cellulose of 55,600 molecular weight, and 20–32 for ethyl cellulose of 98,000 molecular weight.

If a smaller nanoparticle is the objective, then a fast solvent removal process is required. The longer it takes for the solvent to leave, the longer the duration in which the droplets/particles are sticky, giving them a higher probability of coalescing. The concept of utilizing supercritical carbon dioxide to remove the solvent can provide a more controllable and faster method to cause particle hardening (15).

RESIDUAL SOLVENT AND EMULSIFIER

Residual solvent in pharmaceutical preparations, including nanoparticles, is a growing concern because of the toxicological risks associated with such residuals. If proper evaporation and lyphilization is not carried out, then the final nanoparticle may retain the solvent. The limit for the residual solvent is outlined in USP XXIII (16). For example, the limit for methylene chloride is 500 ppm and that for chloroform is 50 ppm. Chattopadhyay et al. (15) utilized supercritical carbon dioxide to extract the solvent from emulsion. Supercritical CO_2 can extract the solvents with a high efficiency in a small contact time, mainly owing to about 100-fold better diffusivity in supercritical fluid than in liquids. The process provides final particles that are very low in the residual solvent. In addition, the particle hardening is expected to be faster.

Residual emulsifiers are also a matter of concern with respect to the toxicological risks, especially for injectable formulations. The most common emulsifier is PVA for PLGA based nanoparticles. A fraction of PVA remains associated with the nanoparticles despite repeated washing because PVA forms an interconnected network with the polymer at the interface (17). Both, the concentration of PVA in the aqueous phase used and the type of organic solvent influence the amount of residual PVA (chap. 6). Other than toxicological concerns, the interfacial PVA influences particle size, zeta potential, polydispersity index, surface hydrophobicity, and drug loading. For example, albumin-loaded nanoparticles with a higher amount of residual PVA had a relatively lower cellular uptake despite their smaller particle size, owing to the higher hydrophilicity (17). Zambaux et al. (18) observed that about 0.1 molecule of PVA adsorbs onto each square nanometer surface of PLA nanoparticles. This amounts to the adsorption of about 20,000 PVA molecules per nanoparticle of about 225 nm.

Both residual solvent and emulsifier can be reduced by cross-flow microfiltration (19). For example, successful elimination of emulsifier (PVA) and solvent (ethyl acetate) was achieved with a concentration step of 40 minutes followed by a diafiltration step of two hours, but the membrane fouling was observed. Cross-flow microfiltration is particularly attractive for the processing of large volumes of nanoparticulate suspension, as the membrane surface can be easily increased. Other methods such as evaporation under reduced pressure or ultracentrifugation usually only treat small batch volumes. When tested for indomethacin-loaded polycaprolactone nanoparticles, the cross-flow microfiltration technique did not alter the nanoparticle size or the drug loading (19).

PROTEIN STABILIZED NANOPARTICLES

Owing to the concerns of residual emulsifier in the final product, several researchers have utilized albumin protein stabilizer because of its complete compatibility with even the injectable formulations. This process is illustrated in Figure 8 for producing albumin-stabilized paclitaxel nanoparticles (20).

The choice of organic solvent and the extent of homogenization can be used to further tailor the nanoparticle size. A variation of the process is shown in Figure 9, in which the aqueous phase was presaturated with the organic solvent and a small amount of ethanol was added to the organic phase. In this variation, smaller nanoparticles, 140–160 nm, are obtained.

The advantage with nanoparticles smaller than 200 nm is that they can be easily sterilized by filtering with standard 0.22 μm filter. Thus, the whole process can be carried out in a nonsterile environment, and the sterilization can be done just before the lyphilization step.

To form a solid and stable layer of albumin onto drug nanoparticles, the protein needs to be cross-linked (or denatured) onto the particle surface. Typically, albumin cross-linking can be achieved by heat, use of cross-linker such as gluteraldehyde, or high shear. Fortunately, in the emulsification solvent evaporation process high shear is already in use, hence it can also be used for cross-linking protein stabilizers. High-shear cross-linking works for the protein-bearing sulfhydryl or disulfide groups (e.g., albumin). The high-shear

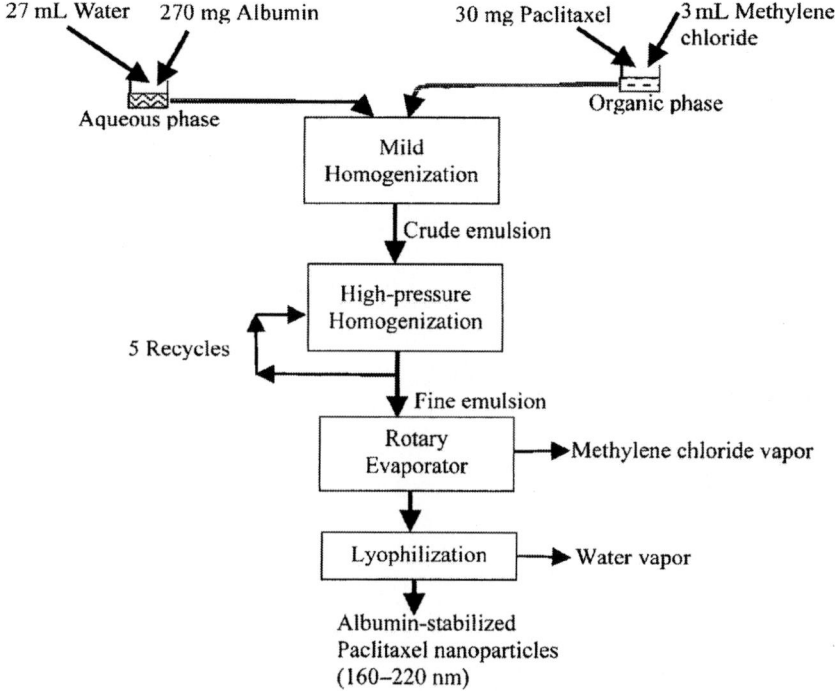

Figure 8   Schematic of the protein-stabilized drug nanoparticle formation.

conditions produce cavitation in the liquid, which causes tremendous local heating and results in the formation of hydroxyl radicals that are capable of cross-linking the polymer, for example, by oxidizing the sulfhydryl residues (and/or disrupting the existing disulfide bonds) to form new, cross-linking disulfide bonds (20–22).

CONCLUSIONS

Polymer- or protein-stabilized drug nanoparticles can be produced by the emulsification solvent evaporation process. With the recent development in the homogenization, very fine emulsions can be created that can yield nanoparticles. The size, zeta potential, hydrophilicity, and drug loading of the

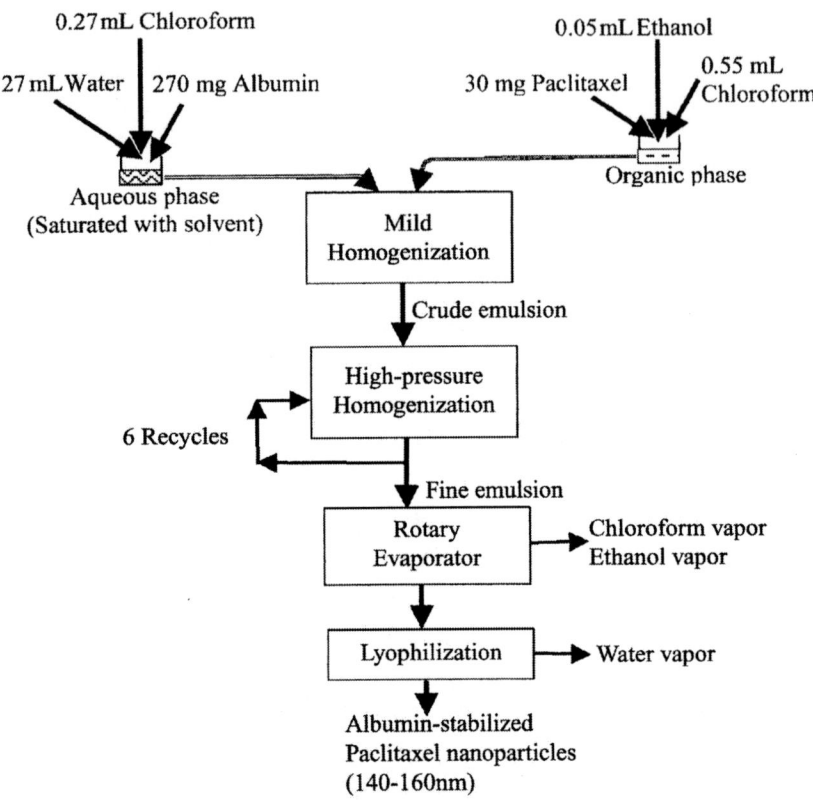

Figure 9 Schematic for the < 200-nm-size protein-stabilized drug nanoparticle formation. Here the aqueous phase is presaturated with the organic solvent, and a small amount of ethanol is added to the organic phase.

nanoparticles can be controlled by various process parameters including the amounts of emulsifier, drug, and polymer, the intensity and duration of homogenization, and the particle hardening profile. Hardening of the polymer particles is achieved by solvent removal, whereas hardening of proteins such as albumin can be done by cross-linking. The single-emulsion method is suitable for hydrophobic drugs, whereas the double-emulsion method is needed for hydrophilic drugs. Because of the relative simplicity of the process, both methods have been widely utilized for a variety of drugs.

REFERENCES

1. O'Donnell PB, McGinity JW. Preparation of microspheres by the solvent evaporation technique. Adv Drug Deliv Rev 1997; 28(1):25–42.

2. Bala I, Hariharan S, Kumar MNVR. PLGA nanoparticles in drug delivery: the state of the art. Crit Rev Ther Drug Carrier Syst 2004; 21(5):387–422.

3. Bibette J, Calderon FL, Poulin P. Emulsions: basic principles. Rep Prog Phys 1999; 62(6):969–1033.

4. Zhao L, Choi P. Molecular dynamics simulation of the coalescence of nanometer-sized water droplets in n-heptane. J Chem Phys 2004; 120(4):1935–1942.

5. Quintanar-Guerrero D, Fessi H, Allemann E, Doelker E. Influence of stabilizing agents and preparative variables on the formation of poly(D,L-lactic acid) nanoparticles by an emulsification-diffusion technique. Int J Pharm 1996; 143(2):133–141.

6. Maa Y-F, Hsu CC. Performance of sonication and microfluidization for liquid-liquid emulsification. Pharm Dev Technol 1999; 4(2):233–240.

7. Mainardes RM, Evangelista RC. PLGA nanoparticles containing praziquantel: effect of formulation variables on size distribution. Int J Pharm 2005; 290(1–2):137–144.

8. Maa YF, Hsu C. Liquid-liquid emulsification by rotor/stator homogenization. J Contr Rel 1996; 38:219–228.

9. Cegnar M, Kos J, Kristl, J. Cystatin incorporated in poly(lactide-co-glycolide) nanoparticles: development and fundamental studies on preservation of its activity. Eur J Pharm Sci 2004; 22(5):357–364.

10. Yoncheva K, Vandervoort J, Ludwig A. Influence of process parameters of high-pressure emulsification method on the properties of pilocarpine-loaded nanoparticles. J Microencapsul 2003; 20(4):449–458.

11. Vandervoort Jo, Yoncheva K, Ludwig A. Influence of the homogenisation procedure on the physicochemical properties of PLGA nanoparticles. Chem Pharm Bull 2004; 52(11): 1273–1279.

12. Dillen K, Vandervoort J, Van den Mooter G, Verheyden L, Ludwig A. Factorial design, physicochemical characterisation and activity of ciprofloxacin-PLGA nanoparticles. Int J Pharm 2004; 275(1–2):171–187.

13. Wang J, Schwendeman SP. Mechanisms of solvent evaporation encapsulation processes: prediction of solvent evaporation rate. J Pharm Sci 1999; 88(10):1090–1099.

14. Desgouilles S, Vauthier C, Bazile D, et al. The design of nanoparticles obtained by solvent evaporation: a comprehensive study. Langmuir 2003; 19(22):9504–9510.

15. Chattopadhyay P, Shekunov BY, Seitzinger JS, Huff RW. Particles from supercritical fluid extraction of emulsion. PCT Int Appl 2004:61 pp. wo2004004862A1.

16. The United States Pharmacopoeia (23rd revision). Organic Volatile Impurities. Rockville, 1995:1747.

17. Sahoo SK, Panyam J, Prabha S, Labhasetwar V. Residual polyvinyl alcohol associated with poly (D,L-lactide-coglycolide) nanoparticles affects their physical properties and cellular uptake. J Contr Rel 2002; 82(1):105–114.

18. Zambaux MF, Bonneaux F, Gref R, et al. Influence of experimental parameters on the characteristics of poly(lactic acid) nanoparticles prepared by a double emulsion method. J Contr Rel 1998; 50(1–3):31–40.

19. Limayem I, Charcosset C, Fessi H. Purification of nanoparticle suspensions by a concentration/diafiltration process. Sep Purif Technol 2004; 38(1):1–9.

20. Desai NP, Tao C, Yang A, et al. Protein stabilized pharmacologically active agents, methods for the preparation thereof and methods for the use thereof. U.S. Patent 6,749,868, Jun 15, 2004.

21. Leucuta SE, Risca R, Daicoviciu D, Porutiu D. Albumin microspheres as a drug delivery system for epirubicin: pharmaceutical, pharmacokinetics and biological aspects. Int J Pharm 1988; 41(3):213–217.

22. Lee TK, Sokoloski TD, Royer GP. Serum albumin beads: an injectable, biodegradable system for the sustained release of drugs. Science 1981; 213(4504):233–235.

5

Physical Characterization of Nanoparticles

ROY J. HASKELL

Pfizer Corporation, Michigan Pharmaceutical Sciences,
Kalamazoo, Michigan, U.S.A.

INTRODUCTION

As discussed elsewhere in this book, the unique qualities and performance of nanoparticles as devices of drug delivery arise directly from their physicochemical properties. Hence, determining such characteristics is essential in achieving a mechanistic understanding of their behavior. A good understanding allows prediction of in vivo performance as well as allowing particle design, formulation development, and process troubleshooting to be carried out in a rational fashion. The following chapter will discuss the means and methods to carry out such determinations on nanoparticles. Many of the tools employed for their characterization are the same as those used for similar analysis of other submicrometer colloids such as micelles, liposomes, and emulsions. Thus, some of the examples are taken from studies in which these species were the focus of study.

Nature of the Analytical Challenge

Before delving into the technical issues, it is worthwhile raising a few questions, consideration of which will help to better define the task ahead. How are nanoparticles different from other analytes such as suspensions or powders? What is it about such objects that suggest different tools will be needed to characterize them? Knowing if and how the analytical challenge is dissimilar from previous experience will assist in making sure the appropriate tools and logic are brought to bear.

Several differences are relevant. The first is certainly the most obvious: size. Clearly, if submicrometer size is a defining characteristic, then accurately quantifying such is important. Many commonly employed methods for determining size will not work in the submicrometer regime and vice versa. For example, the particles are too small for direct imaging using optical microscopy, and some forms of light scattering used for nanoparticles are not suitable for larger objects. In the former case, a reliable tool has become unavailable, and in the latter, methods exist with which practitioners of sizing larger particles may not be familiar. Intermediate scenarios, where conventional methods work but only if applied properly, are of particular concern because one runs the risk of mistaking artifact for reality.

However, size is not brought up for its own sake, i.e., because of the "nano" in nanoparticle. Thus, a second difference emerges: both the behavior and potential uses of such systems vary from those of more conventionally sized particle populations. The resultant low mass of individual nanoparticles means that their kinetic energy is on the same order as the energies involved in interparticle interactions, hence they are observed to behave differently. Their ability to remain suspended under conditions that would lead to the sedimentation of larger particles is an example of this.

Because of their unique properties, nanoparticles are employed in applications uncommon for suspensions, ones for which larger particles would not be used, thus opening a new range of questions. For example, because their small size allows drugs to be delivered via intravenous administration as a solid material, characterizing the upper end of a size distribution becomes important from a size point of view due to safety concerns, i.e., the potential for embolism. Dissolution of the corresponding solid in media other than those typically used to mimic oral conditions is another example.

It should be noted, however, that in some cases the characterization of a nanoparticle system is similar to doing the same with a macroscopic analyte. Determining the state of solid within the particle using thermal or X-ray methods is not much different so long as the possibility of size-induced artifacts is evaluated.

Frame the Question

Any analysis, nano- or otherwise, needs to begin with the end in mind. Why is the analysis being conducted? What is the context of the analysis? For example, in the case of an arrested precipitation process one is looking for the presence of inhomogeneity as particles form from a continuous medium—something from nothing. In the case of comminution, however, one may be looking for the appearance of polydispersity as smaller entities are derived from those that are larger, e.g., at the beginning of the process, or the disappearance of polydispersity due to the residual presence of a few larger particles, e.g., at the end of the process. High-angle light scattering would be a good approach in the former case, whereas particle counting is more appropriate for the latter analysis.

MEASUREMENT OF SIZE

In some cases, usually via precipitation, nanoparticles can be produced with such a high degree of size monodispersity that they become the standard by which distributions of particles in general are measured. More commonly, however, this tends not to be the case, as the effects at play in their formation have a sufficient component of randomness that the result is a nanoparticulate system made from particles of a range of sizes.

As noted above, more than any other the characteristic that defines nanoparticles as such is their small size. Hence, quantifying this value is first on the list of properties to describe. The question, "What is the particle size of this sample?" is deceptively simple and is so for a number of reasons. First, the particle size distribution is defined not only by the size of the average, but by the way in which "average" is defined. Mean, median, and mode are equally valid descriptors. In addition, the population itself can be defined by the number or volume of particles present, and these are only two of the various weighting schemes that can be employed. The choice among these options is best determined by the reason for making the size measurement in the first place (1).

A second issue is the width or shape of the distribution. Is it polydisperse or narrow? Skewed or symmetric? The information content expressed in a complete size distribution usually exceeds greatly that which can be extracted from the available experimental signal. Subtle differences in the experiment can translate to large variations in the result obtained, thus complicating the problem of determining the distribution. The shape of the particle itself is also important as its nature affects the experimental observable directly—a spherical particle will scatter light differently from one that is rectangular, for instance—but it also influences the abscissa of the size distribution, e.g., projected area, Feret's diameter, etc. (1).

Lastly, the results for average and shape of size distributions can depend on how the result was obtained. There is the trivial case of instrumental design, in which data handling and extraction routines, vary among instruments of the same type, i.e., different manufacturers, perhaps. The more significant situation arises when variation results from the differences in physical principles underlying the measurements. For example, a multiangle light scattering experiment relies on the interaction of the photons with the electric field of the particle, whereas dynamic light scattering (DLS) is based on the time-dependent interference pattern generated by particles in motion. So for the same sample, the result reports on the distribution of matter within the particle in the first case, whereas in the latter experiment it is the particles' hydrodynamic nature that is emphasized. The answers will be different; however, both are equally correct. Note also that both are light scattering experiments, so the principles of measurement can vary significantly even among methods that might seem similar to the casual consumer of analytical information.

Sizing methods are frequently classified according to the manner in which they extract information from the sample. In ensemble methods, the collective signal generated by the entire particle population is processed via an appropriate algorithm to produce an estimate of the size distribution. Most spectroscopic methods, such as various forms of light scattering or ultrasonic absorption, are ensemble in nature. The inversion is mathematically ill-defined, thus generating a sensitivity to experimental noise. A common consequence is that different size results that are statistically equivalent can be derived from the same data set. As a result, such methods are not sensitive to small shifts in distribution that may contain valuable information on process or stability. Similarly, results that claim precise measurement of distribution widths, shapes, and number of modes need to be critically evaluated, and confirmed by those of other methods. For these reasons, the differences in instrument and software design can lead to disagreement when comparing results from different manufacturers.

Counting methods, such as microscopy or single-particle optical sensing (SPOS), measure the size of individual particles to compile a histogram reflecting the overall distribution. The effect leading to detection of each particle, e.g., scattering, obscuration, etc., varies among methods. As a result, not all counting techniques should be considered equivalent. These methods are quite sensitive to small changes in the size characteristics of a particle population, but for the same reason are prone to statistical error unless sufficient numbers of particles are counted. This is a particular concern because the probability of detecting one of the few large particles that may be present in a sample can greatly affect the determined volume- or surface area–weighted result.

Separation methods, such as field-flow fractionation or filtration, generate a result by physically ordering the particle population according to size. These methods are valuable because most of them provide an accounting of all the material present in the sample, affording some level of assurance that nothing has been missed. The best detection method is one based on concentration, but an ensemble technique is usually substituted. In the latter case, the effects noted above are less of a concern because the separation presents the detector with more monodisperse "samples," which are less problematic.

Table 1 Methods for Assessing the Properties of Nanoparticles

Property	Relevant analytical method(s)	References
Presence	Dark field optical microscopy	2
Size	Dynamic light scattering, Static light scattering, Ultrasonic spectroscopy, Turbidimetry, NMR, Single particle optical sensing, FFF Hydrodynamic fractionation, Filtration	3–5,6–8,9–41
Morphology	TEM, SEM, Atomic force microscopy	18,19,22–27,42
Surface charge	Electrophoretic light scattering, U-tube electrophoresis, Electrostatic-FFF	7,20,21,43–46
Surface hydrophobicity	Hydrophobic interaction chromatography	20,23,48
Surface adsorbates	Electrophoresis	23,49
Density	Isopycnic centrifugation, sedimentation-FFF	23,50
Interior structure	Freeze-fracture SEM, DSC, X-ray diffraction, NMR	21,40,42,61–63

Abbreviations: DSC, differential scanning calorimetry; FFF, field fractionation; NMR, nuclear magnetic resonance; SEM, scanning electron microscopy; TEM, transmission electron microscopy.

AVAILABLE METHODS

There are a large number of methods available to characterize nanoparticles. Some approaches, such as DLS for size, or nuclear magnetic resonance (NMR) for diffusivity, are unique to the analysis of nanoparticles compared to that of more macroscopic species. Other techniques, such as differential scanning calorimetry (DSC) or X-ray diffraction, are not significantly affected by the submicrometer particle size. Rather, in these cases, it is the interpretation of the results in the context of the problem at hand that renders the corresponding method relevant. The following section will describe various techniques of analysis appropriate to nanoparticles along with references that serve as background or example. Table 1 breaks the classification down orthogonally by summarizing the same information according to the likely properties of interest.

Dynamic Light Scattering (DLS)

DLS, also known as photon correlation spectroscopy (PCS) or quasi-elastic light scattering (QELS) records the variation in the intensity of scattered light on the microsecond time scale (3,4). This variation results from interference of light scattered by individual particles under the influence of Brownian motion, and is quantified by compilation of an autocorrelation function. This function is fit to an exponential, or some combination or modification thereof, with the corresponding decay constant(s) being related to the diffusion coefficient(s). Using standard assumptions of spherical size, low concentration, and known viscosity of the suspending medium, particle size is calculated from this coefficient. The advantages of the method are the speed of analysis, lack of required calibration, and sensitivity to submicrometer particles. Drawbacks include the necessity of significant dilution to avoid artifacts, the need for cleanliness in sample preparation, the mathematical instability of the procedure used to extract decay constants, and the possible influence of interparticle interactions. DLS is a stand-by method for those working in the area of nanoparticles (4,14,15,17–21,23,31,40,41,54,55).

Static Light Scattering/Fraunhofer Diffraction

Static light scattering (SLS) is an ensemble method in which the pattern of light scattered from a solution of particles is collected and fit to fundamental electromagnetic equations in which size is the primary variable (4,5). The method is fast and rugged, but requires more cleanliness than DLS, and advance knowledge of the particles' optical qualities.

Fraunhofer (light, laser) diffraction is frequently employed as a sizing method for nanoparticles, and when appropriately applied, it is not unreasonable to do so with certain caveats. As size drops into the submicrometer regime the differences in the scattering pattern occur primarily at high angles, so collecting such data becomes critical—an ability that varies widely among commercial instruments. The approximations implemented in Fraunhofer theory are acceptable for particles of diameter 2 μm and higher, but full Mie theory is required for smaller sizes (5).

If Fraunhofer calculations are used to extract results from scattered light originating from a population of particles less than 2 μm in diameter, then significant errors will result such as the artifactual presence of particle populations (53). However, using full Mie theory requires knowledge of the values for both the real and the imaginary (absorptive) components of the particle refractive index, the choices for which can profoundly affect the results (54). Values for the real component can be obtained via the Becke method in which fringe patterns arising from the placement of a test particle in a series of oils of varying refractive index are observed in a microscope. This method requires that large enough particles are present to make the measurement or one at least has a macroscopic sample of the material(s) from which the nanoparticles are made (55). A clever approach has been demonstrated by Saveyn et al. (56) whereupon the refractive index of the compound dissolved in a variety of solvents is extrapolated to 100% solute. The method was shown to be simple and straightforward, though it does require solubility in media of widely different polarities.

If the above considerations are taken into account diffraction equipment can be applied to nanoparticle characterization. However the choices, and the reasons for the same, of particle refractive index should be clearly reported. In addition, a complete analysis should include an estimate of the extent to which the sizing results are affected by errors in the refractive index values employed.

Acoustic Methods

Another ensemble approach, acoustic spectroscopy, measures the attenuation of sound waves as a means of determining size through the fitting of physically relevant equations (6). In addition, the oscillating electric field generated by the movement of charged particles under the influence of acoustic energy can be detected to provide information on surface charge. This is termed electroacoustic spectroscopy and can also be reversed so that sound waves generated by the oscillatory motion of charged particles in a varying electric field is the observable (7). Both methods are particularly valuable in that they work with concentrated suspensions and thus can be used to characterize dilution-sensitive systems or for process monitoring.

Turbidimetry

For nonabsorbing particles, turbidity is the complement to light scattering because it represents the amount of incident radiation not reaching a detector, that is, light lost to scattering. Hence the turbidity spectrum is also described by Mie theory and thus can be used to determine particle size as long as the data are normalized for concentration (8). This approach requires tiny amounts of sample and can be easily executed using a spectrophotometer. However, it suffers the ills common to all ensemble methods and the lack of commercial implementation requires the investigator to carry out the appropriate calculations on thier own (51,52,57).

Nuclear Magnetic Resonance

Nuclear magnetic resonance (NMR) can be used to determine both the size and the qualitative nature of nanoparticles. The selectivity afforded by chemical shift complements the sensitivity to molecular mobility to provide information on the physicochemical status of components within the nanoparticle (9).

For example, the mobility of Miglyol 812 within solid lipid nano-particles confirmed the liquid-like nature of the interior, though it was more limited than the same oil in an o/w emulsion (10). Pulsed field gradient methods allow diffusivity of the entire particle to be quantified and compared to produce 2-D, diffusion-ordered plots in which colloidal behavior and chemical speciation are leveraged simultaneously (11). In one case, the diffusion coefficient is used as a surrogate for size of the nanoparticle with results that compare well to separation and DLS, though only NMR could simultaneously detect micellar precursors (12).

Single-Particle Optical Sensing (SPOS)

A particle counting method, SPOS, which is also known as optical particle counting, involves recording the obscuration or scattering of a beam of light that results from the passage of individual particles through a sensor (13). Signal magnitude is translated to the size of the particle via use of a previously determined calibration curve using standards approximating the sample in terms of shape and optical properties. The direct result is a number-based size distribution. SPOS cannot distinguish between a single primary particle and an aggregate (few methods can), and is subject to error at a number of concentrations above which there is a significant chance of multiple particles being present simultaneously in the light beam.

Particles of diameter less than 1 μm are largely unde-tected, thus making SPOS very useful in the determination of the few large particles in a population that may represent a safety concern, indicate a problem in production, or be har-bingers of instability. Count rates of 8000 particles/sec or more are typical, thus thousands to millions of particles are obser-ved in an experiment. Hence, detecting the few large particles present in a distribution is more likely than is the case with microscopy. Drawbacks include the possible dissolution of ana-lyte during analysis, the large dilution required, and the need for low backgrounds. Detecting an interruption in the flow of electrical current through a solution is an analogous method termed electrozone or Coulter counting. This technique sees little recent use because of its need for colloid destabilizing electrolytes and the more complicated instrumentation required, though a novel approach that also determines electrophoretic mobility has been recently reported (14,15).

A great benefit afforded by SPOS is the ability to quantify the large particle population (16). The total volume detected during the experiment can be calculated from the number distribution by assuming a shape, e.g., spherical, and integrating under the resulting volume distribution curve. When compared to the concentration of the suspension, what results is a ratio that describes the fraction of material present as detectable, i.e., large, particles. Hence, a suspension in which all of the particles are larger than the size detection limit of the sensor would yield a recovery of 100%. A recovery of near 0% suggests that, subject to the statistical assurance associated with the number of particles counted, the mass of the distribution resides primarily as particles of less than that size. By comparison, ensemble methods do not measure the absolute amount of material present; only the relative contribution of sizes is determined. Hence, integrating under the corresponding size distribution always sums to 100% regardless of size.

Optical Microscopy

Most nanoparticles are below the resolution limit (ca. 0.5 µm) of direct optical imaging, though microscopy is still useful to get an estimate of size and crystallinity of starting materials, as might be desirable in the instance of comminution or homogenization processing, or other larger particles (17). However, the dark field techniques, in which particles are observed indirectly as bright spots on a dark background because of their scattering under oblique illumination, is extremely valuable in assessing the presence and numbers of nanoparticles (2). Users should be on the lookout for segregation of particles resulting from sample preparation.

Electron Microscopy

Scanning and transmission electron microscopy, SEM and TEM, respectively, provide a way to directly observe nanoparticles, with the former method being better for morphological examination (18–21,42). TEM has a smaller size limit of detection, is a good validation for other methods, and affords structural information via electron diffraction, but staining is usually required, and one must be cognizant of the statistically small sample size and the effect that vacuum can have on the particles. Very detailed images data can result from freeze-fracture approaches in which a cast is made of the original sample (22,23). Sample corruption resulting from the extensive sample preparation is always a possibility, though lower vacuum (environmental- or E-SEM) instrumentation reduces this manipulation, albeit at the loss of some resolution (24).

Atomic Force Microscopy (AFM)

In this technique, a probe tip with atomic scale sharpness is rastered across a sample to produce a topological map based on the forces at play between the tip and the surface. The probe can be dragged across the sample (contact mode), or allowed to hover just above (noncontact mode), with the exact nature of the particular force employed serving to distinguish among the subtechniques. That ultrahigh resolution is obtainable with this approach, which along with the ability to map a sample according to properties in addition to size, e.g., colloidal attraction or resistance to deformation, makes AFM a valuable tool. However, size and shape has been the most common application to date (25,26). The need to raster the probe renders the method very time-consuming and the size of the sample actually observed is small. Nanoparticles are typically presented as an evaporated suspension on a smooth silicon or mica surface, though not without the possibility of deformation (27). Application of various forms of AFM to nanoparticle characterization represents an area of active research.

Other Forms of Microscopy

The size resolution of TEM can be leveraged for morphological studies by rastering the sample across a well-defined electron beam (STEM), and high resolution and some chemical information can be extracted if X-rays are substituted for electrons (STXM) (28). While these methods have not been applied to pharmaceutically relevant nanoparticles, studies of related samples suggest that they may be worth investigating for this purpose (29). The optical analog of AFM is near-field microscopy, which affords nanoscale resolution, and the use of light allows for simultaneous chemical imaging via Raman spectroscopy (30). Confocal microscopy has proved valuable, being used frequently in the study of nanoparticle uptake in biological tissues such as eye, brain, and skin (31–33).

Filtration

A simple, yet effective, approach of determining particle size is filtration, in which the concentration of a suspension is determined before and after passage through filter membranes of various sizes. Subject to the caveats of nonspecific adsorption, aggregation, and particle shape effects, the results give a semiquantitative assessment of the particle size distribution that is not based on instrumentation and algorithms. The practitioner should make sure that if more than one pore size is used, all filters are made of the same material and the same protocol, i.e., the amount of material passed through the filter, is maintained throughout.

Field-Flow Fractionation

Particles are driven toward either the top or the bottom of a thin channel within which eluant is continuously flowing in a direction perpendicular to the driving force. Liquid flow of the eluant is parabolic so that particles spending more time toward the center of the channel where the flow lines are faster emerge first. The nature of the perpendicular force defines the type of field-flow fractionation (FFF) and thus the particle property on which separation occurs: sedimentation (buoyancy, size), flow (hydrodynamic size), electrostatic (charge), or thermal (diffusion) (18,34). FFF necessitates more complicated methodology and the data interpretation is less straightforward than chromatography, but it can provide a wider range of information and can also be used as a preparative method for nanoparticles (18,35).

Hydrodynamic Chromatography

In a sufficiently narrow channel of parabolic flow, particles of different size will on average experience different flow lines because of their differential ability to approach the channel wall (36). The particles will separate based on that property, with those that are smaller eluting later just as they would in flow-FFF. Indeed, hydrodynamic chromatography (HDC) can be thought of as flow-FFF with the narrowness of the channel substituting for the cross flow. Thin capillaries serve as the channels, which can also be created by the interstitial spaces within a packed column (37). The former approach is also known as capillary hydrodynamic fractionation and has been further miniaturized (38,39). The results are highly sensitive to the surfactants employed in the analysis. Size exclusion chromatography is little used for analytical size separation of colloids, though there are examples of its application (51).

Hydrophobic Interaction Chromatography

In this method the analyte is first adsorbed onto a chromatographic stationary phase using a high concentration of an antichaotropic salt (48). Elution occurs using a gradient in which the salt concentration is decreased, so that those materials eluting first are the least hydrophobic because the salt concentration did not need to be decreased much before the analyte desorbed. Originally developed for proteins, hydrophobic interaction chromatography has been pressed into service as a means of characterizing the hydrophobicity of nanoparticle surfaces, a property influenced by the choice of surfactant and/or polymer and also a key parameter in determining their in vivo fate (20,23).

Electrophoresis

The body's response to the introduction of nanoparticles into circulation is such that within a short period of time their surface is festooned with lipoproteins and related species (58). This process will determine the clearance and biodistribution of the colloid, so evaluating the exact nature of the surface coverage is required to achieve a useful understanding. The small size of nanoparticles allows their electrophoretic behavior to be observed using bioanalytical tools such as isoelectric focusing and 2-D polyacrylamide gel electrophoresis (PAGE) (23,49). As with any ex vivo approach, the investigator needs to take into account the effect that sample preparation may have on the experimental observations. Similar information has been derived by electrophoresis of serum proteins desorbed from incubated nanoparticles (59).

Isopycnic Centrifugation

Another bioanalytical method applied to nanoparticles is centrifugation of analyte using a sucrose gradient as the suspending media. Under the influence of Stokes' laws, sedimenting particles will settle until they reach a point where their density matches that of the gradient. This self-focusing separation allows nanoparticle density to be determined, which along with particle size and bulk substituent concentration can in turn be used to calculate a number concentration (23,50). Conventional analytical centrifugation has been employed as well (60). The results can also be used to extract size, rather than buoyancy, information directly from sedimentation FFF.

Zeta Potential

Zeta potential is used as a surrogate for surface change, and is often measured by observing the oscillations in signal that result from light scattered by particles located in an electric field, though there are other approaches (43,44). There are a number of instrumental configurations by which this is achieved, mostly using a Doppler shift, and the user should familiarize themselves with the particular approach implemented in their equipment. Instrumentation concerns aside, the need for dilution begs the question of what is an appropriate diluent, because its choice can profoundly influence the surface chemistry and thus the results. One approach is to use a particle-free supernatant to dilute the sample. This will not account for concentration effects, however, and obtaining such a diluent is nontrivial as the particle size drops. Electroacoustic methods should in principal eliminate or reduce the need for dilution and its inevitable consequences (7). Nonpolar media and the combination of low mobility with high ionic strength are also problematic; however, phase analysis light scattering, a newer method in which a phase delay shift rather than a frequency shift is observed, addresses these issues (45).

X-Ray Diffraction (Power X-ray Diffraction, Small-Angle Neutron Scattering, Small-Angle X-ray Scattering, Electron)

The geometric scattering of radiation from crystal planes within a solid allow the presence or absence of the former to be determined thus permitting the degree of crystallinity to be assessed (21). In one example, the crystallization of interior lipids could be tracked (40). Application of the method is little different from that for bulk powders, though broadening of the diffraction pattern's peaks is observed for particles less than 100 nm in diameter. For nanoparticles, order on the smaller scale can be investigated by reducing the wavelength and angle of incident radiation. Using electron or neutron beams allows reduction of the former parameter due to the shorter DeBroglie wavelengths of such particles (61).

Differential Scanning Calorimetry (DSC)

Another method that is a little different from its implementation with bulk materials, DSC can be used to determine the nature and speciation of crystallinity within nanoparticles through the measurement of glass and melting point temperatures and their associated enthalpies (62,63). A complement to X-ray diffraction, this method is regularly used to determine the extent to which multiple phases exist in the interior or to which the various constituents, including the drug, interact (21,42).

Dissolution Concerns

In some cases nanoparticles are formed to increase the dissolution rate because of the high surface area they afford. This introduces the possibility that particles are dissolving during the analysis. This problem is general in nature and should be carefully considered in any measurement especially when dilution, sometimes significant in extent, is a requisite of the analysis. While an obvious concern when size is to be determined, such dissolution can lead to skewed results in any measurement because the analyte content is not stable. Extrapolating results to initial conditions or using media in which the particles are insoluble are ways of dealing with this problem.

IN VITRO RELEASE

The solubilization of active components from the individual nanoparticles is of obvious interest. This process can involve release of compound from a polymer or lipid matrix, or dissolution of the entire particle. In either case, separation of the ultrasmall particles from the release media is critical so that the nanoparticles are not mistaken for solubilized drug. In the latter case, the high rate of dissolution is frequently an additional complicating factor.

In typical experiments, it is the appearance of solubilized material that signifies that dissolution is taking place. Using conventional filtration to remove undissolved material for in situ experiments presents serious challenges. The nanoparticles can easily pass through most filter membranes typically used for this purpose, if not at the beginning of the experiment, then at the end, when the particle size may have dropped sufficiently. Small filter pore sizes—as low as 0.02 μm—are commercially available, but can be plugged easily. The separation issue can be avoided by using a method, such as polarography, where only solubilized material is detectable (64). In this way, the need for separation is obviated.

Use of dialysis membranes and diafiltration is an option because they are less prone to blockage and the pore size is very small. The nanoparticles can be placed within a dialysis sac and samples taken from the large receiving medium (22). Alternatively, the reverse approach can be used with the nanoparticles dispersed throughout the larger volume and the receiving media located within the sac (21). Diffusion cells have also been used (41). Separation of particles can also be effected by centrifugation, or avoided implicitly by using two immiscible phases with one containing the nanoparticles and the other serving as the receiving medium (19,20).

When nanoparticles are used to increase the dissolution rate, a significant drawback to these approaches is the time it takes for the dissolved material to diffuse across a membrane or boundary. While this transfer function can be determined experimentally, the associated time constant can be on the scale of tens of minutes, if not hours. Such a long lag precludes the deconvolution of the drug release rate from the experimental data when the dissolution occurs within a few minutes or less.

Rather than detecting drug as it appears in a solubilized form, dissolution information can also be derived by observing the disappearance of the undissolved form, i.e., loss of the nanoparticles themselves. Spectroscopic methods such as light scattering or turbidity are good means of making such observations, and are useful because the corresponding measurements are essentially instantaneous in time, thus eliminating the deconvolution problem. Indeed, the limitation on measurable dissolution rate then arises from issues such as mixing times. Deliberately using nonsink conditions is a way of slowing down the process to avoid these problems.

Figure 1A shows results from the author's laboratory in which the disappearance of 300 nm particles of celecoxib is detected, as a function of increasing numbers of particles, via intensity light scattering from a stirred vessel of water using the apparatus described in Figure 3. Rapid dissolution is observed and shown to be occurring under near-sink conditions at the lowest concentration of particles. The behavior of these particles with that of 2 μm particles is presented in Figure 1B. Particularly at longer times, the difference in slopes of the curves is easily seen, though the curves at shorter times overlap with each other. This similarity at short times is due to the presence of smaller particles in the 2 μm sample. Clearly, the time scale over which this experiment was conducted would be difficult to match using dialysis methods.

(A) Dissolution of Celecoxib Measured by Light Scattering

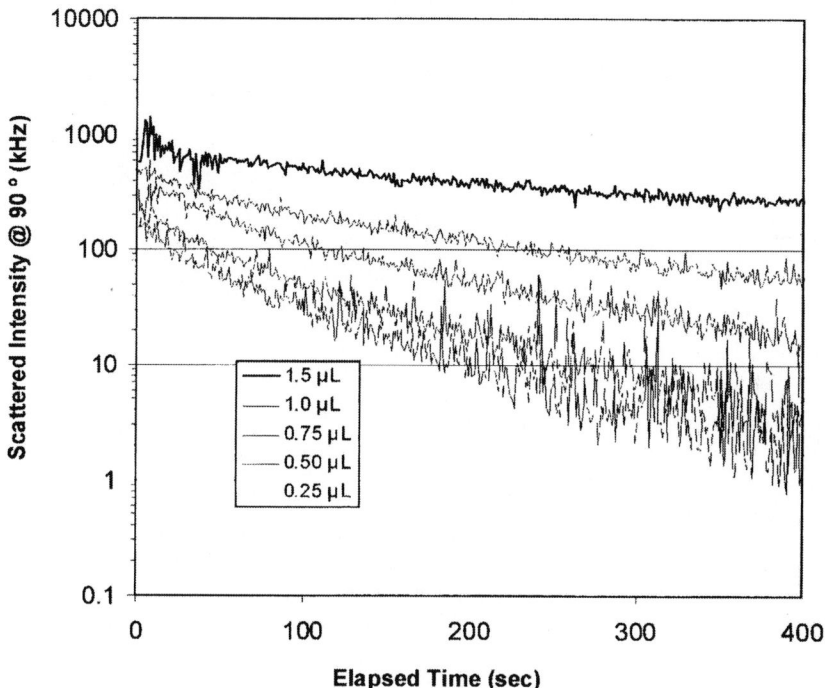

Figure 1 Light observed to scatter at 90-degrees from 300 nm particles of celecoxib stirring in a water-only dissolution medium at ambient temperature using equipment described in Figure 3. The absence of surfactant necessitated the addition of nanoliter volumes of the original suspension in order to maintain sink conditions. **(A)** Five repetitions of the experiment were conducted in which the 2 mg/g nanosuspension was added to 3 mL of water, with a larger aliquot of suspension being used each time. Retardation of the dissolution rate is observed for aliquots greater than 0.50 μL. **(B)** Comparison of 300 nm and 2 μm particles.

EXAMPLE: PARTICLE SIZE

Size is the defining characteristic of nanoparticles even though other properties may be more significant in a given situation. There is great benefit in using multiple analytical methods to characterize size and other properties. The

(B)   **Effect of Particle Size on Dissolution**

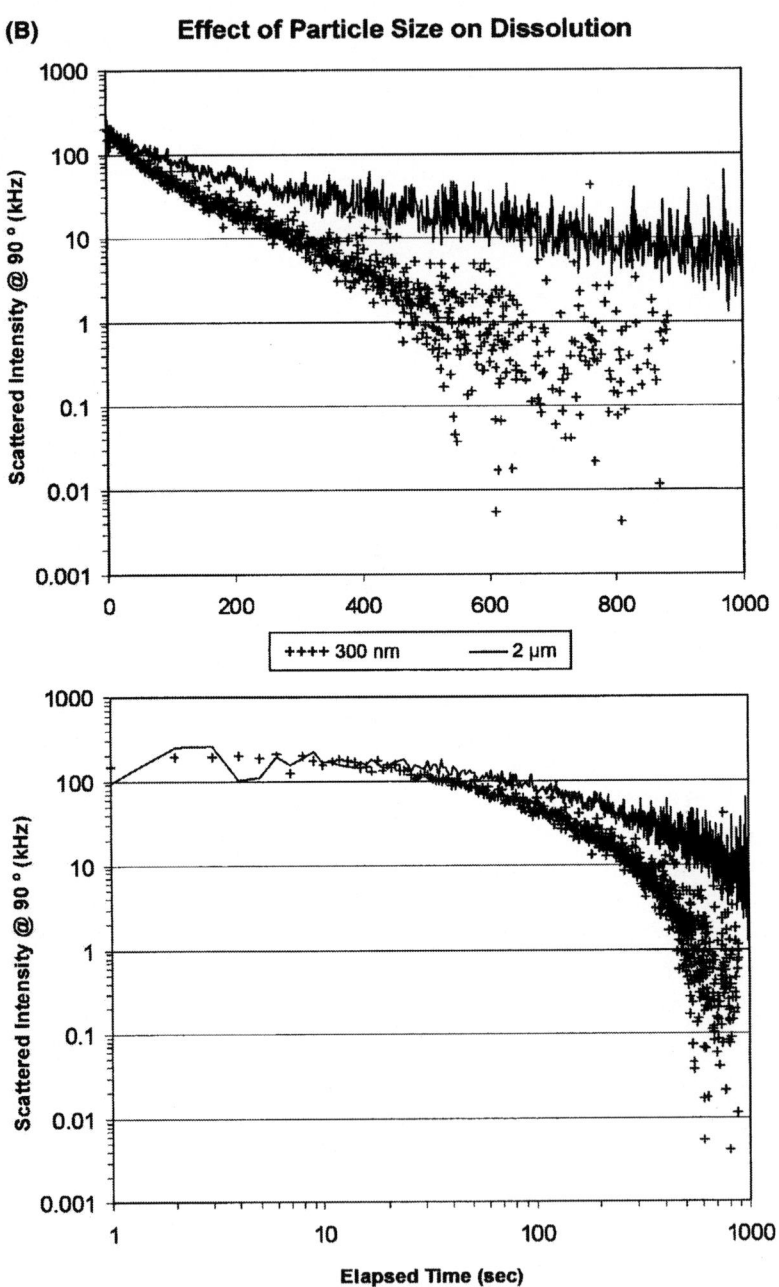

Figure 1   *(Continued)*

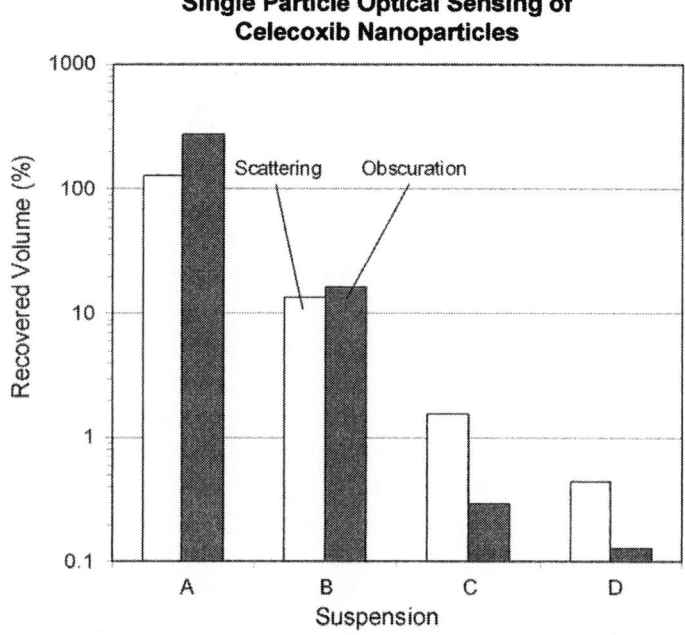

Figure 2 Recovered volume for celecoxib suspensions as detected by both obscuration (Accusizer 770; Particle Sizing Systems; Santa Barbara, California, U.S.A.) and scattering (LE-400–0.5; PSS) SPOS. Suspension aliquots were quantitatively diluted such that the total initial count rate was between 2000 and 8000 counts per second, with a flow rate of 60 mL/min of the diluted suspension being measured for one minute. *Abbreviation*: SPOS, single particle optical sensing.

following size analysis of drug nanoparticles is presented as an example of this approach.

Single-Particle Optical Sensing

Three different aqueous suspensions of celecoxib nanoparticles of decreasing size (Fig. 2B–D) were produced via a laboratory-scale process as a means of evaluating the effect of particle size on the extent and onset rate of oral drug absorption for this compound (65). As noted above, the integrated area under counting-derived size distribution curves is a semiquantitative estimate of the absolute amount of particulate mass detected by the sensor (16). Figure 2 shows this value, as determined by SPOS, for these three suspensions in comparison to that of the initial material, A. As expected, the volume recovered of the unprocessed material approximates 100% because most of the particles are large enough to be detected. Given the optical and shape dissimilarity between the nanoparticles and the monodisperse polystyrene latex standards used as calibrants, the recovery value is reasonable. There is a clear progression to the smallest suspension, whereupon the volume recovered drops from this value to significantly less than 1%, which is clear evidence for the reduction in particle size. A key observation is that even in the first nanosuspension, B, 90% of the particles are less than about 1 μm. Both scattering and obscuration sensors were employed in the analysis, with the detection limits of 0.5–0.7 μm and 1–2 μm, respectively. It is interesting to note that the results for the two approaches are in reasonable agreement until most of the particles drop below the sensitivity limit of the latter detector. At this point the ability of the scattering sensor to detect smaller particles is manifested as a higher value for recovered volume. Note also that there is little difference seen between suspensions C and D suggesting that any differences between the two exist in the submicrometer domain.

Dynamic Light Scattering

Figure 3 presents DLS results for water-diluted samples of the three processed suspensions at low and high scattering angles. The large particles present in suspension A, precluded the corresponding sample from analysis. As noted with SPOS, there is a clear progression of decreasing size, though now the submicrometer particles are observed directly rather than having their nature inferred by the absence of particles. For all samples observed, the size is smaller when the higher scattering angle is employed, which is a consequence of the preferential forward scattering of light from larger particles. This is a good indication of polydispersity in the sample—one that is based directly on the data and not on the algorithm employed in deconvolution of the autocorrelation function.

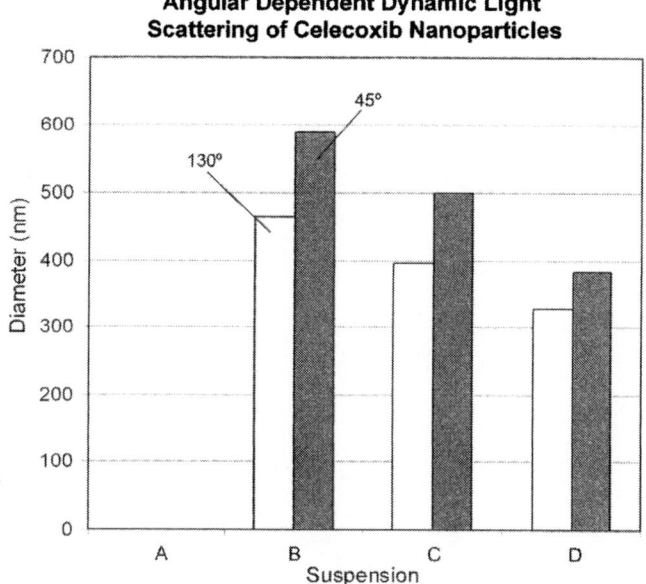

Figure 3 Dynamic light scattering was collected using a goniometer (BI-200 SM; Brookhaven Instruments; Holtsville, New York, U.S.A.) and correlator card (BI-9000AT; BI). Sample time of the correlator was adjusted to account for the change in scattering angle, θ, by normalizing to $\sin^2(\theta)/2$. Particle diameter was calculated from the autocorrelation function via the method of Cumulants, with a third-order fit being used. Samples were diluted in 4 mL of 0.02 μm-filtered water. *Source*: From Ref. 66.

Figure 4 shows DLS, using a fiber optic probe, from the four suspensions for a number of different dilutions in water. Several items are of note. First is the relative insensitivity of the observed size to the extent of dilution, especially at dilutions of 1:20 and greater, suggesting that multiple scattering is not occurring. Second, the sizes of particles in suspensions B–D do not vary significantly. The geometry of the optical probe is such that only light scattered near 180° is observed, so the smallest particles make the largest contribution to the signal. To maintain consistency with the previous data, it is concluded that the process employed leads to particles of a given minimum size near 250 nm.

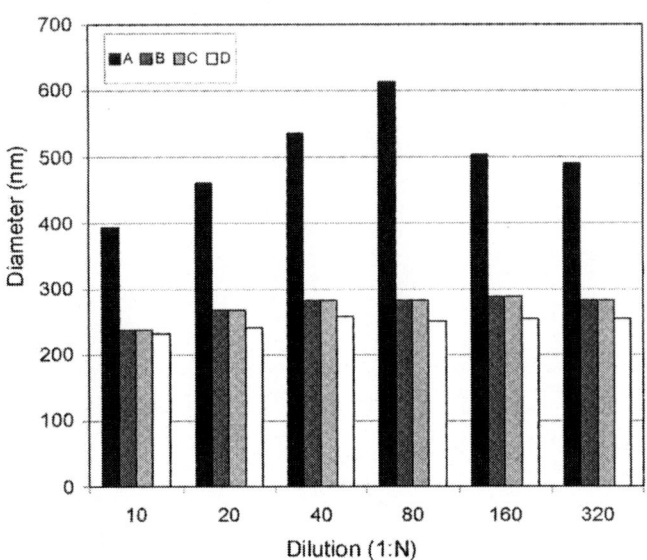

**Fiber Optical Dynamic Light
Scattering of Celecoxib Nanoparticles**

Figure 4 Scattered light was collected using a probe similar to that described by Dhadwal et al. The probe consisted of two adjacent, collinear, single mode optical fibers immobilized within a small length of stainless steel high performance liquid chromatography (HPLC) tubing. Laser light of 514.5 nm was launched into the free end of one fiber using a microscope objective, and backscattered light was delivered to the photomultiplier tube using the free end of the other fiber. Other details of the DLS experiment were as Figure 3. Samples were serially diluted into 0.02 µm-filtered water. *Abbreviation*: DLS, dynamic light scattering. *Source*: From Ref. 67.

Additional processing may add particle population to this size, but it does not lower this size. Finally, it is clear that there are some submicrometer particles (ca. 500 nm) in the starting material, but there are few of these smallest particles present at the outset.

Static Light Scattering

The intensity of scattered light collected with the probe from these same series of diluted samples is presented in Figure 5.

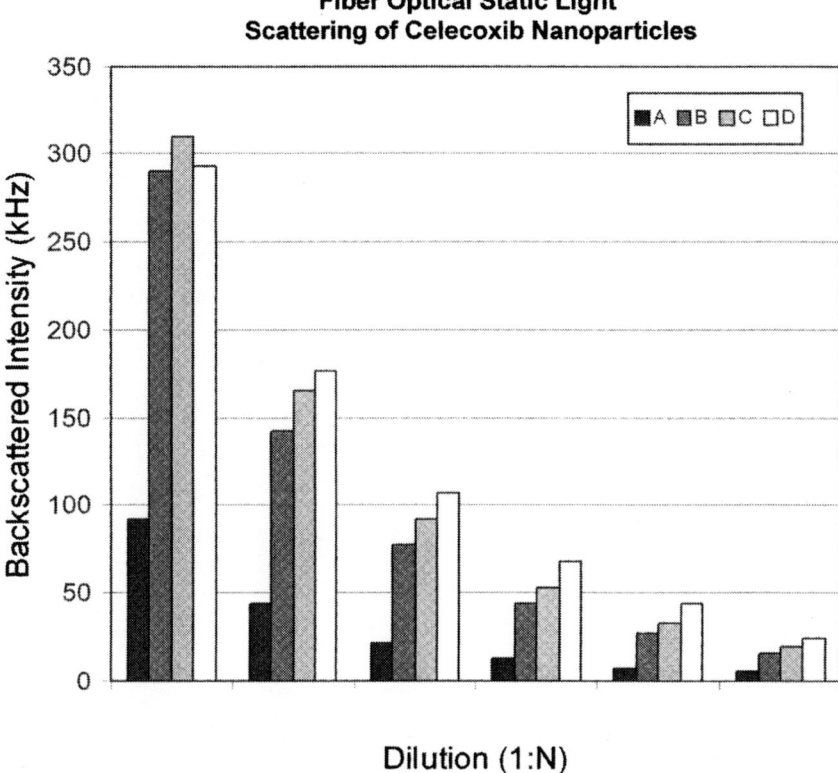

Figure 5 Scattered light intensity corresponding to the data presented in Figure 4.

At intermediate dilutions, the scattered intensity increases for the smaller-sized suspensions, $D > C > B$, thus confirming the above suggestion that additional processing increases the population of the smallest particles, and that there is a difference in this regard among the samples. Consistent with the larger particle size in A, the scattered intensity for this suspension is the lowest because there are few, if any, efficient backscatterers present. The loss of ordering in intensity at the lowest dilution (1:10) suggests the existence of multiple scattering thus accounting for the slightly smaller DLS-derived size for those samples (Fig. 4).

Turbidimetry

Figure 6 shows the size distributions resulting from a fit of the turbidity spectrum using spectra of polystyrene standards collected with the same spectrophotometer as basis functions. Given the errors implicit in comparing spherical latex standards to nonspherical drug nanoparticles, these results should only be used for qualitative purposes. Nonetheless, the conclusions drawn from this data are consistent with those above, and it appears that the size difference between suspensions A and B is greater than that between B and C or C and D.

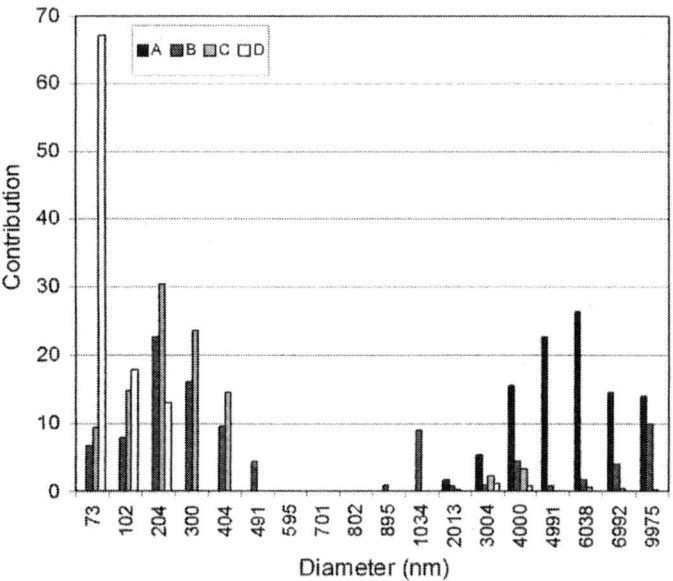

Figure 6 Samples were diluted in water so that a 0.1 cm optical cell produced an "absorbance" of less than 0.5 using a diode-array spectrophotometer (8150A; Hewlett-Packard; Wilmington, Delaware, U.S.A.). Over the wavelength 400–800 nm interval, spectra were fit to a linear combination of spectra collected of individual polystyrene particle standards (NanoSphere®; Duke Scientific; Palo Alto, California, U.S.A.) collected previously. The coefficients of the fit are reported as an estimate of the size distribution.

Microscopy

Finally, scanning electron microscopy (SEM) and polarized optical microscopy are employed with the results presented in Figure 7. The optical micrographs provide a reassuring visual confirmation of the instrumental results, and the SEM validates the light scattering data as well as providing information on the morphology of the processed particles.

Particle Crystallinity

Any optical birefringence that the smallest particles may display is rendered unobservable by polarized microscopy because of their small size. As a result, powder X-ray

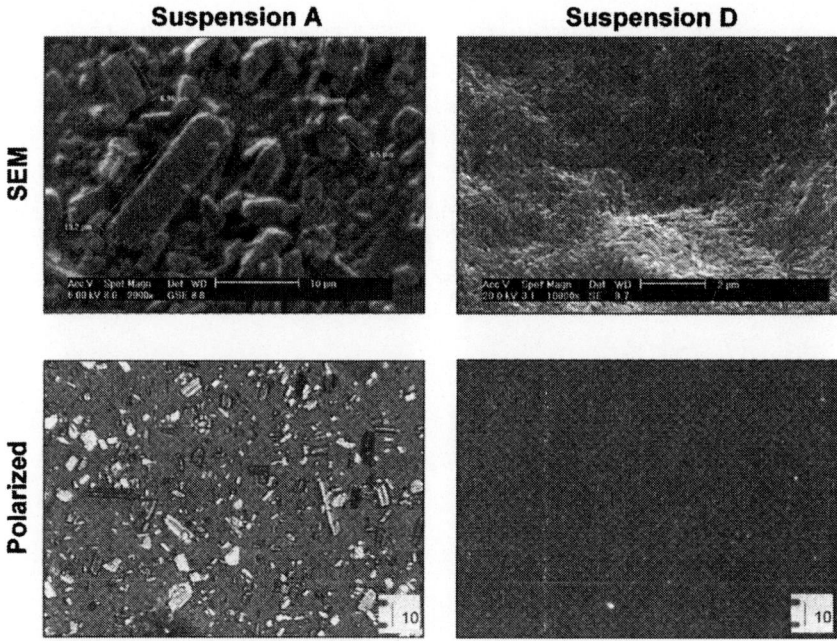

Figure 7 Electron micrographs of Au/Pd-coated samples were collected with an environmental SEM (XL30; Philips; Amsterdam, N.V.) and magnifications of 2000 and 10000×. Magnification for the polarized optical micrographs were 1000×. Bar represents 10 μm. *Abbreviation*: SEM, scanning electron microscopy.

diffraction patterns of dried samples from the three smallest suspensions were compared both to each other and to the unprocessed material, providing assurance that the high level of crystallinity was retained (data not shown). The same conclusion was drawn from scanning calorimetry.

Conclusions

Table 2 summarizes the key findings of the above size evaluation. The use of multiple methods affords a broader understanding of this group of analytes in several ways. First, the results are complementary, e.g., DLS provides direct observation of the submicrometer population, whereas SPOS reports on the large particle population that DLS is poorly suited to determine. Second, confirmatory data is obtained, e.g., turbidimetry, DLS, and SEM all report on the submicrometer population, but rely on different principles to do so: direct scattering, diffusion, and direct observation, respectively. Finally, deficits in the results of one method are addressed by those of another, for example fiber optic-SLS shows that the population of smallest particles, detected but not quantified by fiber optic-quasi elastic light scattering, is increasing with additional processing. Finally, completely orthogonal information is obtained, for example, retention of crystallinity via X-ray versus size from the other investigations.

CONCLUSIONS

Clearly, the ability to characterize nanoparticles goes hand-in-hand with their design and production. Hence, it is as important to advance the appropriate analytical science as it is to progress manufacturing technology. There are several axes along which this can proceed. First, is that of deriving information from concentrated systems when limited quantities of sample are present. This is an analytical challenge due to the strong inter-particle interaction present in such systems. Surface characterization, such as charge and morphology, both in the solid and wetted state, is another area of need. While improvements in instrumentation are always welcome, the real challenge is in

Table 2 Summary of Particle Sizing Results for Celecoxib Nanoparticles

Method	Key findings
Single particle optical sensing (SPOS)	Clear trend to smaller size
	90% of the material is tending submicrometer even for first nanosuspension
	Very little volume present as large particles in final suspension
	Cannot differentiate between the two smallest suspensions
Dynamic light scattering (DLS)	Clear trend to smaller size
	Direct evidence of polydispersity
Fiber optic dynamic light scattering (FO-DLS)	Clear trend to smaller size
	Minimum size near 250 nm
	Submicrometer particles present in starting material
Fiber optic static light scattering (FO-SLS)	Increasing population of the smallest particles as size decreases
	Multiple scattering may affect FO-DLS at least extensive dilutions
Turbidimetry	Clear trend to smaller size
	Large difference between A and B
Microscopy (electron, optical)	DLS data validated
	Particles are tabular though without a high aspect ratio
	Material appears to retain crystallinity
X-ray diffraction	Retention of crystallinity confirmed for all suspensions

separating good data from artifact and in interpretation of the former. Third, is improving the ability to characterize nanoparticles in the solid state, particularly when isolated in a matrix material as is the case with spray drying or lyophilization. Process analytical technology (PAT) is the fourth area deserving research. Though progress has been demonstrated in production environments, the need extends to methods for directly exploring the generation of nanoparticulates and the microscopic processes involved therein (68).

REFERENCES

1. Allen T. Particle Measurement. 5th ed. London: Chapman & Hall, 1997.

2. Amelinckx S. Handbook of Microscopy: Applications in Materials Science, Solid-State Physics, and Chemistry. New York: VCH, 1997.

3. Pecora R. Dynamic light scattering measurement of nanometer particles in liquids. J Nanoparticle Res 2000; 2:123–131.

4. Chu B, Liu T. Characterization of nanoparticles by scattering techniques. J Nanoparticle Res 2000; 2:29–41.

5. Kerker M. The scattering of light and other electromagnetic radiation. New York: Academic Press, 1969.

6. McClements DJ. Principles of ultrasonic droplet size determination in emulsions. Langmiur 1996; 12:3454–3461.

7. Dukhin AS, Goetz PJ. Characterization of aggregation phenomena by means of acoustic and electroacoustic spectroscopy. Colloids Surf A: Physicochem Eng Aspects 1998; 144:49–58.

8. Irache JM, Durrer C, Ponchel G, Duchene D. Determination of particle concentration in latexes by turbidimetry. Int J Pharmaceutics 1993; 90:R9–R12.

9. Westesen K, Bunjes H, Koch MHJ. Physicochemical characterization of lipid nanoparticles and evaluation of their drug loading capacity and sustained release potential. J Controlled Release 1997; 48:223–236.

10. Jenning V, Mader K, Gohla SH. Solid lipid nanoparticles (SLN) based on binary mixtures of liquid and solid lipids: a 1H-NMR study. Int J Pharmaceutics 2000; 205:15–21.

11. Mayer C. Nuclear magnetic resonance on dispersed nanoparticles. Progress in nuclear magnetic resonance spectroscopy 2002; 40:307–366.

12. Valentini M, Vaccaro A, Rehor A, Napoli A, Hubbell JA, Tirelli N. Diffusion NMR Spectroscopy for the characterization of the size and interactions of colloidal matter: The case of vesicles and nanoparticles. J Am Chem Soc 2004; 126:2142–2147.

13. Lieberman A. Particle characterization in liquids. In: Knapp JZ, Barber TA, Lieberman A, eds. Liquid- and Surface-Borne Particle Measurement Handbook. New York: Marcel-Dekker, 1996:1–28.

14. Schwarz C, Mehnert W, Lucks JS, Müller RH. Solid lipid nanoparticles (SLN) for controlled drug delivery. I. Production, characterization and sterilization. J Controlled Release 1994; 30:83–96.

15. Ito T, Sun L, Bevan MA, Crooks RA. Comparison of nanoparticle size and electrophoretic mobility measurements using a carbon-nanotube-based coulter counter, dynamic light scattering, transmission electron microscopy, and phase analysis light scattering. Langmuir 2004; 20:6940–6945.

16. Driscoll DF. The significance of particle/globule-sizing measurements in the safe use of intravenous lipid emulsions. J Dispersion Sci Technol 2002; 23:679–687.

17. Bunjes H, Westesen K, Koch MHJ. Crystallization tendency and polymorphic transitions in triglyceride nanoparticles. Int J Pharmaceutics 1996; 129:159–173.

18. Jores K, Mehnert W, Drechsler M, Bunjes H, Johann C, Maeder K. Investigations on the structure of solid lipid nanoparticles (SLN) and oil-loaded solid lipid nanoparticles by photon correlation spectroscopy, field-flow fractionation and transmission electron microscopy. J Controlled Release 2004; 95:217–227.

19. Chorny M, Fishbein I, Danenberg HD, Golomb G. Study of the drug release mechanism from tyrphostin AG-1295-loaded nanospheres by in situ and external sink methods. J Controlled Release 2002; 83:401–414.

20. Tobio M, Gref R, Sanchez A, Langer R, Alonso MJ. Stealth PLA-PEG nanoparticles as protein carriers for nasal administration. Pharm Res 1998; 15:270–275.

21. Calvo P, Vila-Jato JL, Alonso MJ. Comparative in vitro evaluation of several colloidal systems, nanoparticles, nanocapsules, and nanoemulsions, as ocular drug carriers. J Pharm Sci 1996; 85:530–536.

22. Leo E, Brina B, Forni F, Vandelli MA. In vitro evaluation of PLA nanoparticles containing a lipophilic drug in water-soluble or insoluble form. Int J Pharmaceutics 2004; 278:133–141.

23. Mosqueira VCF, Legrand P, Gulik A, et al. Relationship between complement activation, cellular uptake and surface physicochemical aspects of novel PEG-modified nanocapsules. Biomaterials 2001; 22:2967–2979.

24. Nizri G, Magdassi S, Schmidt J, Cohen Y, Talmon Y. Microstructural characterization of micro- and nanoparticles formed by polymer-surfactant interactions. Langmuir 2004; 20:4380–4385.

25. zur Mühlen A, zur Mühlen E, Niehus H, Mehnert W. Atomic force microscopy studies of solid lipid nanoparticles. Pharm Res 1996; 13:1411–1416.

26. Shi HG, Farber L, Michaels JN, et al. Characterization of crystalline drug nanoparticles using atomic force microscopy and complementary techniques. Pharm Res 2003; 20:479–484.

27. Montasser I, Fess H, Coleman AW. Atomic force microscopy imaging of novel type of polymeric colloidal nanostructures. Eur J Pharmaceutics Biopharmaceutics 2002; 54:281–284.

28. Liu J. Scanning transmission electron microscopy of nanoparticles. In: Wang ZL, ed. Characterization of Nanophase Materials. Weinheim: Wiley-VCH Verlag, 2000:81–132.

29. Hitchcock AP, Morin C, Heng YM, Cornelius RM, Brash JL. Towards practical soft X-ray spectromicroscopy of biomaterials. J Biomaterials Sci Polymer Edition 2002; 13:919–937.

30. De Serio M, Zenobi R, Deckert V. Looking at the nanoscale: scanning near-field optical microscopy. TrAC, Trends Anal Chem 2003; 22:70–77.

31. de Campos AM, Diebold Y, Carvalho ELS, Sanchez A, Alonso MJ. Chitosan nanoparticles as new ocular drug delivery systems: in vitro stability, in vivo fate, and cellular toxicity. Pharm Res 2004; 21:803–810.

32. Ramge P, Unger RE, Oltrogge JB, et al. Polysorbate-80 coating enhances uptake of polybutylcyanoacrylate (PBCA)-nanoparticles by human and bovine primary brain capillary endothelial cells. Eur J Neurosci 2000; 12:1931–1940.

33. Shim J, Kang HS, Park WS, Han SH, Kim J, Chang IS. Transdermal delivery of minoxidil with block copolymer nanoparticles. J Controlled Release 2004; 97:477–484.

34. Colfen H, Antonietti M. Field-flow fractionation techniques for polymer and colloid analysis. Adv Polym Sci 2000; 150(New Developments in Polymer Analytics I):67–187.

35. Tan JS, Butterfield DE, Voycheck CL, Caldwell KD, Li JT. Surface modification of nanoparticles by PEO/PPO block copolymers to minimize interactions with blood components and prolong blood circulation in rats. Biomaterials 1993; 14:823–833.

36. DosRamos JG, Silebi CA. The determination of particle size distribution of submicrometer particles by capillary hydrodynamic fractionation (CHDF). J Colloid Interface Sci 1990; 135:165–177.

37. Williams A, Varela E, Meehan E, Tribe K. Characterization of nanoparticulate systems by hydrodynamic chromatography. Int J Pharmaceutics 2002; 242:295–299.

38. DosRamos JG. Recent developments on resolution and applicability of capillary hydrodynamic fractionation (CHDF). Polymeric Mat Sci Eng 2002; 87:338.

39. Blom MT, Chmela E, Oosterbroek RE, Tijssen R, Van den Berg A. On-chip hydrodynamic chromatography separation and detection of nanoparticles and biomolecules. Anal Chem 2003; 75:6761–6768.

40. Westesen K, Siekmann B, Koch MHJ. Investigations on the physical state of lipid nanoparticles by synchrotron radiation X-ray diffraction. Int J Pharmaceutics 1993; 93:189–199.

41. Panyam J, Williams D, Dash A, Leslie-Pelecky D, Labhasetwar V. Solid-state solubility influences encapsulation and release of hydrophobic drugs from PLGA/PLA nanoparticles. J Pharmaceutical Sci 2004; 93:1804–1814.

42. Molpeceres J, Aberturas MR, Guzman M. Biodegradable nanoparticles as a delivery system for cyclosporin: preparation and characterization. J Microencapsulation 2000; 17:599–614.

43. Hunter RJ, ed. Colloid Science: Zeta Potential in Colloid Science: Principles and Applications. London: Academic Press, 1981.

44. Yang SC, Zhu JB. Preparation and characterization of camptothecin solid lipid nanoparticles. Drug Dev Ind Pharm 2002; 28:265–274.

45. McNeil-Watson F, Tscharnuter W, Miller J. A new instrument for the measurement of very small electrophoretic mobilities using phase analysis light scattering (PALS). Colloids Surf, A: Physicochemical and Engineering Aspects 1998; 140:53–57.

46. Schwarz C, Mehnert W. Solid lipid nanoparticles (SLN) for controlled drug delivery II. Drug incorporation and physicochemical characterization. J Microencapsulation 1999; 16:205–213.

47. Yang SC, Zhu JB. Preparation and characterization of camptothecin solid lipid nanoparticles. Drug Dev Ind Pharm 2002; 28:265–274.

48. Müller RH. Hydrophobic interaction chromatography (HIC) for determination of the surface hydrophobicity of particulates. In: Particle and Surface Characterisation Methods, Based on the Invited Lectures presented at the Colloidal Drug Carriers Expert Meeting, 2nd, Mainz, Mar. 6, 1997. Müller RH, Mehnert W, Hildebrand GE, eds. 1997.

49. Goeppert TM, Müller RH. Alternative sample preparation prior to two-dimensional electrophoresis protein analysis on solid lipid nanoparticles. Electrophoresis 2004; 25:134–140.

50. Vauthier C, Schmidt C, Couvreur P. Measurement of the density of polymeric nanoparticulate drug carriers by isopycnic centrifugation. J Nanoparticle Res 1999; 1:411–418.

51. Huve P, Verrecchia T, Bazile D, Vauthier C, Couvreur P. Simultaneous use of size-exclusion chromatography and photon correlation spectroscopy for the characterization of poly(lactic acid) nanoparticles. J Chromatogr A 1994; 675:129–139.

52. Khlebtsov BN, Kovler LA, Bogatyrev VA, Khlebtsov NG, Shchyogolev SY. Studies of phosphatidylcholine vesicles by spectroturbidimetric and dynamic light scattering methods. J Quant Spectr Rad Transfer 2003; 79–80:825–838.

53. Teipel U. Problems in characterizing transparent particles by laser light diffraction spectrometry. Chem Eng Technol 2002; 25:13–21.

54. Zhang H, Xu G. The effect of particle refractive index on size measurement. Powder Technol 1992; 70:189–192.

55. Delly JG. Microscopy: The setup and operation of the polarized-light microscopy lab for particle identification. In: Knapp JZ, Barber TA, Lieberman A, eds. Liquid- and Surface-Borne Particle Measurement Handbook. New York: Marcel-Dekker, 1996:29–60.

56. Saveyn H, Mermuys D, Thas O, van der Meeren, P. Determination of the refractive index of water-dispersible granules for use in laser diffraction experiments. Part Part Syst Charact 2002; 19:426–432.

57. Gulari E, Bazzi G, Gulari E, Annapragada A. Latex particle size distributions from multiwavelength turbidity spectra. Part Charact 1987; 4:96–100.

58. Moghimi SM, Szebeni J. Stealth liposomes and long circulating nanoparticles: critical issues in pharmacokinetics, opsonization and protein-binding properties. Prog Lipid Res 2003; 42:463–478.

59. Olivier JC, Vauthier C, Taverna M, Puisieux F, Ferrier D, Couvreur P. Stability of orosomucoid-coated poly(isobutyl cyanoacrylate) nanoparticles in the presence of serum. J Controlled Release 1996; 40:157–168.

60. Bootz A, Vogel V, Schubert D, Kreuter J. Comparison of scanning electron microscopy, dynamic light scattering and analytical ultracentrifugation for the sizing of poly(butyl cyanoacrylate) nanoparticles. Eur J Pharmaceutics Biopharmaceutics 2004; 57:369–375.

61. Lukowski G, Pflegel P. Electron diffraction of solid lipid nanoparticles loaded with aciclovir. Pharmazie 1997; 52:642–643.

62. Lacoulonche F, Gamisans F, Chauvet A, Garcia ML, Espina M, Egea MA. Stability and in vitro drug release of flurbiprofen-loaded poly(ε-caprolactone) nanospheres. Drug Dev Ind Pharm 1999; 25:983–993.

63. Cavalli R, Caputo O, Carlotti E, Trotta M, Scarnecchia C, Gasco MR. Sterilization and freeze-drying of drug-free and drug-loaded solid lipid nanoparticles. Int J Pharmaceutics 1997; 148:47–54.

64. Kontoyannis CG, Douroumis D. Release study of drugs from liposomic dispersions using differential pulse polarography. Analytica Chimica Acta 2001; 449:135–141.

65. Kararli TT, Kontny MJ, Desai S, Hageman MJ, Haskell RJ. Cyclooxygenase-2 inhibitor compositions having rapid onset of therapeutic effect. PCT Int. Appl. 2001; WO 2001041760 A2 20010614.

66. Koppel DE. Analysis of macromolecular polydispersity in intensity correlation spectroscopy: The method of Cumulants. J Chem Phys 1972; 57:4814–4820.

67. Dhadwal HS, Ansari RR, Meyer WV. A fiber-optic probe for particle sizing in concentration suspensions. Rev Sci Inst 1991; 62:2963–2968.

68. Higgins JP, Arrivo SM, Thurau G, Green RL, Bowen W, Lange A, Templeton AC, Thomas DL, Reed RA. Spectroscopic approach for on-line monitoring of particle size during the processing of pharmaceutical nanoparticles. Anal Chem 2003; 75:1777–1785.

6

Nanoparticle Interface: An Important Determinant in Nanoparticle-Mediated Drug/ Gene Delivery

SANJEEB K. SAHOO and
VINOD LABHASETWAR

Department of Pharmaceutical Sciences,
University of Nebraska Medical Center, Omaha,
Nebraska, U.S.A.

INTRODUCTION

There is significant interest in recent years in developing biodegradable nanoparticles as a drug/gene delivery system (1–5). Nanoparticles are colloidal particles that range in size from 10 to 1000 nm in diameter, and are formulated using biodegradable polymers in which a therapeutic agent can be entrapped, adsorbed, or chemically coupled (2,4). The advantages of using nanoparticles for drug delivery applications result from their three main basic properties. First, nanoparticles, because of their small size, can penetrate through smaller capillaries, which could allow efficient drug accumulation at the target sites (6,7). Second, the use of biodegradable materials for nanoparticle preparation can allow sustained drug release within the target site over a period of days or even weeks (8–10). Third, the nanoparticle surface can be modified to alter biodistribution of drugs or can be conjugated to a ligand to achieve target-specific drug delivery (11,12). Although a number of different polymers have been investigated for formulating biodegradable nanoparticles, poly(D,L-lactide-co-glycolide) (PLGA) and poly lactic acid (PLA) are the most extensively studied polymers for controlled drug delivery applications (13,14). The lactide/glycolide polymers chains are cleaved by hydrolysis into natural metabolites (lactic and glycolic acids), which are eliminated from the body by the citric acid cycle (14). Further, these polymers are approved by the U.S. Food and Drug Administration for human use.

The interface of nanoparticles can significantly influence various physical as well as biological properties of nanoparticles. Various factors such as the emulsifier used for their stabilization, the polymer material and its composition, or adsorption of certain polymers can influence the interfacial properties of nanoparticles (15). This chapter reviews various aspects of nanoparticle interface and its effect on physical properties of nanoparticles, cellular uptake and drug/gene delivery, and in vivo biodistribution.

INFLUENCE OF EMULSIFIER ON PHARMACEUTICAL PROPERTIES OF NANOPARTICLES

The emulsion–solvent evaporation method, as described in chapter 4, is commonly used to formulate PLA and PLGA nanoparticles (15,16). In general, the method involves emulsifying a polymer dissolved in an organic solvent (e.g., chloroform, methylene chloride, etc.) into a nonsolvent (mostly water) which contains an emulsifier(s) to form an oil-on-water (o/w) emulsion. Nanoparticles are formed once the organic solvent from the emulsion is evaporated. Lipophilic drugs can be incorporated into nanoparticles by dissolving them in the organic solvent along with the polymer prior to emulsification. To encapsulate a hydrophilic drug, it is first dissolved in water, and then emulsified into the polymer solution to form water-in-oil (w/o) emulsion. This emulsion is further emulsified into an aqueous solution containing an emulsifier to form (w/o/w) double emulsion.

The emulsifier added in the aqueous phase stabilizes the emulsion and plays an important role in particle formation. Because it is present at the boundary layer between the water phase and the organic phase during particle formation, the stabilizer can get incorporated into the nanoparticle polymer matrix at the interface or is adsorbed because of ionic or hydrophobic interactions, thus modifying the nanoparticle properties such as their size, zeta potential, hydrophilicity/hydrophobicity, surface charge, adhesion, etc. (17,18). Both size and zeta potential are important determinants as they can influence the physical stability as well as the biopharmaceutical properties of nanoparticles. Further, the interfacial properties could influence the drug release rate, biodistribution of nanoparticles, and/or their cellular/tissue uptake (15,19).

Different types of emulsifiers are used for the formulation of nanoparticles; however, poly(vinyl alcohol) (PVA) is the most commonly used emulsifier because it forms particles that are relatively smaller in size and uniform in size distribution (20). We and others have shown that a fraction of PVA remains associated with the nanoparticle surface despite repeated washing because PVA forms an interconnected network with the polymer matrix at the interface (20,21). This occurs because the hydrophobic portion of PVA, polyvinyl acetate, anchors into the nanoparticle matrix during their formulation (22). The commercial PVA contains some unhydrolyzed segments as polyvinyl acetate. It is estimated that PVA forms about six multilayer deposits around the nanoparticle surface (20,21).

Particle Size

Interfacial property directly affects the size and size distribution of nanoparticles formed. We have shown that the mean particle size of nanoparticles formed is a function of PVA concentration and is used as an emulsifier. The particle size decreases from 520 to 380 nm with an increase in PVA concentration from 0.5% to 5% (w/v) (Fig. 1), and also the polydispersity index is reduced, thus forming more uniform particles (15). Mainardes and Evangelista (23) also have observed a decrease in particle size (345–242 nm) with increase in PVA concentration from 0.15% to 0.7% (w/v). This drop in particle size with increase in PVA concentration is probably due to the differences in the stability of the emulsion formed. At

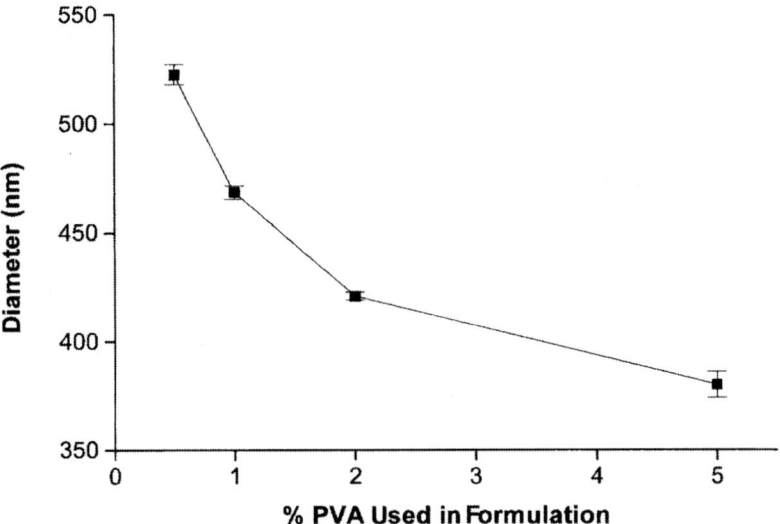

Figure 1 Mean particle size of nanoparticles formulated with different concentrations of PVA emulsifier. Particle size was determined using PCS. A dilute suspension of (100 μg/mL) nanoparticles was prepared in double-distilled water and sonicated on an ice bath for 30 seconds. The sample was subjected to particle size analysis in the ZetaPlus™ particle size analyzer (Brookhaven Instrument Corp. Holtsville, New York, U.S.A.). Data as mean ± s.e.m. (*n* = 5). *Abbreviations*: PVA, poly(vinyl alcohol); PCS, photon correlation spectroscopy.

concentrations lower than 2.5% (w/v), PVA exists as single molecules in solution and at higher concentrations it exists in an aggregated form and has an enhanced surfactant activity. In addition, the viscosity of the aqueous solution increases by increasing the PVA concentrations (e.g., 2.1 cps for 2% to 5.7 cps for 5%), which could also help in the stabilization of the emulsion, leading to the formation of smaller-sized nanoparticles with low polydispersity index (15). Similar to the effect of PVA, human serum albumin (HSA), when used as an emulsifier, also demonstrated a reduction in particle size with increase in its concentration (24). HSA in solution can exist as a monomer but at 3% and higher concentration it forms trimer, tetramer, pentamer, and hexamer, and helps in the formation of smaller particles with a lower polydispersity index (25).

The size of a drug carrier system is an important parameter as it could affect the cellular and tissue uptake. It has been shown that in some cell lines, only smaller-sized nanoparticles are taken up (26,27). Thus, there is a size-dependent cutoff for cellular and tissue uptake of nanoparticles, with the exception of macrophages in which larger-size particles are also taken up efficiently (27). In our previous studies, we have shown a size-dependent uptake of particles by the gastrointestinal tissue in a rat intestinal loop model as well as by the arterial wall in an ex vivo model; with smaller-size nanoparticles showing significantly greater uptake. For example, the uptake of nanoparticles increased exponentially with the decrease in particle size in the arterial wall (28,29). Thus, the size becomes an important parameter to achieve higher drug localization in the target tissue.

Zeta Potential

Zeta potential, a measure of surface charge, can influence particle stability as well as cellular uptake and intracellular trafficking (9). Higher zeta potential values, either positive or negative, are necessary to ensure stability and avoid aggregation of particles. Zeta potential can be altered by varying the stabilizer concentration or by surface modification. Zeta potential of PLGA nanoparticles formulated without PVA in neutral buffer is about $-45\,mV$ (30). This high negative charge is attributed to the presence of uncapped end carboxyl groups of the polymer at the particle surface. However, zeta potential becomes relatively less negative (-6 to $-10\,mV$) when nanoparticles are formulated using PVA as an emulsifier (15). This occurs because of coating of emulsifier, thus masking the surface groups. We have shown that the increase in PVA used for emulsification increases the amount of PVA associated with nanoparticles, which also affects the zeta potential of nanoparticles, especially with pH of the medium. Nanoparticles formulated using 2% (w/v) PVA had about 3% (w/w) nanoparticles surface-associated PVA whereas those formulated using 5% (w/v) PVA had about 5% (w/w) PVA. Although all the formulations of nanoparticles demonstrated negative surface charge at pH 7, the formulation prepared with a higher amount of PVA demonstrated less positive charge in the acidic pH or less negative charge in the basic pH (Fig. 2). The surface charge reversal of nanoparticles from negative in neutral or basic pH to positive in acidic pH can be attributed to the transfer of protons from the bulk solution onto their surface (31,32). Hydroxyl groups at the surface of nanoparticles can become $-OH_2{}^+$ by protonation. A similar charge reversal with the change in pH has been observed for polystyrene nanoparticles with carboxyl functional groups on the surface and was attributed to a positive charge acquired by hydrogen bonding of hydronium ions to the carboxylic group (30). Coating of nanoparticles with some amphiphilic polymers normally decreases the zeta potential because the coating layers shield the surface charge and move the shear plane outward from the particle surface (33). Redhead et al. (34) have reported a similar reduction in the zeta potentials of PLGA nanoparticles after coating with amphiphilic polymers such as poloxamer 407 and poloxamine 908.

Surface Hydrophobicity/Hydrophilicity

The fate of nanoparticles upon intravenous injection mainly depends on their size and surface properties. Nanoparticles, like other colloidal carriers after intravenous administration, are normally taken up mainly by the reticuloendothelial system (RES). Smaller-size nanoparticles have a relatively long circulation time because they can avoid the RES uptake and also can penetrate deep into tissues through fine capillaries (35). In general, the strategy that is followed to avoid the uptake of nanoparticles by the RES is to sterically stabilize

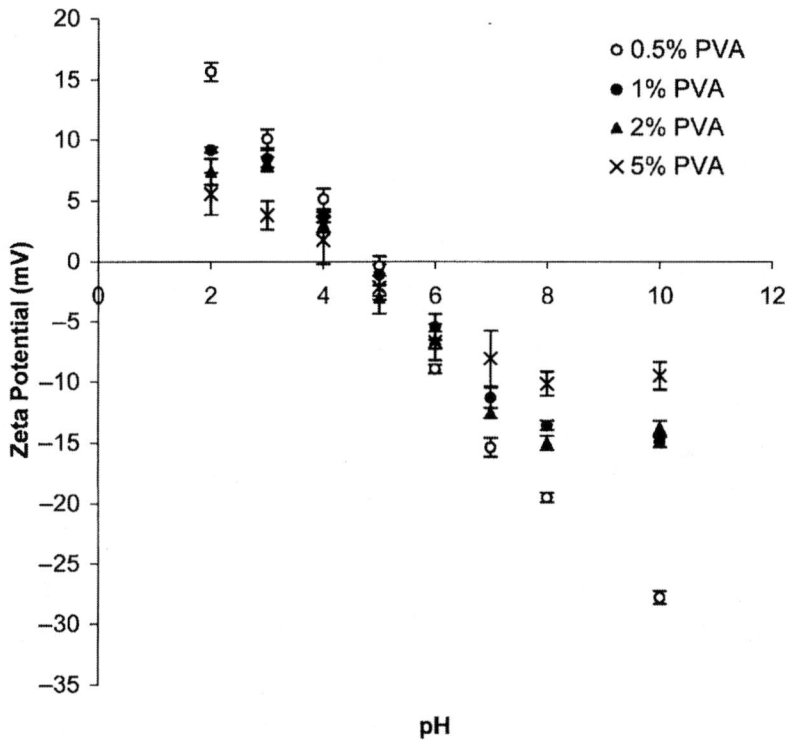

Figure 2   Effect of pH on the zeta potential of nanoparticles formulated with 0.5%, 1%, 2%, and 5% PVA concentrations. To measure the zeta potential of nanoparticles as a function of pH, a suspension of nanoparticles was prepared as above in 0.001 M HEPES buffer of different pH (pH adjusted either with 0.1 M HCL or 0.1 M NaOH). The zeta potential was measured immediately using the ZetaPlus™ zeta potential analyzer. Data as mean ± s.e.m. ($n = 5$). *Abbreviations*: PVA, poly(vinyl alcohol); HEPES, *N*-2-hydroxyethylpiperazine-*N'*-2-ethanesulfonic acid. *Source*: From Ref. 15.

them with a layer of amphiphilic polymer chains like polyethylene oxide (PEO), poloxamers, poloxamines, etc. (19,36,37). The presence of these poloxamers and poloxamines on the surface decreases protein adsorption (opsonization) and the subsequent phagocytosis of the nanoparticles by the Kupffer cells in the liver. In our studies, we have demonstrated that nanoparticles formulated with 5% PVA are more hydrophilic compared to those formulated with 0.5% PVA (15). This effect could be compared to the effect of coating a hydrophilic polymer such as polyethylene glycol (PEG) or poloxamer on to hydrophobic nanoparticle surface (34,38).

Drug Release

Drug release from nano- and microparticles is a complex process, and is generally assumed to involve several steps, such as (i) diffusion through the polymer matrix, (ii) release by polymer degradation, and (iii) solubilization and diffusion of the drug through fine channels that exist in the polymer matrix or are formed as a result of polymer erosion (39). The initial rapid drug release occurs owing to the release of the drug deposited at the interface, which is generally referred to as burst effect. The subsequent release occurs via diffusion followed by flow through the water channels that are created as a result of erosion of the polymer matrix (40). Drug release not only is influenced by the molecular weight of drug molecules but also seems to depend on the interfacial properties of nanoparticles. In our earlier studies, we have shown that the cumulative release of the encapsulated bovine serum albumin (BSA) in vitro from PLGA nanoparticles formulated using 0.5% PVA as an emulsifier is higher than that from the nanoparticles formulated with 5% PVA. Although the release profiles were biphasic and the initial release rates were similar for both the formulations, the release rate was faster during the later phase for the formulation, which was prepared with a lower concentration of PVA (15). The possibility is that the presence of PVA at a higher concentration at the interface slows down the degradation of the polymer, thus affecting the release rate at a later stage. PVA at the interface could form a gel-like boundary that could affect drug diffusion as well as the polymer degradation. Recently, we have determined that the cumulative release of doxorubicin from the nanoparticles formulated with PVA is slower than that from the nanoparticles formulated without it (Fig. 3). Further,

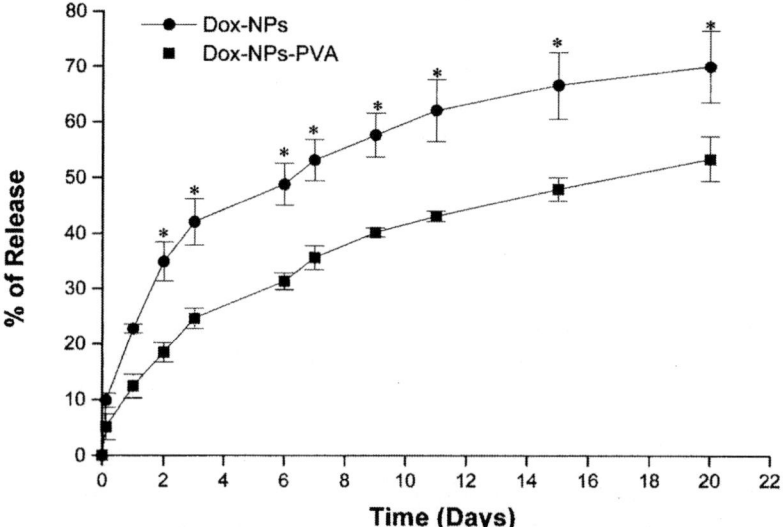

Figure 3 Release of doxorubicin from NPs formulated with PVA Dox-NPs-PVA) and without PVA (Dox-NPs). The in vitro release of the drug was determined in phosphate-buffered saline (0.15 M, pH 7.4) containing 0.1% (w/v) Tween® 80 at 37°C utilizing double-chamber diffusion cells placed on a shaker at 100 rpm (Environ®, Lab Line, Melrose Park, Illinois, U.S.A.). The donor chamber was filled with a 2.5 mL suspension of nanoparticles (2 mg/mL) and the receiver end was filled with the buffer. A Millipore® hydrophilic low protein binding membrane (Millipore Co., Bedford, Massachusettes, U.S.A.) with 0.1 μm pore size was placed between the two chambers. Tween-80 was used in the buffer to maintain sink condition during the release study. At a predetermined time interval, the solution in the receiver end was collected and doxorubicin concentration was determined by fluorescence spectrophotometer (Varian, Cary Eclipse, Walnut Creek, California, U.S.A.) measuring the fluorescence intensity at $l_{ex} = 485$ nm and $l_{em} = 591$ nm). Data as mean ± s.e.m. ($n = 3$). *$p < 0.05$. *Abbreviations*: NPs, nanoparticles; PVA, poly(vinyl alcohol).

the emulsifier used can also affect drug loading in nanoparticles. For example, using D-α-tocopheryl PEG 100 succinate (Vitamin E TPGS) as an emulsifier, Feng's group has shown that paclitaxel loading in PLGA nanoparticles is higher than in nanoparticles formulated using PVA (41,42).

IMPLICATION ON CELLULAR UPTAKE/ TOXICITY/GENE DELIVERY

Cellular Uptake

Intracellular uptake of nanoparticles is affected by a number of factors including particle size, surface characteristics, hydrophilicity, and zeta potential. We have demonstrated that nanoparticles fractionated into greater than 100 nm and less than 100 nm sizes have different levels of gene expression. The smaller-size fraction of nanoparticles demonstrated 27-fold higher transfection in COS-7 cells than the larger–particle size fraction (43). Effect of particle size on gene transfection has been reported for other systems such as polyplex and DNA–lipid complexes, with smaller-size complexes demonstrating better transfection than larger size complexes or aggregate (44).

Because the nanoparticle interface comes in direct contact with the cell surface, it is anticipated that their interfacial properties would influence the cellular uptake as well as gene expression. In our studies, the surface-associated PVA has been shown to influence the cellular uptake of nanoparticles. For example, the cellular uptake of nanoparticles formulated with 0.5% PVA was about threefold higher than the uptake of nanoparticles formulated with 5% of PVA in vascular smooth muscle cells (Fig. 4). We attributed the reduced cellular uptake of nanoparticles with an increase in surface associated PVA to the increase in their higher hydrophilicity and hence the reduced interaction with the cell surface (15). Further, we have demonstrated that the surface-associated PVA also influences the intracellular distribution of nanoparticles. Nanoparticles with a higher amount of surface-associated PVA demonstrated lower nanoparticle levels in the cytoplasmic fraction in MCF-7 cells as compared

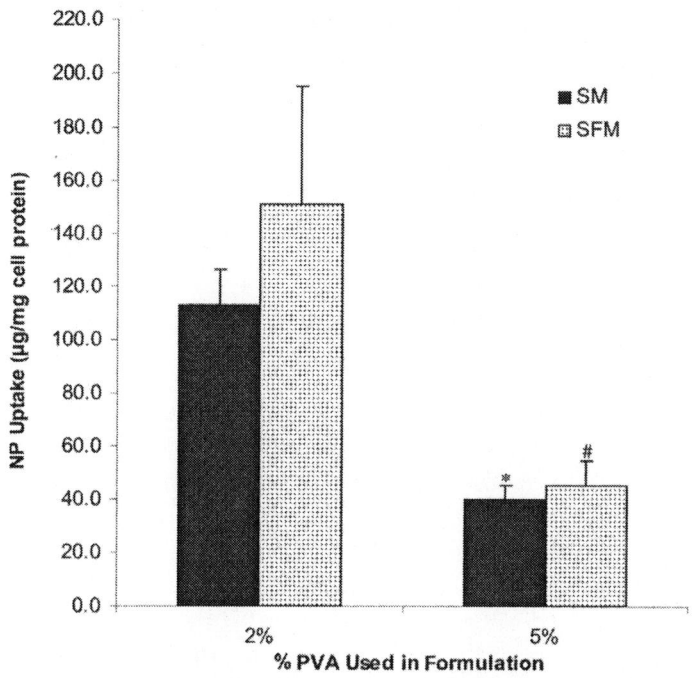

Figure 4 Effect of surface associated PVA on nanoparticle uptake in VSMCs. Nanoparticles were formulated using 2% and 5% (w/v) PVA as an emulsifier. VSMCs (50,000 cells per plate in a 24-well plate) were allowed to attach for 24 hours, the medium was changed with a suspension of nanoparticles (100 µg/mL) prepared either in SM or SFM. Cells were incubated with particles for one hour, washed, and the nanoparticle levels in the cell were determined by HPLC. Nanoparticles formulated with 2% PVA demonstrated greater uptake than the uptake of nanoparticles formulated with 5% PVA despite their similar particle size. *, #, $p < 0.05$ compared to uptake of corresponding 2% PVA nanoparticle group. *Abbreviations*: PVA, poly(vinyl alcohol); VSMC, vascular smooth muscle cell; SM, serum medium; SFM, serum-free medium; NP, nanoparticles, HPLC, high-performance medium liquid chromatomediumgraphy. *Source*: From Ref. 15.

to the levels for the formulation with a lower amount of surface associated PVA. It is suggested that the PVA present on the nanoparticle surface shields the charge reversal of nanoparticles in the endo-lysosomal compartment, resulting in a lower number of nanoparticles escaping into the cytoplasm (45). Our studies with doxorubicin-loaded nanoparticles formulated without PVA were seen to demonstrate greater cellular drug uptake than that with the nanoparticles formulated with PVA. This was evident from their confocal microscopic pictures (data not shown) and also from the flow cytometry data that showed a twofold difference in the uptake (Fig. 5).

In Vitro Cytotoxicity of Doxorubicin-Loaded Nanoparticles

To evaluate the effect of interfacial property of nanoparticles on drug effect, we determined the cytotoxicity of doxorubicin-loaded nanoparticles, which were formulated either with or without PVA. At the lowest dose of the drug studied (50 ng/mL), the drug in solution demonstrated 25% inhibition in cell proliferation whereas drug-loaded nanoparticles formulated with or without PVA demonstrated 15% inhibition in cell proliferation. However, at higher doses (500 or 1000 ng/mL) doxorubicin in solution and doxorubicin-loaded nanoparticles formulated without PVA demonstrated similar antiproliferative activity, whereas doxorubicin-nanoparticles formulated with PVA showed lower antiproliferative activity (Fig. 6). The IC_{50} calculated from dose–response study was $5.1 \times 10^{-1} \mu M$ doxorubicin-loaded nanoparticles formulated without PVA whereas it was $7.1 \times 10^{-1} \mu M$ for the nanoparticles formulated with PVA. The difference in the antiproliferative effect of the two formulations of nanoparticles could have been the combined effect of the difference in their drug release rates and cellular uptake.

Gene Transfection

Recently, we studied the effect of surface-associated PVA on gene expression. Nanoparticles formulated using 2% and 5% (w/v) PVA concentrations were used for comparison as these formulations had almost similar DNA loading and particle size. Nanoparticles with a lower amount of surface-associated PVA demonstrated 12- to 20-fold higher gene transfection in

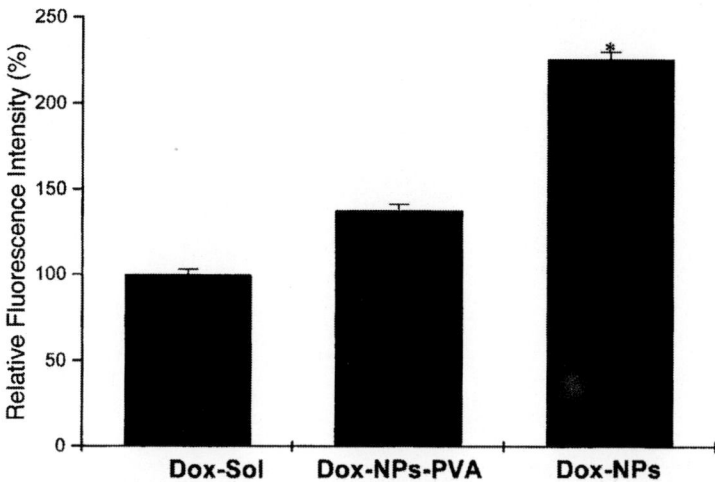

Figure 5 Relative fluorescence intensity of drug as determined by flow cytometry in MCF-7 cells incubated with doxorubicin-loaded NPs formulated with PVA (Dox-NPs-PVA) and without PVA (Dox-NPs) and free doxorubicin. Cells were seeded into a 100 mm culture dish at a cell density of 500,000 cells per dish in 10 mL growth medium and were allowed to attach overnight. The medium from each dish was replaced with a suspension of 50 μM doxorubicin solution or nanoparticle suspension. Cells treated with empty nanoparticles and plain medium were used as controls. At the end of three days, cells were washed with phosphate-buffered saline and then detached by trypsinization. The harvested cells were analyzed using flow cytometry. The gates were arbitrarily set for the detection of green fluorescence (FL1-H > 200, 535 nm, linear scale). The relative fluorescence intensity was calculated from control cells treated either with medium or with control nanoparticles. *, $p < 0.05$ compared to uptake of doxorubicin solution or Dox-NPs-PVA. *Abbreviations*: NPs, nanoparticles; PVA, poly(vinyl alcohol).

MCF-7 cells than nanoparticles having a higher amount of surface associated PVA. Similar higher transfection was observed in PC-3 cell line for the nanoparticles formulated using 2% (w/v) PVA; however, the difference in the transfection was only twofold in this cell line (45). In the human bronchial cell line Calu-3, Bivas-Benita et al. (46) have

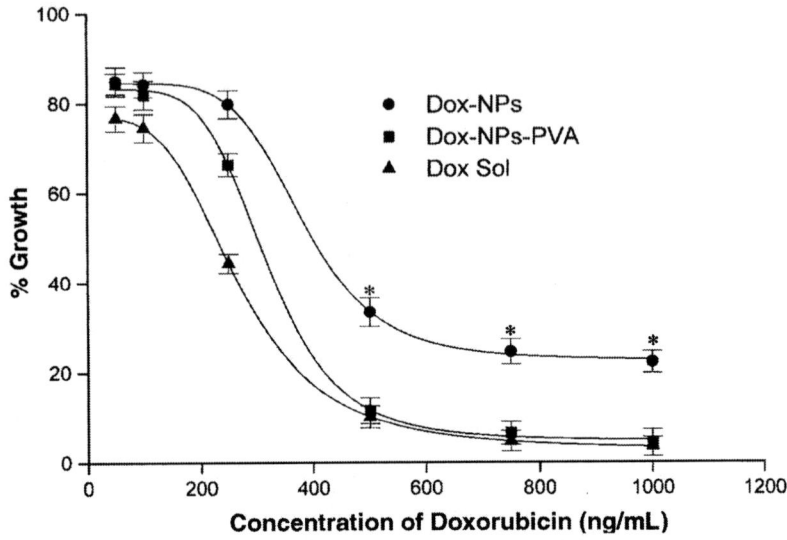

Figure 6 Dose-dependent cytotoxicity of doxorubicin in MCF-7 cells. Cells were seeded at a density of 4000 per well in 96-well plates and allowed to attach for 24 hours. Different concentrations of (1–1000 ng/mL) doxorubicin in solution or loaded in NPs with PVA (Dox-NPs-PVA) and without PVA (Dox-NPs) were used with medium or control nanoparticles serving as respective controls. After 120 hours, cell viability was determined using a standard MTS assay (CellTiter 96® AQueous, Promega). The effect of each drug was calculated as a percentage of control cell growth obtained from vehicle-treated cells grown in the same culture medium. Data as mean ± s.e.m. ($n = 6$). *, $p < 0.05$ compared to uptake of doxorubicin solution or Dox-NPs-PVA *Abbreviations*: NPs, nanoparticles; PVA, poly(vinyl alcohol).

demonstrated that gene transfection efficiency of PLGA nanoparticles bearing polyethyleneimine (PEI) depends on the ratio of PEI–DNA. Ravi Kumar et al. (47) demonstrated enhanced DNA binding to PLGA nanoparticles that were formulated using PVA-chitosan blend as a stabilizer. The nanoparticles were tested for their ability to transport across the nasal mucosa in vivo in mice but also as a gene expression vector. The results show that both modified nanoparticles facilitate gene delivery and expression in vivo with increased efficiency and without causing inflammation, as measured by IL-6. These results indicate that surface charge affects DNA binding, cytotoxicity, and gene transfection.

BIODISTRIBUTION

Rapid clearance of intravenously injected colloidal carrier systems including nanoparticles from blood circulation by the tissues of the mononuclear phagocyte system (MPS) is the major obstacle to the delivery of drugs to organs or cells other than MPS (48–50). Different strategies have been proposed to modify the body distribution of polymeric nanoparticles, most of them are based on the modification of the hydrophobic particle surface by physical adsorption of a hydrophilic polymer. One of the most commonly used polymers for particle hydrophilization is the series of linear or branched copolymers of poly (ethylene oxide) and poly (propylene oxide) (Pluronic Tetronic™ or Poloxamer/Poloxamine) (51–54). Another approach includes the synthesis of amphiphilic copolymer in which the hydrophobic block itself is able to form a solid phase (particle core), while the hydrophilic part remains as a surface-exposed "protective cloud." Examples are the block-copolymer of PEG and poly(lactide/glycolide) (PEG PLGA) or PEG and *n*-hexadecylcyanoacrylate (PEG-PHDCA) (38,53). With such types of hydrophilic coatings, the natural blood opsonization process of the particles is reduced; hence, a relative avoidance of the recognition by macrophages in the liver and spleen is conferred, increasing particle blood half-life and therefore, their extravasation to non-RES tissues.

Among the different surfactants used to modify particle biodistribution, coating with polysorbate-80 has been shown to also cross the blood–brain barrier (BBB) (55). Several therapeutic agents such as dalargin, loperamide, tubocurarinc, and MKZ 2/576, a novel N-methyl-D-aspartate (NMDA) receptor antagonist and doxorubicin have been tested for brain delivery using the modified nanoparticles (56–60). The study demonstrates that the brain concentration of systemically administered doxorubicin can be enhanced by over

60-fold by binding to biodegradable poly(butyl cyanoacrylate) nanoparticles which were overcoated with polysorbate-80. The pharmacokinetics of another drug, amitriptyline, a tricyclic antidepressant that normally cannot penetrate the BBB also showed an improvement in brain AUC (area under the curve) following intravenous injection of polysorbate-80 coated nanoparticles (61).

CONCLUSIONS

Nanoparticle interface thus not only affects the physicochemical properties of nanoparticles but also their biological properties. Therefore, a critical analysis of various parameters and their influence on interfacial properties of nanoparticles, and how these properties affect the nanoparticle-mediated drug and gene delivery is important. It is possible that at the cellular level, the nanoparticle interface influences the cell signaling and hence, the uptake pathway that could influence the intracellular distribution of nanoparticles. Therefore, in addition to studying the effect of interfacial properties of nanoparticles on their biodistribution, a better understanding of the molecular mechanism of intracellular trafficking of nanoparticles and sorting pathways as a function of surface properties of nanoparticles would be useful in developing formulations that can target therapeutic agents at specific intracellular compartments.

REFERENCES

1. Brannon-Peppas L, Blanchette JO. Nanoparticle and targeted systems for cancer therapy. Adv Drug Deliv Rev 2004; 56: 1649–1659.

2. Panyam J, Labhasetwar V. Biodegradable nanoparticles for drug and gene delivery to cells and tissue. Adv Drug Deliv Rev 2003; 55:329–347.

3. Kubik T, Bogunia-Kubik K, Sugisaka M. Nanotechnology on duty in medical applications. Curr Pharm Biotechnol 2005; 6:17–33.

4. Sahoo SK, Labhasetwar V. Nanotech approaches to drug delivery and imaging. Drug Discov Today 2003; 8:1112–1120.

5. Ravi Kumar M, Hellermann G, Lockey RF, Mohapatra SS. Nanoparticle-mediated gene delivery: state of the art. Expert Opin Biol Ther 2004; 4:1213–1224.

6. Sahoo SK, Ma W, Labhasetwar V. Efficacy of transferrin-conjugated paclitaxel-loaded nanoparticles in a murine model of prostate cancer. Int J Cancer 2004; 112:335–340.

7. Panyam J, Lof J, O'Leary E, Labhasetwar V. Efficiency of Dispatch and Infiltrator cardiac infusion catheters in arterial localization of nanoparticles in a porcine coronary model of restenosis. J Drug Target 2002; 10:515–523.

8. Panyam J, Labhasetwar V. Sustained cytoplasmic delivery of drugs with intracellular receptors using biodegradable nanoparticles. Mol Pharm 2004; 1:77–84.

9. Panyam J, Zhou WZ, Prabha S, Sahoo SK, Labhasetwar V. Rapid endo-lysosomal escape of poly(DL-lactide-*co*-glycolide) nanoparticles: implications for drug and gene delivery. FASEB J 2002; 16:1217–1226.

10. Prabha S, Labhasetwar V. Nanoparticle-mediated wild-type p53 gene delivery results in sustained antiproliferative activity in breast cancer cells. Mol Pharm 2004; 1:211–219.

11. Moghimi SM, Hunter AC. Capture of stealth nanoparticles by the body's defences. Crit Rev Ther Drug Carrier Syst 2001; 18:527–550.

12. Moghimi SM, Hunter AC, Murray JC. Long-circulating and target-specific nanoparticles: theory to practice. Pharmacol Rev 2001; 53:283–318.

13. Bala I, Hariharan S, Kumar MN. PLGA nanoparticles in drug delivery: the state of the art. Crit Rev Ther Drug Carrier Syst 2004; 21:387–422.

14. Shive MS, Anderson JM. Biodegradation and biocompatibility of PLA and PLGA microspheres. Adv Drug Deliv Rev 1997; 28:5–24.

15. Sahoo SK, Panyam J, Prabha S, Labhasetwar V. Residual polyvinyl alcohol associated with poly(D,L-lactide-*co*-glycolide)

nanoparticles affects their physical properties and cellular uptake. J Control Release 2002; 82:105–114.

16. Davda J, Labhasetwar V. Characterization of nanoparticle uptake by endothelial cells. Int J Pharm 2002; 233:51–59.

17. Scholes PD, Coombes AG, Illum L, et al. Detection and determination of surface levels of poloxamer and PVA surfactant on biodegradable nanospheres using SSIMS and XPS. J Control Release 1999; 59:261–278.

18. Feng S, Huang G. Effects of emulsifiers on the controlled release of paclitaxel (Taxol) from nanospheres of biodegradable polymers. J Control Release 2001; 71:53–69.

19. Stolnik S, Daudali B, Arien A, et al. The effect of surface coverage and conformation of poly(ethylene oxide) (PEO) chains of poloxamer 407 on the biological fate of model colloidal drug carriers. Biochim Biophys Acta 2001; 1514:261–279.

20. Zambaux MF, Bonneaux F, Gref R, et al. Influence of experimental parameters on the characteristics of poly(lactic acid) nanoparticles prepared by a double emulsion method. J Control Release 1998; 50:31–40.

21. Murakami H, Kobayashi M, Takeuchi H, Kawashima Y. Preparation of poly(DL-lactide-co-glycolide) nanoparticles by modified spontaneous emulsification solvent diffusion method. Int J Pharm 1999; 187:143–152.

22. Shakesheff KM, Evora C, Soriano II, Langer R. The adsorption of poly(vinyl alcohol) to biodegradable microparticles studied by x-ray photoelectron spectroscopy (XPS). J Colloid Interface Sci 1997; 185:538–547.

23. Mainardes RM, Evangelista RC. PLGA nanoparticles containing praziquantel: effect of formulation variables on size distribution. Int J Pharm 2005; 290:137–144.

24. Verrecchia T, Spenlehauer G, Bazile DV, Murry-Brelier A, Archimbaud Y, Veillard M. Non-stealth (poly(lactic acid/albumin)) and stealth (poly(lactic acid-polyethylene glycol)) nanoparticles as injectable drug carriers. J Control Release 1995; 36:49–61.

25. Zini R, Barre J, Bree F, Tillement JP, Sebille B. Evidence for a concentration-dependent polymerization of a commercial human serum albumin. J Chromatogr 1981; 216:191–198.

26. Desai MP, Labhasetwar V, Walter E, Levy RJ, Amidon GL. The mechanism of uptake of biodegradable microparticles in Caco-2 cells is size dependent. Pharm Res 1997; 14:1568–1573.

27. Tabata Y, Ikada Y. Effect of the size and surface charge of polymer microspheres on their phagocytosis by macrophage. Biomaterials 1988; 9:356–362.

28. Desai MP, Labhasetwar V, Amidon GL, Levy RJ. Gastrointestinal uptake of biodegradable microparticles: effect of particle size. Pharm Res 1996; 13:1838–1845.

29. Song C, Labhasetwar V, Cui X, Underwood T, Levy RJ. Arterial uptake of biodegradable nanoparticles for intravascular local drug delivery: results with an acute dog model. J Control Release 1998; 54:201–211.

30. Stolnik S, Garnett MC, Davies MC, et al. The colloidal properties of surfactant-free biodegradable nanospheres from poly(*b*-malic acid-*co*-benzyl malate)s and poly(lactic acid-*co*-glycolide). Colloids Surf A 1995; 97:235–245.

31. Makino K, Ohshima H, Kondo T. Transfer of protons from bulk solution to the surface of poly(L-lactide) microcapsules. J Microencapsul 1986; 3:195–202.

32. Corkill JM, Goodman JF, Wyer J. Nuclear magnetic resonance of aqueous solutions of alkylpolyoxyethylene glycol monoethers. Trans Faraday Soc 1969; 65:9–18.

33. Barnes TJ, Prestidge CA. PEO-PPO-PEO block copolymers at the emulsion droplet-water interface. Langmuir 2000; 16: 4116–4121.

34. Redhead HM, Davis SS, Illum L. Drug delivery in poly(lactide-*co*-glycolide) nanoparticles surface modified with poloxamer 407 and poloxamine 908: in vitro characterisation and in vivo evaluation. J Control Release 2001; 70:353–363.

35. Vinogradov SV, Bronich TK, Kabanov AV. Nanosized cationic hydrogels for drug delivery: preparation, properties and interactions with cells. Adv Drug Deliv Rev 2002; 54:135–147.

36. Illum L, Church AE, Butterworth MD, Arien A, Whetstone J, Davis SS. Development of systems for targeting the regional lymph nodes for diagnostic imaging: in vivo behaviour of

colloidal PEG-coated magnetite nanospheres in the rat following interstitial administration. Pharm Res 2001; 18:640–645.

37. Lin W, Garnett MC, Davis SS, Schacht E, Ferruti P, Illum L. Preparation and characterization of rose Bengal-loaded surface-modified albumin nanoparticles. J Control Release 2001; 71:117–126.

38. Gref R, Minamitake Y, Peracchia MT, Trubetskoy V, Torchilin V, Langer R. Biodegradable long-circulating polymeric nanospheres. Science 1994; 263:1600–1603.

39. Panyam J, Dali MM, Sahoo SK, et al. Polymer degradation and in vitro release of a model protein from poly(D,L-lactide-*co*-glycolide) nano- and microparticles. J Control Release 2003; 92: 173–187.

40. Lin SY, Chen KS, Teng HH, Li MJ. In vitro degradation and dissolution behaviors of microspheres prepared by three low molecular weight polyesters. J Microencapsul 2000; 17:577–586.

41. Mu L, Feng SS. Vitamin E TPGS used as emulsifier in the solvent evaporation/extraction technique for fabrication of polymeric nanospheres for controlled release of paclitaxel (Taxol). J Control Release 2002; 80:129–144.

42. Mu L, Feng SS. A novel controlled release formulation for the anticancer drug paclitaxel (Taxol): PLGA nanoparticles containing vitamin E TPGS. J Control Release 2003; 86:33–48.

43. Prabha S, Zhou WZ, Panyam J, Labhasetwar V. Size-dependency of nanoparticle-mediated gene transfection: studies with fractionated nanoparticles. Int J Pharm 2002; 244:105–115.

44. Cherng JY, van de Wetering P, Talsma H, Crommelin DJ, Hennink WE. Effect of size and serum proteins on transfection efficiency of poly ((2-dimethylamino)ethyl methacrylate)-plasmid nanoparticles. Pharm Res 1996; 13:1038–1042.

45. Prabha S, Labhasetwar V. Critical determinants in PLGA/PLA nanoparticle-mediated gene expression. Pharm Res 2004; 21: 354–364.

46. Bivas-Benita M, Romeijn S, Junginger HE, Borchard G. PLGA-PEI nanoparticles for gene delivery to pulmonary epithelium. Eur J Pharm Biopharm 2004; 58:1–6.

47. Ravi Kumar MN, Bakowsky U, Lehr CM. Preparation and characterization of cationic PLGA nanospheres as DNA carriers. Biomaterials 2004; 25:1771–1777.

48. Gaur U, Sahoo SK, De TK, Ghosh PC, Maitra A, Ghosh PK. Biodistribution of fluoresceinated dextran using novel nanoparticles evading reticuloendothelial system. Int J Pharm 2000; 202:1–10.

49. Maitani Y, Nakamura K, Kawano K. Application of sterylglucoside-containing particles for drug delivery. Curr Pharm Biotechnol 2005; 6:81–193.

50. Torchilin VP. Polymer-coated long-circulating microparticulate pharmaceuticals. J Microencapsul 1998; 15:1–19.

51. Moghimi SM. Mechanisms regulating body distribution of nanospheres conditioned with pluronic and tetronic block copolymers. Adv Drug Deliv Rev 1995; 16:183–193.

52. Moghimi SM. Mechanisms of splenic clearance of blood cells and particles: towards development of new splenotropic agents. Adv Drug Deliv Rev 1995; 17:103–115.

53. Peracchia MT, Fattal E, Desmaele D, et al. Stealth PEGylated polycyanoacrylate nanoparticles for intravenous administration and splenic targeting. J Control Release 1999; 60:121–128.

54. Stolnik S, Illum L, Davis SS. Long circulating microparticulate drug carriers. Adv Drug Deliv Rev 1995; 16:195–214.

55. Troster SD, Kreuter J. Influence of the surface properties of low contact angle surfactants on the body distribution of 14C-poly(methyl methacrylate) nanoparticles. J Microencapsul 1992; 9:19–28.

56. Kreuter J, Alyautdin RN, Kharkevich DA, Ivanov AA. Passage of peptides through the blood-brain barrier with colloidal polymer particles (nanoparticles). Brain Res 1995; 674:171–174.

57. Schroder U, Sabel BA. Nanoparticles, a drug carrier system to pass the blood-brain barrier, permit central analgesic effects of i.v. dalargin injections. Brain Res 1996; 710:121–124.

58. Alyautdin RN, Petrov VE, Langer K, Berthold A, Kharkevich DA, Kreuter J. Delivery of loperamide across the blood-brain

barrier with polysorbate 80-coated polybutylcyanoacrylate nanoparticles. Pharm Res 1997; 14:325–328.

59. Alyautdin RN, Tezikov EB, Ramge P, Kharkevich DA, Begley DJ, Kreuter J. Significant entry of tubocurarine into the brain of rats by adsorption to polysorbate 80-coated polybutylcyanoacrylate nanoparticles: an in situ brain perfusion study. J Microencapsul 1998; 15:67–74.

60. Gulyaev AE, Gelperina SE, Skidan IN, Antropov AS, Kivman GY, Kreuter J. Significant transport of doxorubicin into the brain with polysorbate 80-coated nanoparticles. Pharm Res 1999; 16:1564–1569.

61. Kreuter J, Ramge P, Petrov V, et al. Direct evidence that polysorbate-80-coated poly(butylcyanoacrylate) nanoparticles deliver drugs to the CNS via specific mechanisms requiring prior binding of drug to the nanoparticles. Pharm Res 2003; 20: 409–416.

7

Toxicological Characterization of Engineered Nanoparticles

PAUL J. A. BORM

Centre of Expertise in Life Sciences,
Zuyd University, Heerlen,
The Netherlands

ROEL P. F. SCHINS

Institut fur Umweltmedizinische
Forschung (IUF), University of
Düsseldorf, Düsseldorf, Germany

INTRODUCTION

Nanotechnology is expected to bring a fundamental change in manufacturing in the next few years and will have an enormous impact on life sciences, including drug delivery, diagnostics, nutraceuticals and production of biomaterials (1,2). Engineered nanoparticles (NP) (<100 nm) are an important tool to realize a number of these applications. The reason why these NP are attractive for such purposes is based on their important and unique features, such as their surface to mass ratio, which is much larger than that of other particles, their quantum properties and their ability to adsorb and carry other compounds. NP on the one hand have a large (functional) surface which is able to bind, adsorb and carry other compounds such as drugs, probes and proteins. On the other hand, NP have a surface that might be chemically more reactive as compared to their fine (>100 nm) analogs. Many of these special purpose engineered NP are produced in small quantities. In 2003, single-walled and multiwalled nanotubes had a worldwide production of 2954 kg. However, the Carbon Nanotechnology Research Institute (Japan) plans on expanding their production from ~1000 kg in 2003 to 120,000 kg per year within the next five years. Although current production of engineered nanomaterials is small, it is evident that production rates will accelerate exponentially in the next few years (3).

In addition to these specifically engineered nanomaterials, nano-sized particles are also being produced non-intentionally in diesel exhaust and other combustion processes. It is estimated that 50,000 kg/year of nano-sized materials are being produced through these un-intended anthropogenic sources. These combustion NP are included in particulate matter (PM) which is measured by mass and related to adverse effects in patients with lung and cardiovascular disease. Combustion NP have also been denominated as ultrafine particles, and are primary particles or agglomerates with a diameter <100 nm. These ultrafine particles are a small mass fraction of total anthropogenic particulate emissions, described with total suspended particles, PM or PM beyond a specific size in micrometers (PM_{10}, $PM_{2.5}$, PM_1). The first publication on this topic was the so-called Six Cities study (4) that described an association between mortality in six United States cities and the annual mean of particulate mass sampled by convention with a 50% cutoff at 2.5 μm ($PM_{2.5}$). From this and later studies it is estimated that per 10 μg/m^3 increase in the concentration of $PM_{2.5}$, overall mortality increases by 0.9%, while deaths from specific respiratory diseases can increase by as much as 2.7%. There is ample evidence that a small proportion of the mass but a large proportion of the number of the particles in ambient air are ultrafine in size. Numerous toxicological studies have now forwarded these ultrafine particles to be responsible for adverse effects (5,6), but so far few human studies have been able to investigate this (7–9).

Interestingly most of the toxicological work on NP has been generated with a small set of bulk NP, that have been around in industry for some decades and are produced in quantities that currently exceed many tons per year (Table 1). According to the National Nanotechnology Initiative (United States), the largest production volume in 2004 was for colloidal silica, titanium dioxide (TiO_2), and various iron-oxides (Table 2). All these bulk NP were considered to be so-called nuisance

Table 1 Various Denominations of Particles in Inhalation Toxicology and Drug Delivery in Relation to Their Source (Ambient, Bulk, Engineered)

Particle type	Description
PM_{10}, $PM_{2.5}$	Particle mass fraction in ambient air with a mean diameter of 10 or 2.5 µm, respectively. Basis of current standards for ambient particles in Europe and United States
Coarse particles	The mass fraction of PM_{10}, which is bigger than 2.5 µm
Ultrafine particles ($PM_{0.1}$)	The fraction of PM_{10} with a size cutoff at 0.1 µm. Contains primary particles and agglomerates <100 nm
PSP	Poorly soluble particles with low specific toxicity. Terminology used in relation to bulk synthetic ultrafine particles
Nanoparticles	Primary particles of any material <100 nm
Liposomes	Particles, not strictly NP, consisting of fatty acids and derivatives
Carriers-conjugates	Polymer-protein or polymer-drug conjugates with a size below 100 nm used in drug delivery

Abbreviations: PSP, poorly soluble particles; PM, particulate matter; NP, nanoparticles.

Table 2 Different Sources and Applications of NP

Source of NP	Examples	Application and use
Combustion NP	Diesel exhaust particles	Environmental exposure
	Fly-ashes	
Bulk synthetic NP	Titanium dioxide (TiO_2)	Cosmetics
	Carbon blacks	Pigments, tires, toner
	Amorphous silica	Paints, fillers
	Iron oxides (Fe_2O_3)	Color pigments
	Zinc oxides (ZnO)	UV absorber
	Vitamins	Food
	Pd/Pt	Hydrogenation catalyst
	Azodyes	Color pigments
Engineered NP	Organic	
	Liposomes	Drug delivery
	Polycyanoacrylates	
	Inorganic	
	Gold, dendrimers	Drug delivery
	Zeolites, silver	Quantum dots (imaging)
	Iron oxides	Diagnostics

Abbreviation: NP, nanoparticles.

dusts until it was observed that upon prolonged exposure in rats inflammation and lung tumors can occur (5,12,13) A schematic summary of key studies on toxicological effects of NP is given in Table 3, and this is considered as both direct and indirect evidence that NP are important components in the adverse effects of PM_{10}. The question now is whether in this triangle of different applications and sources of NP (Fig. 1) the different pieces of toxicological evidence can be mutually used or whether a more sophisticated approach is necessary. In this chapter the different parts of evidence and know-how are listed and used to suggest a multidisciplinary design for toxicological testing of NP engineered for drug delivery.

INHALATION OF PARTICLES

The inhalation exposure of particles is mainly relevant in environmental or occupational exposure to combustion and bulk NP and historically particle toxicology has developed in

Table 3 Important Findings on the Biological Activity and Key Publications in the Toxicity of Combustion and Bulk Nanoparticles (NP) Between 1990 and Now

Description of finding	References
NP TiO$_2$ causes pulmonary inflammation. Later studies show that inflammation is mediated by surface area dose	10,11
NP cause more lung tumors than fine particles in rat chronic studies. Effect is surface area mediated	12,13
NP inhibit macrophage phagocytosis, mobility and killing	14–16
NP affect immune response to common allergens	17
NP are related to lung function decline in asthmatics	
NP cause oxidative stress in vivo and in vitro, by inflammatory action and generation of surface radicals	18–20
NP exposure adversely affects cardiac function and vascular homeostasis	9,21–25
NP have access to systemic circulation upon inhalation and instillation	26–28
NP interfere with Ca-transport and cause increased binding of pro-inflammatory transcription factor NF-κB	29,30
NP cause progression of plague formation	31
NP can affect mitochondrial function	32
NP can translocate to the brain from the nose	33
NP do affect rolling in hepatic tissue	34

this area (35). Inhalation studies with particles have led to the understanding of the effects of particles and are therefore outlined in the next section of this chapter before discussing applications of engineered NP in intravenous (i.v.) or oral drug delivery. In addition, some of the anticipated drug delivery systems do consider inhalation as a port of entry.

General Paradigms in Particle Toxicology

For the interpretation of inhaled particle effects, the following five parameters have to be taken into account: dose, deposition, dimension, durability, and defense. First of all the dose at a specific site (in the lungs) determines the potential toxicity. This deposited dose is of course dependent on the concentration and the dimensions of the particle. Interestingly, the deposition probability of NP increases steeply in the

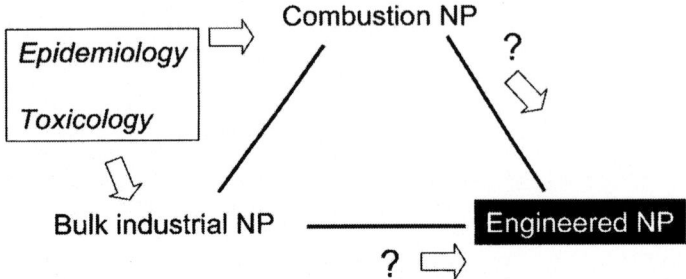

Figure 1 Schematic illustration of the different sources and applications of nanoparticles (NP) and the evidence for their relation with adverse effects in humans or animals. Epidemiology and toxicology have demonstrated acute effects of combustion NP in humans, as well as chronic effects of NP in animals. It remains an open issue whether the hazards and risks found with those types of NP can be extrapolated to engineered NP, which is illustrated by the question marks.

respiratory tract the smaller the particles are. Moreover, a major fraction will be deposited on the fragile epithelial structures of the terminal airways and gas exchange region (36). If a particle is neither soluble nor degradable in the lung it has a high durability and there will be rapid local accumulation upon sustained exposure. The lung, however, has extensive defense systems such as mucociliary clearance (upper airways) and macrophage clearance (lower airways, alveoli) to remove deposited particles. Although the above concept is simple, most of these parameters are interrelated and dimension—as in the case of fibers or nanotubes—may have profound effects on defense and thereby chronic dose. Long (>20 μm) fibers are not taken up by alveolar macrophages, and therefore have a longer half-life in the lung when compared with the same material with shorter fibers and consequently has a higher toxic potency. In addition, particle transport by macrophages from the alveolar region toward the larynx is slow in man even under normal conditions, thus, eliminating only about a third of the deposited particles in the lung periphery; i.e., the other two-thirds accumulate in the lungs without clearance unless they are biodegradable and cleared by other mechanisms (37). If particles are reactive or present at sufficient dose, macrophages and epithelial cells can be activated or damaged leading to inflammation which drives most pathogenic effects of particles.

Pulmonary Deposition and Translocation of NP

Although the deposition of inhaled NP in the respiratory tract follows largely the same distribution as fine particles, the underlying mechanisms are different. NP (< 100 nm) have a size dimension that makes them less subject to gravity and turbidometric forces and therefore their deposition occurs mostly by diffusion (36). In addition, their size makes them interact with other potential targets than conventional fine particles. As a result of their small size, defense is less efficient as recognition by macrophages is suggested to be impaired or less effective. In addition for drug delivery, particle surfaces have been treated to behave as "stealth" particles and remain unrecognized by phagocytosing cells (38). Because of their low uptake by macrophages and their diffusion behavior, NP are suggested to be taken up by endothelial cells and they have access to cells in the epithelium, the interstitium and the vascular walls. However, after instillation of massive doses of NP into the lungs of experimental animals, most particles are located in the interstitium and do not reach the blood stream (Fig. 2). It is only after increasing endothelial or epithelial permeability that particles do translocate to the blood. This may be achieved by mediators released during an inflammatory response such as hydrogen peroxide or histamine (39,40). Wherever they deposit or translocate to, NP has properties such as a large surface that can carry and absorb many endogenous substances such as proteins. It has been shown that particle recognition and distribution can be dramatically affected upon coating with plasma proteins such as ceruloplasmin or cations such as aluminum (41,42). Drug delivery uses this phenomenon termed "stealth particles" to create particle surfaces that are not recognized by the reticulo-endothelial system, which is usually achieved by coating the particle surface with polyethylene glycol (43).

(A)

(B)

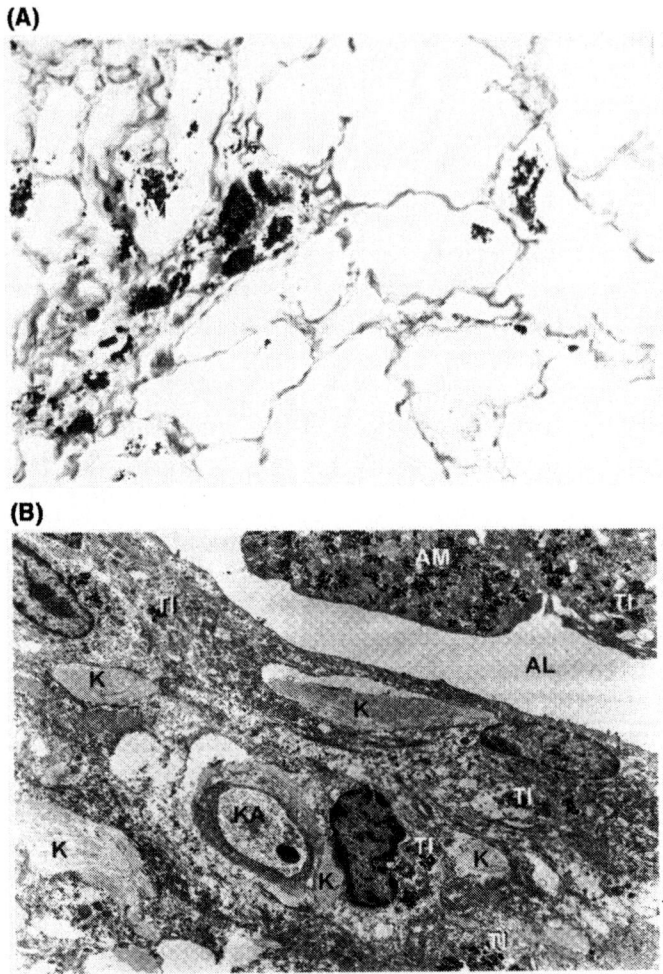

Figure 2 (**A**) Interstitial localization of ultrafine TiO$_2$ particles two years after intratracheal instillation of a high dose (30 mg) of TiO$_2$ (20 nm) in female Wistar rats. The black particle agglomerates are located either in the interstitium or the alveolar lumen. The red staining indicates areas with collagen formation. (**B**) Cellular and subcellular distribution of ultrafine TiO$_2$ (20 nm) two years after in vivo pretreatment (*as in panel A*). The TEM picture shows TiO$_2$ particles (Ti) in an epithelial cell adjacent to an alveolar macrophage (AM). Magnification of the lower panel is 12,800. *Abbreviations*: TiO$_2$, titanium dioxide; TEM, transmission electron microscopy. *Source*: Photo courtesy of Welf Mahlke (**A**) and Dr. Doris Höhr (**B**).

Pulmonary Inflammation and Immune Defense

The toxicological profile of (bulk and combustion) NP has only emerged during the past decade. An early key study demonstrated that ultrafine TiO_2 (20 nm) caused more inflammation in rat lungs than exposure to the same airborne mass concentration of fine TiO_2 (250 nm) (10). Until then TiO_2 had been considered a nontoxic dust and indeed had served as an inert control dust in many studies on the toxicology of particles. Therefore, this report was highly influential in highlighting that a material was low in toxicity in the form of fine particles but could be toxic in the form of ultrafine particles. Later studies have demonstrated that the pulmonary inflammation, usually measured as the number of neutrophilic granulocytes (PMN) in bronchoalveolar lavage (BAL), is related to the instilled or inhaled surface area of particles (11) although at similar surface some ultrafines seem to be more inflammatory than others (18). Among mechanisms by which NP could cause an enhanced inflammatory response, direct effects have been reported on alveolar macrophages such as inward leaching of Ca^{2+}(29), impairment of phagocytosis (14,15) and cytoskeletal changes (16). Epithelial and nerve cells may also contribute to airway inflammation by producing pro-inflammatory cytokines such as interleukin-8 (Donaldson, 2004) or pharmacologically active compounds such as capsacein (30). In this neurogenic inflammation, stimulation of sensory nerve endings releases neurotransmitters which may affect many types of white blood cells in the lung, as well as epithelial and smooth muscle cells. Another potential consequence of exposure to NP may be their effect on the capacity to defend against microorganisms or, in contradiction, an augmentation of allergic immune response to common allergens (17).

Pulmonary Carcinogenicity

Poorly soluble particles (PSP) without specific toxicity such as carbon black and titanium dioxide are known to cause fibrosis, neoplastic lesions and lung tumors in the rat (13). NP (TiO_2, carbon black) can induce lung tumors in rats at considerably lower gravimetric lung burdens than their

larger-sized analogs and actually the retained particle surface metric has been used to describe the lung tumor rate
in chronic inhalation studies (13). It is now generally
accepted that the continued presence of high levels of particle
surface leads to impairment of alveolar macrophage clearance, culminating in rapid buildup of particles, chronic
inflammatory response, fibrosis and tumorigenesis, known
as the so-called rat lung overload. The overall pattern is
one of chronic inflammation that occurs upon saturation of
lung clearance by overloading of macrophages at which point
particle accumulation starts and inflammatory cell influx
increases sharply (5,11,13). The inflammatory cell influx is
held responsible for the lung tumors after chronic particle
exposure to PSP due to their mutagenic activity and actions
on cell proliferation (44,45). The importance of particle surface is illustrated by a graph that summarizes findings on
lung tumors in chronic animal studies using PSP, including
NP (Fig. 3). The graph shows that both inhalation and instillation of particles cause induction of tumors that is related to
the deposited particle surface. As NP have a larger specific
surface area, at similar gravimetric dose, NP cause higher
tumor doses at similar mass dose. Still this surface dose concept is probably an oversimplification for several reasons.
First, ultrafine particles at similar surface area appear to
exhibit significant differences in inflammatory activity (18).
Secondly, it is unclear whether ultrafine particles following
inhalation have a different lung distribution between alveolar spaces, macrophages and interstitium and how relevant
this is for tumor formation (Fig. 2). Thirdly at high local concentrations of NP, these particles should be considered to
penetrate target cells and enter the mitochondria and
nucleus exerting direct effects to DNA (32,46).

EFFECTS OF NANOPARTICLES

Studies with Inhaled NP

Studies with inhaled NP have forwarded several major mechanisms by which the ultrafine component of PM may cause

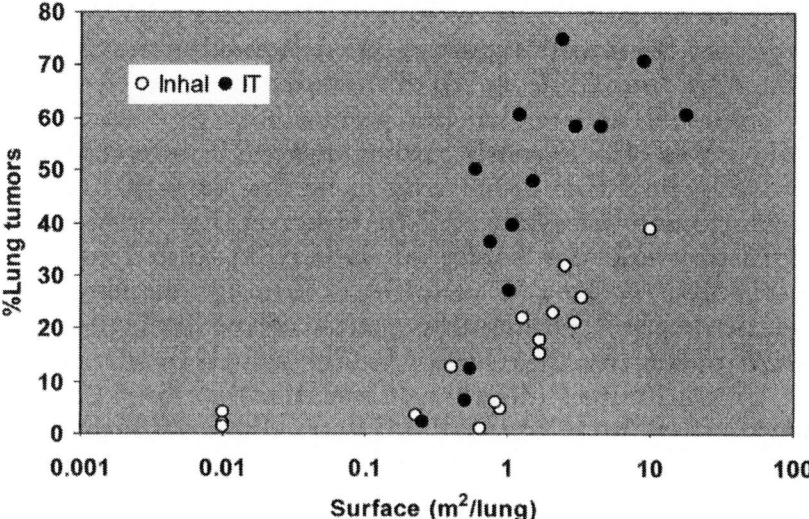

Figure 3 Association between lung tumor response and the particle surface area. Various poorly soluble low toxicity particles (PSP) including nanoparticles have been shown to cause lung tumors after pulmonary administration in rats. The open circles represent different inhalation studies done over the past 15 years. The closed circles are taken from a study where fine and ultrafine particles were administered by intratracheal instillation in rats and lung tumors were evaluated after 129 weeks. In both cases a straight line is obtained with a threshold between 0.2 and 0.3 m^2 surface dose per rat lung, which suggests a no-effect level. *Source*: From Refs. 12, 13.

responses that explain the mortality in those with existing pulmonary and cardiovascular diseases (47). Mechanisms to explain these effects can be discriminated into direct and indirect pathways, as effects by particles themselves or processes induced by particles (mainly in the lung). As a mechanism for direct effects of NP a series of studies have addressed the issue whether NP can translocate from the lung to the circulation, and exert their effects when being in the systemic circulation. However, quantitative estimates of translocation range between 50% of ^{13}C NP (26 nm size) within 24 hours in a rat model to <1% using 18 nm iridium particles in vivo or in isolated perfused lung (27,28,40). This wide variation shows that apart from particle size, particle surface chemistry and particle charge may be important parameters determining the translocation of NP from the lung. Apart from particle characteristics, also epithelial and endothelial permeability are considered to play a role (48). Recently, carbonaceous NP were shown to translocate from the nasal cavity along the same pathway to the central nervous system (CNS), based on their presence in the olfactory bulb of rats after inhalation (33). Such a mechanism was first reported for polio virus (30 nm) in monkeys and was later described for nasally deposited colloidal gold particles (50 nm) moving into the olfactory bulb of squirrel monkeys (49).

Among indirect effects, inflammation has been considered to affect target organs by lung mediators that become systemically available. However, inhalation studies with NP at particle numbers found in the general environment did not demonstrate pulmonary inflammation as described at higher doses. Two mechanisms have been supposed that could be considered as indirect mechanisms:

- Seaton et al. (50) suggested that in susceptible individuals, exposure to NP will invoke alveolar inflammation, and that the release of inflammatory mediators can trigger systemic hypercoagulability of the blood thereby increasing the risk for cardiovascular events.
- A second mechanism is the progression and destabilization of atheromatous plaques by inhalation of PM (31). Although this mechanism remains to be investigated using NP, NP properties should be able to invoke the same destabilization mechanisms (inflammation, LDL oxidation, lipid peroxidation) as the PM used in earlier studies.

A large series of molecular epidemiological studies have supported aspects of the plausibility of the above mechanisms. A large multinational trial on cardiovascular risks (MONICA) performed between 1984 and 1988, reported a higher blood viscosity (51) and C-reactive protein (24) during an air pollution episode that coincided with the survey in 1985. Recent studies from the same research group in Erfurt (Germany) have identified combustion NP as an important variable explaining cardiac deaths due to increased ambient particle exposure (8). In fact the association increased with the smaller particles and individuals with cardiovascular diseases were more likely to die than others. Clearly further research is needed, but the research reported to date has direct relevance to public-health policy, as both coal-burning and traffic emissions continue to be major sources of NP exposure worldwide. Recent cohort and intervention studies in the Netherlands (52) and Ireland (53) have demonstrated the importance of regulation combustion-derived particle emissions.

Studies with NP upon Intravenous Injection

Studies with NP in systemic delivery have been performed with particles developed for therapeutic purposes such as polymers, liposomes and engineered inorganic NP. For i.v. administration the choice of an appropriate NP is crucial with regard to many chemical and biological properties that determine biocompatibility. The toxicology of NP that are used in drug delivery is now well understood thanks to extensive studies on cytotoxicity, hemotoxicity, complement activation, and cellular or humoral immunogenecity of many candidate NP (54). Polycations are in general cytotoxic, hematolytic, and can activate complement. On the contrary, polyanions are less cytotoxic, but can cause anti-coagulant activity and can also stimulate cytokine release from lymphocytes and mononuclear cells. Polymeric macromolecules (including polyamino-acids or polysaccharides) can elicit a humoral response characterized by increased total or specific IgE or IgM levels. All biological responses seem to be molecular-weight-dependent for polymers and also have a size component when considering (in)organic NP. It has to be realized that the potential biological interactions of the NP can change when drugs or complementary antibodies are attached. Both kinetics, distribution, and metabolism can change when new chemical entities are attached or molecular weight and properties are modified. Ideally, the NP or the skeleton should be biodegradable or soluble. If the skeleton is not degradable at all the molecular weight of the co-polymers should be limited to

<40,000 Da to ensure renal elimination as a back up for clearance. All these elements and toxicological concepts are developed and demonstrated in the use of surface engineered iron oxide NP, which have been used to pioneer the biocompatibility of nanomaterials. Magnetite due to its strong magnetic properties was used first in biology and then in medicine for cell separation and magnetic guidance of particles for site-specific drug delivery. Nowadays a whole series of compounds is known (e.g., polyethylene glycol) that can be used for coating of iron oxide particles to allow application in drug delivery (55).

Gastrointestinal Uptake and Effects of NP

Nanoparticles (<0.1 μm) and microparticles (0.1–3 μm) are ingested at high levels per person per day and it is estimated that 10^{12}–10^{14} particles are ingested per person per day in the Western world (56), and concerns mainly silicates and titanium dioxide. They are scavenged by M cells overlying the intestinal mucosa and in this way circumvent active uptake by intestinal epithelium. A gastrointestinal (GI) route of translocation of ingested ultrafine particles to the blood is supported by studies in rats and humans that have shown that TiO_2 particles (150–500 nm) taken in via food can translocate to the blood and are taken up by liver and spleen (57,58). Earlier studies described a mechanism of persorption in epithelial cells of the GI tract by which even larger particles are taken up into lymphatic and blood circulation and translocate to the liver and other organs (59). Recently, nanocrystals have become the subject of intense investigation for oral administration of drugs and functional food components. Drugs or food constituents are produced in 100% pure form in nanocrystals, by precipitation or other processes (60). As they are easy to produce and are very efficient in vivo, their production for oral application is expected to increase considerably. Interestingly, studies with ultrafine metal particles did not show a significant translocation from the GI tract to other organs via the blood circulation (26). In the latter study after esophageal administration of 18 nm ^{192}Ir particles in suspension, virtually the whole amount of ^{192}Ir was found in fecal excretion within 2–3 days.

During the 6-day observation period no detectable ^{192}Ir in urine was observed at any day. Six days after administration there was no detectable ^{192}Ir in any organ or tissue of the body. Hence, it was concluded that for these particles there was no uptake and/or absorption from the GI tract. Some studies have shown that the NP (titanium dioxide, silicates) can accumulate locally in the GI tract in so-called Peyer's patches and have suggested that this may be related to exacerbation of inflammation in Crohn's disease, but the evidence is weak (61).

Dermal Uptake and Effects of NP

With a surface of well over $2\,m^2$ the skin is one of the major exposure routes for NP. Particles with a size of approximately 50–500 nm are widely used in cosmetic products, in order to improve the homogeneity of the distribution of the formulations on the skin surface, or to act as a UV filter against sun radiation. The smallest particles act as "nanomirrors" on the skin and partly reflect the sunlight. Because of their scattering properties, they increase the optical pathway of UV photons entering the upper part of the horny layer, and energy is absorbed by the stratum corneum and by the applied organic filter substances. On the contrary, one of the mostly handled NP in cosmetics, i.e., TiO_2, has considerable photocatalytic activity. In order to prevent potential adverse effects caused by this property, the titanium dioxide used in cosmetic preparations is often coated. The usual concentration of the NP in the formulations is significantly $< 3\%$. Sunscreens are usually applied on to the skin at a concentration of $2\,mg/cm^2$, which means that if $1\,m^2$ of skin is treated, the total external amount of NP is $0.03 \times 2 \times 10,000$ mg, i.e., 600 mg of NP. There is considerable discussion about the uptake of NP through the skin. In principle there are three possible penetration pathways of topically applied substances through the skin: the intercellular penetration, the intracellular penetration, and the follicular penetration (Fig. 4). In the past, the penetration was described as a diffusion process through the lipid layers of the stratum corneum. Liposomes with a diameter between 20 and 200 nm have been found to

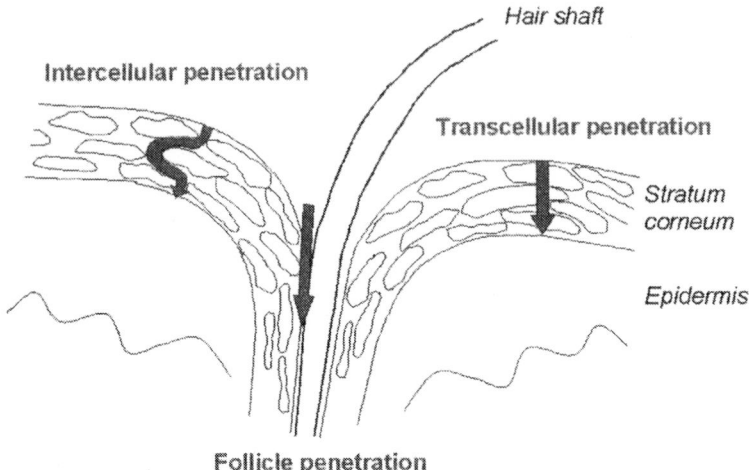

Figure 4 Different routes of penetration through the skin. With a surface of 2 m² the skin is a major potential route of uptake of nanoparticles. In principle there are three penetration pathways through the skin: intercellular, intracellular, and follicular penetration.

be active carriers of topically applied drugs into the living epidermis via the intercellular penetration route (62).

Lademann et al. (63) investigated the penetration of coated titanium dioxide NP into the stratum corneum of living human skin by tape stripping and biopsies in combination with spectroscopic measurements. A penetration of NP through the horny layer could not be detected by X-ray fluorescence in the other histological sections of the biopsy. These data also confirmed the results of an earlier study which detected no percutaneous absorption of particles in skin samples from humans treated with a microfine titanium oxide-containing sunscreen (64). A more recent study on the percutaneous penetration of two different micronized titanium dioxide preparations used in sunscreens: (i) particle size of 10–15 nm, which aggregated to particles of approximately 100 nm, and (ii) particle size of approximately 100 nm, revealed that these particles remain on the surface of the stratum corneum, and do not penetrate into the stratum corneum or living compartments of the skin (65). The absence of skin penetration of

NP is also consistent with the results of a recent study that measured in vitro the percutaneous penetration of micronized zinc oxide (mean particle size: 40 nm) through pig skin. The study found no measurable penetration of zinc oxide NP into the viable parts of the skin (BASF Study No 52H0546/032193, 2004, unpublished data). Although it cannot be excluded that the physical properties of NP may enhance the absorption/penetration of certain substances applied to the skin, such as reported for methanol or octanol (66), the results of available studies suggest that, although small particles may be deposited on the follicle orifice, they do not penetrate the skin via the follicle. This was confirmed by the results of a recent study, which showed that polystyrene NP (20–200 nm) accumulated in the follicle orifices but did not penetrate into the skin or the follicle (67).

Effects of NP in the Central Nervous System

Nanoparticles can get access to the brain by two different mechanisms:

- Trans-synaptic transport after inhalation through the olfactory epithelium.
- Uptake through the blood–brain barrier (BBB).

The first pathway has been studied primarily with model particles such as carbon, Au and MnO_2 in experimental inhalation models. The second pathway has been the result of extensive research and particle surface manipulation in drug delivery, as an approach to try and get drugs to the brain (60,68). The latter studies suggest that the physiological barrier may limit the distribution of some proteins and viral particles after transvascular delivery to the brain, suggesting that the healthy BBB contains defense mechanisms protecting it from blood borne NP exposure. A number of pathologies, including hypertension and allergic encephalomyelitis, however have been associated with increased permeability of the BBB to NP in experimental set ups. Reversely, the NP surface charges have been shown to alter blood–brain integrity (69) and need consideration for brain toxicity and brain distribution profiles.

The use of paramagnetic NP for magnetic resonance imaging of different cell types within neural tissue has proved useful experimentally, and it has been suggested that this might be useful in humans to track, for example, the development of stem cell grafts used to treat neurodegenerative diseases (70). However, the potential impact of NP on human neuronal tissue is as yet not investigated in detail. As NP have been shown to induce the production of reactive oxygen species (ROS) and oxidative stress, and oxidative stress has been implicated in the pathogenesis of neurodegenerative diseases such as Parkinson's and Alzheimer's, it is conceivable that the long-term effects might include a decrease in cognitive function (71). Evidence for such effects is presented by studies in biopsies from city dwellers and Alzheimer's like pathology that have demonstrated increased markers of inflammation and AB42-accumulation in frontal cortex and hippocampus in association with the presence of NP (72). Recently, also inhalation exposure of BALB/c mice to PM showed the activation of pro-inflammatory cytokines in the brain of exposed mice (73). Whether this is due to the fraction of combustion NP remains to be investigated.

SCREENING ENGINEERED NP FOR TOXICOLOGICAL HAZARDS

Engineered NP are increasingly used as devices to target drugs to specific tissues, to increase their biological half life, or for imaging purposes such as extravasation and tumor vascularization. The testing of engineered NP for these purposes follows guidelines such as defined in the European Union or Food and Drug Administration medical device regulations, which are based on their biocompatibility as measured by platelet adhesion and activation, neutrophil attachment, angiogenesis, and cell spreading (54,55).

As discussed earlier on, one of the crucial questions is whether the hazards and risks of *inhaled* bulk and combustion NP can be extrapolated to engineered NP for application in drug delivery (Fig. 1). NP are being advocated for exploration in atherosclerosis, inflammatory lung diseases, diabetes, brain tumors and hemorrhagic disorders (55,68,74). When knowing the effects of inhaled combustion NP or PM_{10} in these patient groups, a striking discrepancy emerges between the anticipated therapy and the observation that these patients are the primary targets at air pollution episodes. Development of drug delivery systems based on NP should attain the utmost care not to use particles that aggravate symptoms or conditions of these patients. Secondly, as the epidemiology of combustion NP (PM) has identified those with COPD, asthma and cardiovascular disease as the risk groups (47,52,53), it needs careful consideration whether animal models for these diseases should be used and developed to test hazards of engineered NP.

Surface Modification and Coatings

Whatever test will be used or developed, it needs to be considered that most suppliers apply postsynthetic strategies to modify NP to prevent aggregation or stimulate disaggregation. This part is discussed extensively in the first part of this book (A. technologies for NP manufacturing). In summary, postsynthetic routes open a variety of possible surface modifications which can be adjusted to any application, using chemicals such as 4-dimethylaminopyridine, various thiols, fluoroalkanes, alkoxysilanes, and phosphorous containing substances. There is now a body of evidence from drug delivery and toxicological literature that surface modification as well as surface charge can have major impact on biological response to the particles, including phagocytosis, genotoxicity, and inflammation. Particle coating with polyethylene glycol or dextran are common treatments to prevent recognition by the reticulo-endothelial system and to increase the half-life of the particle-conjugated drugs (55). A clear example from particle toxicology demonstrating the crucial role of particle surface comes from work with respirable, nonultrafine quartz samples. Coating with aluminum lactate or the polymer poly-4-vinyl pyridine-N-oxide (PVNO) has a dramatic beneficial impact on the various adverse effects of the native quartz, including phagocytosis/endocytosis, oxidative DNA damage and inflammation upon intratracheal instillation in rat lung (75,76). In sunscreens NP are often used as "nanomirrors" on the skin and partly reflect the sunlight. Because of their scattering properties, they increase the optical pathway of UV photons entering the upper part of the horny layer. In this way, more photons are absorbed by the stratum corneum and by the applied organic filter substances. Coated titanium dioxide NP are commonly used as UV filter substances in commercial sunscreen products. Concern has been raised about a possible photocatalytic activity of titanium dioxide on living tissues and to reduce potential adverse effects, the titanium dioxide used in cosmetic preparations is often coated. Surface-modified TiO_2 has been the subject of considerable toxicological investigation and has shown that the hydrophobic coatings usually tend to lower the inflammatory response after inhalation or instillation (Table 4). However, one study reported a very high acute toxicity after instillation of doses around 1 mg per rat (77). With this regard it is crucial to know how the surface modification has been achieved and if this can be released from the NP in biological media (low pH in macrophages). In the case of sunscreen-grade-coated titanium NP

Table 4 Intratracheal Instillation Studies with Surface-Modified Ultrafine TiO_2 in Rats

Material tested	Dose-exposure	Result	References
T805, silanized TiO_2 (P25, 21 nm, 45 m²/g)	0.1–120 mg, 24 hrs	High toxicity of coated TiO_2. Lethal effects	77
T805, P25 and quartz	0.15, 0.3, 0.6 and 1.2 mg, 90 days	No difference in inflammation and proliferation	78
T805 versus P25	50 and 500 µg	Less inflammation with T805	5
Methylated TiO_2 (fine and NP) versus native form	1 and 6 mg	Less inflammation with methylated TiO_2 at 1 mg dose	79

Abbreviations: TiO_2, titanium dioxide; NP, nanoparticles.

the stability of the coating was investigated by laser-induced plasma spectroscopy (63). No changes in the mechanical stability of the coated microparticles could be detected during the manufacturing and penetration of the sunscreen.

Other studies have indicated that blood coagulation by latex particles, when infused into the jugular vein of hamsters was dependent on the surface charge (80). Studies on nasal translocation showed that surface charge and chemistry affected the rate of translocation to the blood (81). Uptake of lipid particles through the BBB was only achieved successfully when using a specific (Tween®80) surface coating, which mediates its binding to the ApoE receptor (68). Most likely, but unknown, surface chemistry also plays a role in the uptake of NP through the olfactory epithelium into the brain (33). Therefore, it is recommended that before testing an NP formulation, the surface modification procedure and its effects on typical surface properties as zeta potential and surface reactivity should be known.

Tests for Toxicity of Engineered NP

Considering the large amount of research on effects and mechanisms of (combustion) NP it is surprising to note that little of this work has been done to screen engineered NP to prevent adverse biological effects. There are different opinions on the statement whether existing tests may pick up all of the hazards. Existing tests may pick up the toxicological hazards of NP but not sensitive enough, or hazards are not seen at all, because insensitive models are being used. The latter is underscored by the negative outcomes of animal research with combustion NP trying to reproduce the effects of PM seen in epidemiological studies. This underscores the need to develop and validate new test models as well as to evaluate and validate existing methods for testing of engineered NP.

Currently, there is no alternative to the approach used in the pharmaceutical industry, i.e., a case-by-case approach. There is definitely a need to develop concepts of testing, which can be done by bridging studies with the right dosimetry. The dosimetry should be related to the anticipated application of the nanostructured materials and to the metric which is chosen or investigated. Another reason for the right dosimetry is that often only small amounts of nanomaterials are available, and for instance (chronic) inhalation studies are virtually impossible. With regard to testing of the toxicity of engineered NP, a variety of tests may be used that were applied to particles of highly contrasting size and dimension, such as asbestos and man-made fibers, quartz and coal mine dust, fly ashes, and diesel exhaust particles. Particularly, in view of the "NP hypothesis," a number of in vitro and in vivo tests have been introduced for comparative toxicity testing at equal mass of commercial particles of fine versus ultrafine size (e.g., carbon black, TiO_2). Tests for NP toxicity can be arbitrarily subdivided into four levels:

1. Testing of the (re)activity of NP in acellular or subcellular systems (e.g., dissolution, radical generation, protein/DNA oxidation, lipid peroxidation, enzyme inactivation/immobilization, action on isolated mitochondria, etc.).
2. Testing of NP in vitro, using intact cells or cell systems (e.g., lung epithelial cells, tracheal explants, vascular endothelium, macrophages, etc.).
3. Testing of NP on isolated organ (culture systems) (e.g., intact skin models, whole blood, isolated perfused lung, heart, etc.).
4. In vivo testing.

Short-Term Tests for Toxicity and
Inflammatory Potential

Very rapid and basic acellular approaches that have been used to predict particle reactivity include testing of plasmid DNA, unwinding, or oxidation of calf thymus DNA (19,82). Electron paramagnetic resonance (EPR) combined with a spin-trap has been used to determine the radical generation properties of particulate materials well above the nano-size range such as quartz and asbestos in relation to (surface) modification as well as of ambient PM. In general, these EPR studies showed positive associations with toxicity in vitro and/ or in vivo toxicity (42,75,82).

Cellular tests are very common in particle toxicology as little material is required, they have short-term read outs and microscopical techniques are easy to apply concomitantly. A classic test is the hemolysis assay, which is based on properties of reactive constituents to elicit hemoglobin leakage upon red blood cell membrane damage (83). Nowadays, toxicity testing of particles in vitro mainly uses primary cells and/or immortalized cell lines and has been performed on the basis of membrane damaging properties (e.g., LDH leakage) or for instance changes in mitochondria-associated metabolic competence measurements (e.g., MTT assay, ATP), reflecting necrosis and/or apoptosis (83). As such, various cells or cell types can be tested, e.g., in relation to the route of entrance and/or target tissue or organ of concern (e.g., pulmonary type II cells, colon cells, keratinocytes, endothelial cells). Notably, alveolar epithelial cells are often used to screen for markers of pro-inflammatory pathways and/or toxicity of inhaled NP (84) but other assays can be chosen based on the application of the NP. Differential cell adhesion and toxicity (fibroblast) are often used as a screening test to optimize surface modification in NP for drug delivery of coatings of biocompatible materials (85).

It is important to realize that many of the assays require molecular biological tools and methods, and are often of semi-quantitative nature, and it is therefore recommended to use clear-cut basic endpoints. One of these is the production of the key inflammatory protein tumor necrosis factor-alpha (TNF-α). A test may be a combination of several endpoints in one target cell such as alveolar macrophages to evaluate the inflammogenic potential of engineered NP. The outcomes are projected as vectors in a multidimensional matrix including pathological stimulation (TNF-α and ROS secretion) and impairment of cellular functions (LDH- and PMA-stimulated ROS secretion) (86,87). These various parameters turned out to be at least partly independent and the vector model has been used for the characterization of particles from different origin including commercial particles such as metal NP, toner particles, particles from sinter technology (also NP), species of quartz powders of different deposits and particles from dusty worksites with specific endangerment (86). Most of the in vitro tested particle classes have been analyzed also in vivo (subchronic, intratracheal instillation) on representative samples for inflammation and genotoxicity (e.g., 70). The most common in vivo system for testing of NP is represented by bronchoalveolar lavage analysis of lungs and determination of specific tissue markers, e.g., for genotoxicity and proliferation of rats or mice following particle application, e.g., using whole-body inhalation, intratracheal inhalation, or intratracheal instillation (5,78,79,88).

Short-Term Tests for Immunogenic Potential

Various ultrafine particles have been shown to act as an adjuvant in mice when co-exposed via subcutaneous injection, nasal instillation or inhalation to common allergens (e.g., Ovalbumin, house dust mite, pollen), and this was associated with the physicochemical properties of particles such as surface area or soluble metals (17,89). The allergic potential of particles may also be screened for with other typical immunological assays such as the mouse popliteal lymph node (PLN) assay (90). In general these mouse models are excellent tools to test for the possible adjuvant effects of NP. However, contrasting observations in the literature indicate that it will be a major challenge to determine the treatment order and interval for the particles and the allergen. Particles may also be contaminated or carriers of endotoxin or pollen allergens. For instance, endotoxin has been shown to be a potent inflammatory compound on specific PM samples and was demonstrated to act as a priming agent for pulmonary inflammation by particles (5,91). The "pyrogenic" properties of contaminants of NP (e.g., endotoxin, glycans) can be tested using short-term in vivo instillation assays as well as in vitro assays such as described for the testing of the inflammatory potential. In this regard, incubation of NP with whole blood for subsequent cytokine production could be envisaged as a rapid initial screening tool and possible replacement of animal studies (91,92). Finally, NP have been shown to impair macrophage function (14,15).

Cardiovascular Effects and Hematocompatibility

Patients with cardiovascular diseases are the main vulnerable group with respect to the effects of inhaled combustion NP. Therefore, it is recommended that toxicological testing of engineered NP will also include cardiovascular endpoints and models. Broadly, the cardiovascular effects of NP can be classified into (1) effects on clotting homeostasis, and (2) effects on neural control of cardiovascular function, and therefore testing procedures and models should follow these principles.

An overview of endpoints and models that can be used is given in Table 5. A number of animal models have been used to investigate the potential cardiovascular effects of combustion NP. These models include both healthy animals and compromised models such as Watanabe heritable hyperlipidemic rabbits, ApoE knockout mice with hyperlipoproteinemia that develop atheromatous plaques, and spontaneous hypertensive rats (24,31,93). Men, dogs, and rats have also been used to study heart rate variability after inhalation of combustion NP (21,94–96). Both in vivo and ex vivo experiments have been conducted after in vivo exposure to NP. For example, artery diameters and blood pressure was measured in vivo, while isolated aorta rings or hearts were used to study autonomic innervation and function of the heart and vascular tissue (22–24,94,97). Telemetric procedures can be applied in most animals nowadays to measure cardiac function and innervation in aware, unanesthetized animals (94,97).

Several toxicological studies have demonstrated that combustion and model NP can gain access to the blood following inhalation or instillation and can enhance experimental thrombosis (80). Ligand-coated engineered NP have been explored for decades as agents for molecular imaging or drug delivery tools. This has led to a considerable understanding of particle properties that cause low hematocompatibility. In

Table 5 Test Methods and Models That Can Be Used to Explore Potential Hazards of NP Particles for the Cardiovascular System

Test system	Endpoint	Interpretation
Cardiomyocytes		
Primary	Beating frequency, Ca^{2+}-response Ca-dependent channels, cytoskeletal behavior	Basic effects on cardiac metabolism
Cell lines	Ca^{2+}-response, ATP response, apoptosis	
Aorta rings (in vitro)	Contraction–relaxation with epinephrin–carbachol	Autonomic innervation
Langendorf perfusion(ex vivo)	Coronary flow, left ventricular developing pressure, heart rate	Heart function
Telemetry (whole animal)	Heart rate variability, ECG, blood pressure, vessel diameter	Cardiac and vascular function
Biomarkers in plasma/blood	Fibrinogen, CRP	Acute phase response
	Factor VII Viscosity clotting time Plasminogen activator inhibitor	Blood coagulation
	Endothelin, ACE	Endothelial damage/ activation

Abbreviations: ECG, electrocardiogram; CRP, C-reactive protein; ACE, angiotensin-converting enzyme, ATP, adenosine triphosphate.

general cationic NP, including gold and polystyrene, have been shown to cause hemolysis and blood clotting, while usually anionic particles are quite nontoxic. This conceptual understanding may be used to prevent potential effects of unintended NP exposure. Similarly, drug-loaded NP have been used to prolong half-life or reduce side-effects and have shown which particle properties need to be modified to allow delivery, while being biocompatible (55). Also this know-how can help to further develop engineered NP for other applications that are with low hazard.

Tests for Uptake and Effects in CNS

Only recently, it has been demonstrated that inhaled ultra-fine particles may translocate into the CNS via the olfactory epithelium nerve, and NP drugs have been shown to cross the BBB under certain conditions (33,68). Routine tests to allow testing of uptake into the CNS by various NP, as well assays to predict their possible adverse consequences for this organ need to be developed (e.g., neuroimaging techniques in exposed animals, in vitro tests with glia cells).

CONCLUSION

Presently, it is unknown whether the hazards, vulnerable groups, and mechanism of action induced by combustion NP in epidemiological studies are applicable to hazard and risk estimation of the immense variety of engineered NP. In addition, it is unsure whether inhalation studies with bulk NP (carbon black, TiO_2) can be used for the same purpose. Nevertheless, it seems important that nanomedicine should learn from these observations to double check and to update its testing strategies for NP used in drug delivery. Whereas toxicology is trying to understand the mechanisms of NP translocation and how these minute amounts of NP might invoke systemic response, pharmacology intends to use NP for systemic delivery of drugs. In this review we have indicated what effects of (inhaled) NP have been found by toxicologists and epidemiologists, and how this know-how could be used to develop new screening procedures for safe NP for drug delivery.

Communication and open minds are needed for exchange of know-how and testing methods between inhalation toxicologists and those active in drug delivery. They do have at least one mutual question of interest that is what material surface properties determine its acute and chronic interaction with biological systems where they deposit or interact. Interactions between cells and NP are mediated by the surface characteristics of both the material and the target cells. On the one hand, proteins, extracellular matrix and cell recognition play an important role. On the other hand, the physicochemical properties of NP including their size play an important role in the pharmaceutical and dynamical phase of the NP-drug conjugate. However, NP may elicit a biological response in one tissue (e.g., bone) but not in another (e.g., blood). In addition, inhalation toxicology tells us that NP usually invoke responses in those with existing diseases. As drugs are primarily used in those with diseases, it should be stressed that toxicological testing of NP should be done in various models that reflect human diseases. Therefore, it is recommended that a close interaction between both areas of research should be established, which will lead to screening methods that can be used to develop both safe NP for drug delivery and a better understanding of NP toxicology after inhalation.

ACKNOWLEDGMENTS

The authors are indebted to many collaborators in the development of their know-how in this area. In particular, collaboration with Ken Donaldson (ELEGI, Edinburgh), Wolfgang Kreyling (GSF, Munich) and Detleff Müller-Schulte (Magnamedics GmbH) has generated a lot of the thinking that is included in this chapter. In addition, we thank Catrin Albrecht, Doris Hoehr, and Welf Mahlke for the use of their work on lung microscopy.

REFERENCES

1. Duncan R. The dawning era of polymer therapeutics. Nat Rev 2003; 2:347–360.

2. Wagner V. Nanobiotechnologie II: Anwendungen in der Medizin und Pharmazie. Dusseldorf, Germany: Zukunftige technologien Nr 50, 2004:184. ISSN 1436–5928.

3. http://www.corporate.basf.com/basfcorp/img/innovationen/ felder/nanotechnologie/ e/c_021206_distler.pdf.

4. Dockery DW, Pope CA III, Xu X, et al. An association between air pollution and mortality in six U.S. cities. N Engl J Med 1993; 329:1753–1759.

5. Oberdörster G. Pulmonary effects of ultrafine particles. Int Arch Occup Environ Health 2001; 74:1–8.

6. Donaldson K, Brown D, Clouter A, Duffin R, MacNee W, Renwick L, Stone V. The pulmonary toxicology of ultrafine particles. J Aerosol Med 2002; 15:213–220.

7. Peters AE, Wichmann HE, Tuch T, Heinrich J, Heyder J. Respiratory effects are associated with the number of ultrafine particles. Am J Respir Crit Care Med 1997; 155:1376–1383.

8. Ibald-Mulli A, Wichmann HE, Kreyling W, Peters A. Epidemiological evidence on health effects of ultrafine particles. J Aerosol Med 2002; 15:189–201.

9. de Hartog JJ, Hoek G, Peters A, et al. Effects of fine and ultrafine particles on cardiorespiratory symptoms in elderly subjects with coronary heart disease. Am J Epidemiol 2003; 157:613–623.

10. Ferin J, Oberdörster G, Penney DP. Pulmonary retention of fine and ultrafine particles in rats. Am J Respir Cell Mol Biol 1992; 6:535–542.

11. Tran CL, Buchanan D, Cullen RT, Searl A, Jones AD, Donaldson K. Inhalation of poorly soluble particles. II. Influence of particle surface area on inflammation and clearance. Inhal Toxicol 2000; 12(12):1113–1126.

12. Morrow PE, Haseman JK, Hobbs CH, Driscoll KE, Vu V, Oberdorster G. The maximum tolerated dose for inhalation bioassays: toxicity vs overload. Fundam Appl Toxicol 1996; 29(2): 155–167.

13. Borm PJA, Schins RPF, Albrecht CA. Inhaled particles and lung cancer. Part B: Paradigms and risk assessment. Int J Cancer 2004; 110(1):3–14 (Review).

14. Renwick LC, Donaldson K, Clouter A. Impairment of alveolar macrophage phagocytosis by ultrafine particles. Toxicol Appl Pharmacol 2001; 172:119–127.

15. Lundborg M, Johard U, Lastbom LP, Gerde P, Camner P. Human alveolar macrophage phagocytic function is impaired by aggregates of ultrafine carbon particles. Environ Res 2001; 86:244–253.

16. Möller W, Hofer T, Ziesenis A, Karg E, Heyder J. Ultrafine particles cause cytoskeletal dysfunctions in macrophages. Toxicol Appl Pharmacol 2002; 182:197–207.

17. Granum B, Lovik M. The effect of particles on allergic immune responses. Toxicol Sci 2002; 65:7–17.

18. Zhang G, Kusaka Y, Sato K, Nakakuki K, Kohyama N, Donaldson K. Differences in the extent of inflammation caused by intratracheal exposure to three ultrafine metals: role of free radicals. J Toxicol Environ Health 1998; 53:423–438.

19. Dick CAJ, Brown DM, Donaldson K, Stone V. The role of free radicals in the toxic and inflammatory effects of four different ultrafine particle types. Inhal Toxicol 2003; 15:39–52.

20. Donaldson K, Stone V, Borm PJA, et al. Oxidative stress and calcium signalling in the adverse effects of environmental particles (PM_{10}). Free Radic Biol Med 2002; 34:1369–1382.

21. Stone PH, Godleski JJ. First steps toward understanding the pathophysiologic link between air pollution and cardiac mortality. Am Heart J 1999; 138:804–807.

22. Brook RD, Brook JR, Urch B, Vincent R, Rajagopalan S, Silverman F. Inhalation of fine particulate air pollution and ozone causes acute arterial vasoconstriction in healthy adults. Circulation 2002; 105:1534–1536.

23. Batalha JRF, Saldiva PHN, Clarke RW, et al. Concentrated ambient air particles induce vasoconstriction of small pulmonary arteries in rats. Environ Health Perspect 2002; 12:1191–1197.

24. Peters A, Frohlich M, Doring A, et al. Particulate air pollution is associated with an acute phase response in men; results from the MONICA-Augsburg Study. Eur Heart J 2001; 22(14):1198–1204.

25. Bagate K, Meiring JJ, Gerlofs-Nijland ME, Vincent R, Cassee FR, Borm PJ. Vascular effects of ambient particulate matter instillation in spontaneous hypertensive rats. Toxicol Appl Pharmacol 2004; 197(1):29–39.

26. Nemmar A, Vanbilloen H, Hoylaerts MF, Hoet PH, Verbruggen A, Nemery B. Passage of intratracheally instilled ultrafine particles from the lung into the systemic circulation in hamster. Am J Respir Crit Care Med 2001; 164(9):1665–1668.

27. Kreyling WG, Semmler M, Erbe F, et al. Translocation of ultrafine insoluble iridium particles from lung epithelium to extrapulmonary organs is size dependent but very low. J Toxicol Environ Health A 2002; 65(20):1513–1530.

28. Oberdorster G, Sharp Z, Atudorei V, et al. Extrapulmonary translocation of ultrafine carbon particles following wholebody inhalation exposure of rats. J Toxicol Environ Health 2002; 65(20):1531–1543.

29. Stone V, Tuinman M, Vamvakopoulos JE, et al. Increased calcium influx in a monocytic cell line on exposure to ultrafine carbon black. Eur Respir J 2000; 15:297–303.

30. Veronesi B, de Haar C, Roy J, Oortgiesen M. Particulate matter inflammation and receptor sensitivity are target cell specific. Inhal Toxicol 2002; 14:159–183.

31. Suwa T, Hogg JC, Quinlan KB, Ohgami A, Vincent R, van Eeden SF. Particulate air pollution induces progression of atherosclerosis. Am Coll Cardiol 2002; 39:943–945.

32. Li N, Sioutas C, Cho A, et al. Ultrafine particulate pollutants induce oxidative stress and mitochondrial damage. Environ Health Perspect 2003; 111(4):455–460.

33. Oberdorster G, Sharp Z, Atudorei V, et al. Translocation of inhaled ultrafine particles to the brain. Inhal Toxicol 2004; 16(6–7):437–445.

34. Khandoga A, Stampfl A, Takenaka S, et al. Ultrafine particles exert prothrombotic but not inflammatory effects on the hepatic microcirculation in healthy mice in vivo. Circulation2004; 109(10):1320–1325.

35. Borm PJA. Particle toxicology: from coal mining to nanotechnology. Inhal Toxicol 2002; 14:311–324.

36. Bair WJ, Bailey MR, Cross FT, et al. ICRP Publication 66. Ann ICRP 1994; 24:1–3.

37. Kreyling WG, Scheuch G. In: Gehr P, Heyder J, eds. Particle Lung Interactions. New York-Basel: Marcel Dekker, 2000:323–376.

38. Moghimi SM, Hunter AC. Capture of stealth nanoparticles by the body's defences. Crit Rev Ther Drug Carrier Syst 2001; 18:527–550.

39. Nemmar A, Nemery B, Hoet PH, Vermylen J, Hoylaerts MF. Pulmonary inflammation and thrombogenicity caused by diesel particles in hamsters: role of histamine. Am J Respir Crit Care Med 2003; 168(11):1366–1372.

40. Meiring JJ, Borm PJA, Bagate K, et al. The role of endothelial permeability in translocation of nanoparticles in the isolated perfused rat lung. Particle Fibre Toxicol 2005; 2:3.

41. Gupta AK, Curtis AS. Lactoferrin and ceruloplasmin derivatized superparamagnetic iron oxide nanoparticles for targeting cell surface receptors. Biomaterials 2004; 25(15):3029–3040.

42. Schins RPF, Duffin R, Höhr D, et al. Surface modification of quartz inhibits its ability to cause radical formation, particle uptake and oxidative DNA damage in human lung epithelial cells. Chem Res Toxicol 2002; 15:1166–1173.

43. Harris JM, Chess RB. Effect of pegylation on pharmaceuticals. Nat Rev Drug Discov 2003; 2:214–221.

44. Driscoll KE, Deyo LC, Carter JM, Howard BW, Hassenbein DG, Bertram TA. Effects of particle exposure and particle-elicited inflammatory cells on mutation in rat alveolar epithelial cells. Carcinogenesis 1997; 18(2):423–430.

45. Knaapen A, Borm PJA, Albrecht CA, Schins RPF. Inhaled particles and lung cancer. Part I; mechanisms. Int J Cancer 2004; 109(6):799–809 (Review).

46. Kneuer C, Sameti M, Bakowsky U, et al. A nonviral DNA delivery system based on surface modified silica-nanoparticles can efficiently transfect cells in vitro. Bioconjug Chem 2000; 11(6):926–932.

47. Pope CA III, Burnett RT, Thurston GD, et al. Cardiovascular mortality and long-term exposure to particulate air pollution: epidemiological evidence of general pathophysiological pathways of disease. Circulation 2004; 109(1):71–77 (Epub 2003; Dec 15).

48. Hamoir J, Nemmar A, Halloy D, et al. Effect of polystyrene particles on lung microvascular permeability in isolated perfused rabbit lungs: role of size and surface properties. Toxicol Appl Pharmacol 2003; 190(3):278–285.

49. De Lorenzo A, Darin J. Wolstenholme GEW, Knight J, eds. Taste and Smell in Vertebrates. London: Churchill, 1970:151–176.

50. Seaton A, MacNee W, Donaldson K, Godden D. Particulate air pollution and acute health effects. Lancet 1995; 345:176–178.

51. Peters A, Doring A, Wichmann HE, Koenig W. Increased plasma viscosity during an air pollution episode: a link to mortality? Lancet 1997; 349:1582–1587.

52. Hoek G, Brunekreef B, Goldbohm S, Fischer P, van den Brandt PA. Association between mortality and indicators of traffic-related air pollution in the Netherlands: a cohort study. Lancet 2002; 360(9341):1203–1209.

53. Clancy L, Goodman P, Sinclair H, Dockery DW. Effect of air-pollution control on death rates in Dublin, Ireland: an intervention study. Lancet 2002; 360(9341):1210–1214.

54. Rihova B. Biocompatibility of biomaterials: haemocompatibility, immunocompatibility and biocompatibility of solid polymeric materials and soluble targetable polymeric carriers. Adv Drug Delivery Rev 1996; 21:157–176.

55. Gupta AK, Gupta M. Synthesis and surface engineering of iron oxide nanoparticles for biomedical applications. Biomaterials 2005; 26:3995–4021.

56. Lomer MCE, Thompson RPH, Powell JJ. Fine and ultrafine particles of the diet: influence on the mucosal immune response and association with Crohn's disease. Proc Nutr Soc 2002; 61:123–130.

57. Jani PU, McCarthy DE, Florence AT. Titanium dioxide (rutile) particles uptake from the rat GI tract and translocation to systemic organs after oral administration. Int J Pharm 1994; 105:157–168.

58. Bockmann J, Lahl H, Eckert T, Unterhalt B. Titanium blood levels of dialysis patients compared to healthy volunteers. Pharmazie 2000; 55(6):468.

59. Volkheimer G. Passage of particles through the wall of the gastrointestinal tract. Environ Health Perspect 1974; 9:215–225.

60. Muller RH, Keck CM. Drug delivery to the brain—realization by novel drug carriers. J Nanosci Nanotechnol 2004; 4:471–483.

61. Lomer MC, Hutchinson C, Volkert S, et al. Dietary sources of inorganic microparticles and their intake in healthy subjects and patients with Crohn's disease. Br J Nutr 2004; 92(6):947–955.

62. Egbaria K, Weiner N. Liposomes as a topical drug delivery system. Adv Drug Del Rev 1990; 5:287–300.

63. Lademann J, Weigmann HJ, Rickmeier CH, et al. Penetration of titanium dioxide microparticles in a sunscreen formulation into the horny layer and the follicular orifice. Skin Pharm Appl Skin Phys 1999; 12:247–256.

64. Tan MH, Commens CA, Burnett L, Snitch PJ. A pilot study on the percutaneous absorption of microfine titanium dioxide from sunscreens. Australas J Dermatol 1996; 37(4):185–187.

65. Pflücker F, Wendel V, Hohenberg H, et al. The human stratum corneum layer: an effective barrier against dermal uptake of different forms of topically applied micronised titanium dioxide. Skin Pharmacol Appl Skin Physiol 2001; 14(suppl 1):92–97.

66. Cappel MJ, Kreuter J. Effect of nanoparticles on transdermal drug delivery. J Microencapsul 1991; 8(3):369–374.

67. Alvarez-Roman R, Naik A, Kalia YN, Guy RH, Fessi H. Skin penetration and distribution of polymeric nanoparticles. J Contr Rel 2004; 99(1):53–62.

68. Kreuter J, Shamenkov D, Petrov V, et al. Apolipoprotein-mediated transport of nanoparticle-bound drugs across the blood–brain barrier. J Drug Target 2002; 10(4):317–325.

69. Lockman PR, Koziara JM, Mumper RJ, Allen DD. Nanoparticle surface charges alter blood–brain barrier integrity and permeability. J Drug Target 2004; 12:635–641.

70. Jendelova P, Herynek V, Urdzikova, et al. Magnetic resonance tracking of transplanted bone marrow embryonic stem cells labelled by iron oxide nanoparticles in rat brain and spinal cord. J Neurosci Res 2004; 76:232–243.

71. Kedar NP. Can we prevent Parkinson's and Alzheimer's disease? J Postgrad Med 2003; 49(3):236–245.

72. Calderon-Garciduenas L, Reed W, Maronpot RR, et al. Brain inflammation and Alzheimer's-like pathology in individuals

exposed to severe air pollution. Toxicol Pathol 2004; 32(6): 650–658.

73. Campbell A, Oldham M, Becaria A, et al. Particulate matter in polluted air may increase biomarkers of inflammation in mouse brain. Neurotoxicology 2005; 26:133–140.

74. Buxton DB, Lee SC, Wickline SA, Ferrari M. National Heart, Lung, and Blood Institute Nanotechnology Working Group. Recommendations of the National Heart, Lung, and Blood Institute Nanotechnology Working Group. Circulation 2003; 108(22):2737–2742.

75. Knaapen AM, Albrecht C, Becker A, Höhr D, Winzer A, Haenen GR, Borm PJA, Schins RPF. DNA damage in lung epithelial cells isolated from rats exposed to quartz: role of surface reactivity and neutrophilic inflammation. Carcinogenesis 2002; 23:1111–1120.

76. Albrecht C, Schins RP, Hoehr D, et al. Inflammatory time course following quartz instillation: role of TNFalpha and particle surface. Am J Respir Cell Mol Biol 2004; 31(3):292–301.

77. Pott F, Althoff G-H, Roller M, Höhr D, Friemann J. High acute toxicity of hydrophobic ultrafine titanium dioxide in an intratracheal study with several dusts in rats. In: Dungworth DL et al., eds. Relationships between Respiratory Disease and Exposure to Air Pollution. Washington DC: ILSI Press, 1998:270–272.

78. Rehn B, Seiler F, Rehn S, Bruch J, Maier M. Investigations on the inflammatory and genotoxic lung effects of two types of titanium dioxide: untreated and surface treated. Toxicol Appl Pharmacol 2003; 189:84–95.

79. Höhr D, Steinfartz Y, Martra G, Fubini B, Borm PJA. The surface area rather than the surface coating determines the acute inflammatory response after instillation of fine and ultrafine TiO$_2$ in the rat. Int J Hyg Environ Health 2002; 205:239–244.

80. Nemmar A, Hoylaerts MF, Hoet PH, Vermylen J, Nemery B. Size effect of intratracheally instilled particles on pulmonary inflammation and vascular thrombosis. Toxicol Appl Pharmacol 2003; 186(1):38–45.

81. Brooking J, Davis SS, Illum L. Transport of nanoparticles across the rat nasal mucosa. J Drug Target 2001; 9(4):267–279.

82. Knaapen AM, Shi T, Borm PJA, Schins RPF. Soluble metals as well as the insoluble particle fraction are involved in cellular DNA damage induced by particulate matter. Mol Cell Biochem 2002; 234/235:317–326.

83. Fubini B, Aust AE, Bolton RE, et al. Non-animal tests for evaluating the toxicity of solid xenobiotics—the report and recommendations of ECVAM Workshop 30. Altern Lab Anim 1998; 26:579–617.

84. Baulig A, Garlatti M, Bonvallot V, et al. Involvement of reactive oxygen species in the metabolic pathways triggered by diesel exhaust particles in human airway epithelial cells. Am J Physiol Lung Cell Mol Physiol 2003; 285:L671–L679.

85. Gupta AK, Curtis ASG. Surface modified superparamagnetic nanoparticles for drug delivery: interaction studies with human fibroblasts in culture. J Mater Sci 2004; 15:493–496.

86. Bruch J, Rehn S, Rehn B, Borm PJA, Fubini B. Variation of biological responses to different respirable quartz flours determined by a vector model. Int J Hyg Environ Health 2004; 207(3):1–14.

87. Rehn B, Seiler F, Rehn S, Bruch J. A new in-vitro-testing concept (vector model) in biological screening and monitoring the lung toxicity of dusts. Presentation of the concept and testing the method with dust of known lung toxicity. Reinhaltung der Luft 2001; 61(7/8):301–312.

88. Li XY, Brown D, Smith S, MacNee W, Donaldson K. Short-term inflammatory responses following intratracheal instillation of fine and ultrafine carbon black in rats. Inhal Toxicol 1999; 11:709–731.

89. van Zijverden M, van der Pijl A, Bol M, et al. Diesel exhaust, carbon black, and silica particles display distinct Th1/Th2 modulating activity. Toxicol Appl Pharmacol 2000; 168:131–139.

90. Dybing E, Lovdal T, Hetland RB, Lovik M, Schwarze PE. Respiratory allergy adjuvant and inflammatory effects of urban ambient particles. Toxicology 2004; 198:307–314.

91. Schins RPF, Lightbody J, Borm PJA, Donaldson K, Stone V. Inflammatory effects of coarse and fine particulate matter in

relation to chemical and biological constituents. Toxicol Appl Pharmacol 2004; 195:1–11.

92. Fennrich S, Fischer M, Hartung T, et al. Detection of endotoxins and other pyrogens using human whole blood. Dev Biol Stand 1999; 101:131–139.

93. Williams H, Johnson JL, Carson KG, Jackson CL. Characteristics of intact and ruptured atherosclerotic plaques in brachiocephalic arteries of apolipoprotein E knockout mice. Arterioscler Thromb Vasc Biol 2002; 22:788–792.

94. Kodavanti UP, Schladweiler MC, Ledbetter AD, et al. Pulmonary and systemic effects of zinc-containing emission particles in three rat strains: multiple exposure scenarios. Toxicol Sci 2002; 70(1):73–85.

95. Riediker M, Cascio WE, Griggs TR, et al. Particulate matter exposure in cars is associated with cardiovascular effects in healthy young men. Am J Respir Crit Care Med 2004; 169(8):934–940.

96. Wellenius GA, Batalha JR, Diaz EA, et al. Cardiac effects of carbon monoxide and ambient particles in a rat model of myocardial infarction. Toxicol Sci 2004; 80(2):367–376.

97. Chang CC, Hwang JS, Chan CC, Wang PY, Hu TH, Cheng TJ. Effects of concentrated ambient particles on heart rate, blood pressure, and cardiac contractility in spontaneously hypertensive rats. Inhal Toxicol 2004; 16(6–7):421–429.

8

Injectable Nanoparticles for Efficient Drug Delivery

BARRETT RABINOW and MAHESH V. CHAUBAL

BioPharma Solutions, Baxter Healthcare, Round
Lake, Illinois, U.S.A.

INTRODUCTION: MEDICAL NEEDS ADDRESSABLE BY NANOPARTICULATE DRUG DELIVERY

Development of nanoparticulates arose in response to broad medical needs, common to a number of therapeutic areas and targets. Earlier work on liposomes and emulsions had established iconic examples of enhancements that drug delivery could confer on established agents such as doxorubicin and amphotericin. These involved improvements in pharmacokinetics as well as in targeting to certain organs, cell types, or organelles. Although its in vivo disposition was altered, the molecular structure of the drug was not changed, thus facilitating overall development. For broader applicability,

199

nanosuspensions offer additional features, compelling their current popularity. As contrasted with microparticulates, nanoparticulates are sufficiently small to avoid embolism associated with intravenous (i.v.) delivery, and can also be used for the less invasive parenteral routes.

A large proportion of i.v. drugs in development are antineoplastic agents or antiinflammatory compounds. While they are fewer in number, there is a need for improved antimicrobial agents as well, although many companies are exiting this area. Opportunities for enhancement in these specific therapeutic areas will be considered from a biological barrier perspective. Additionally, medical benefits arising from the ability to target to specific organs will also be shown. The limitations of predicate dosage form platforms will be noted, which define the opportunities of nanosuspensions to address unmet needs.

Disease Process Perspective

Many antineoplastic agents suffer from a very narrow therapeutic index. Clinically significant efficacy is attained often at the expense of systemic side effects, afflicting the relatively fast growing cells of the bone marrow, mucosa, etc. A major effort is focused on achieving greater specificity for tumor tissue by tailoring the molecular structure to target specific tumor cells, receptors, signal pathways, etc. However, it has also been demonstrated that conventional molecules may increase their therapeutic index by retaining their native structure while enhancing targeting specificity. Pharmacokinetic improvements can also be realized for cell cycle-specific antineoplastic agents, which utilize the drug more efficiently upon prolonged, rather than acute, exposure (1).

Cancer

Specific delivery of intravenously administered drugs to solid tumors can be significantly enhanced by considering the factors involved in increased vascular permeability of tumors. Endemic to the tumorigenic process is angiogenesis, which leads to a high vascular density. Compared to normal tissue, tumor vasculature is leaky, exhibiting a discontinuous endothelial cell lining, which creates transvascular pores, permitting the entry of macromole-cules and particles that normally have very limited access to normal tissue. Furthermore, a wide variety of permeability-enhancing factors is elaborated by tumor cells, including bradykinin, nitric oxide, prostaglandins, vascular endothelial growth factor (VEGF), kallikrein and cytokines such as tumor necrosis factor. These all act to increase the permeability of tumor vasculature. Additionally, matrix metalloproteinases (collagenases) effect disintegration of the matrix tissue sur-rounding blood vessels, increasing its apparent leaky nature (2). Because lymphatic drainage is impaired as well, this constel-lation of effects selectively concentrates macromolecules larger than 40 kDa and particles less than about 300 nm in tumors (3,4). Such entities are able to permeate through the vascular defects, but cannot subsequently re-equilibrate with systemic concentrations, as can smaller diffusible molecules, because they are dependent upon lymphatic drainage. This phenomenon of tumor circulation was first elaborated by Maeda et al. (5) as the enhanced permeability and retention (EPR) effect.

Infection and Inflammation

Tumor vascular physiology resembles that for sites involving infection and inflammation (6). Indeed, the mutually causa-tive inter-relationship between inflammation and cancer has been remarked upon for more than a century (7). In inflam-mation and infection, without cancer, the major feature appears to be an inflammatory cascade involving bradykinin, triggered either by microbial products or the host's own upregulated immune system. Microbial proteases and host macrophage and polymorphonuclear (PMN) cell cytokine and protease elaboration have been implicated. This leads to a series of events involving vasodilatation and permeability, facilitating entry of the host immune cells, and also of inflam-matory mediators, which aggravate the process in a positive feedback system. The major difference between the vascular physiology of cancer versus that for noncancer-involved infection and inflammation is the presence of a functioning lymphatic system in the latter. Thus while macromolecules and particles may gain entry into the compromised tissue served by the defective vasculature, they are cleared faster than in the case for tumor, reducing the possibility for sustained drug release at the disease site (8).

Biological Barriers Imposed by the Monocyte Phagocytic System

Based upon an understanding of compromised vasculature, the requirements of a drug delivery system intended for targeting to sites of tumor, infection, or inflammation can be specified. There is an upper limit placed upon the size of the particle, permitting diffusion through the vascular pores (9). The range of pore sizes is 300–700 nm, depending upon the tumor type, and therefore targeting particles should be substantially smaller, preferably <250 nm. The particles should be designed to target the pores rather than suffer less productive competitive encounters, the major one being that of entrapment by the monocyte phagocytic system (MPS).

Entrapment: Phagocytosis

The MPS system consists of fixed macrophage cells in key tissues, such as liver, kidney, lung, bone marrow, and spleen, as well as circulating monocytes, macrophages, and PMN cells. These are designed to rid the body of bacterial, viral, and particulate waste. The first step in the MPS removal process involves deposition of specific circulating blood proteins onto the particle, termed opsonization, which subsequently signal receptors on the macrophages and PMN for particle uptake. Following opsonin docking on the receptors, an intracytoplasmic process is activated, reorganizing actin filaments, causing the extension of pseudopodia to project from the phagocyte, surrounding the particle. The pseudopodia follow the contours of the particle as guided by further receptor docking onto the opsonized particle. Provided the particle is smaller than approximately 8 µm, the spreading pseudopodia will eventually meet, totally engulfing the particle. The particle is then encased in an intracytoplasmic vacuole, termed a phagosome, formed from a remnant of the spreading pseudopodia (10).

While this process of phagocytosis is applicable to particles as small as 500 nm, a similar receptor-mediated endocytosis is more generally available to many different kinds of cells. This extends to particles as small as 100 nm and probably smaller (11). Non–receptor-mediated pincocytosis also becomes more prominent as particle size decreases from 1100 down to 100 nm (12).

Escape

Over the course of 15–30 minutes, the pH of the phagosome decreases from 7.4 to 4–5, as digestive enzymes are also added by docking vacuoles. Eventually, the phagosome unites with a lysosome, emptying its contents into the low pH environment (13). If the particle is not metabolizable or soluble, it will remain in the phagocyte (14).

There are several ways in which phagocytized particles may escape the lysosome to enter the cytoplasm, and from there, the extracellular milieu. If the pH–solubility characteristic of the particle is such that it simply dissolves in the low pH environment of the lysosome, then the particle will dissolve. If additionally, the solubilized constituents are soluble in phospholipid membranes, they may then dissolve into the lysosomal membrane and enter the cytoplasm, diffusing down a concentration gradient. By the same process, the dissolved constituents may dissolve into the cytoplasmic membrane and diffuse into the extracellular space. Itraconazole nanosuspension exhibits this behavior, and is able to vacate the phagolysosomal compartment, as from a depot, to provide sustained release to the systemic circulation (15). Alternatively, the particle coating may feature endosomolytic agents, which cause the lysosomal membrane to rupture, thus emptying the contents of the lysosome, including the particle, into the cytosol (16,17).

Targeting or Evasion

Depending upon the pharmacokinetic and targeting needs, nanoparticulate dosage forms may be engineered to either target or evade the MPS. Targeting may be accomplished passively, simply by ensuring that the nanoparticulate remains intact to be phagocytized minutes after i.v. infusion. Alternatively, targeting motifs may be intentionally added to the coating of the particle, for the purpose of actively docking with particular macrophage receptors, thus triggering phagocytosis (18). Evasion of the MPS is most commonly performed by inhibiting the initial opsonization process. This is accomplished by coating the nanoparticles with a molecular layer that prevents deposition of the opsonizing proteins (19). The result is a significantly prolonged circulation time, than would otherwise occur. This affords sufficient time for the particle to encounter and diffuse through vascular pores, resulting in higher ratios of drug concentration in sites of tumor, infection, or inflammation, relative to normal tissue. This increases the therapeutic index by increasing local site efficacy and decreasing systemic toxicity.

Avoidance

Optimization of coating for minimizing MPS uptake has been exceedingly well studied, and utilizes predominantly hydrophilic polymers that are attached by various means to the particle surface. There is precedent for this from nature, where a strain of *Pseudomonas aeruginosa* is known to elaborate a viscous polyuronic acid polysaccharide, which interferes with phagocytosis by virtue of its hydrophilicity (20). The coating most often used in drug delivery applications features polymers of ethylene oxide. These may be adsorbed onto preexisting nanoparticulates, by using triblock copolymers, containing a central hydrophobic polyoxypropylene segment, flanked by hydrophilic polyoxyethylene chains on either side. The hydrophobic portion permits physical adsorption onto hydrophobic surfaces of nanoparticles enabling the hydrophilic chains to project into the aqueous medium (21). The steric barrier inhibits opsonic protein deposition. Consistent with this concept, it has been found that the hydrophilic chains should be sufficiently long (98 or more units of ethylene oxide) to create a corona of sufficient thickness to prevent protein deposition. And the hydrophobic section should be sufficiently long (greater than 67 propylene oxide units) to resist shear detachment following administration in the blood (22,23). Certain inconsistencies with the brush-like theory have been raised, namely that the experimentally effective grafting density, polymer chain length, and poly(ethylene oxide) (PEO) molecular weight are too low compared with required theoretical values. It is argued that surface bonding is at least as important as steric barrier effects, as shown by studies with phenoxy-substituted dextran polymers (24).

Despite success in this area, much remains to be done. The polymers that have proved most effective for prolonging circulation time, poloxamine-908, poloxamer-407, etc., are not approved for use in injectable drugs. Furthermore, although they work well with polystyrene model nanoparticles, poloxamers and poloxamines do not prolong circulation time for a wide variety of nanoparticles with more hydrophilic surfaces such as albumin and PLGA. For this reason graft copolymers, primarily involving poly(ethylene glycol) (PEG), have been studied. PEG coating employs the same ethylene oxide repeat unit found to be effective in poloxamer, but is covalently bonded to the polymer comprising the bulk of the nanoparticle. Because it is tethered to the surface of the nanoparticle it is therefore expected to avoid the desorption issues found with the free surfactants. PEG–PLGA copolymer was found to extend the half-life of incorporated albumin from 14 minutes, found with non-PEGylated PLGA nanoparticles, to 4.5 hours (25). The systematic variation of both components of the polymer was studied. The PEG moiety was shown to repel the deposition of the opsonizing protein complement, as shown with Western blot using antiopsonin antibodies, but was less effective in repelling Immunoglobulin G (IgG) (26).

Biological Barrier Considerations in Targeting Organs

Monocyte Phagocytic System (MPS)

For diseases that solely afflict the MPS, nanoparticles unadorned with stealth coatings are optimal for targeting these organs. For example, intracellular bacteria such as *Brucella, Listeria, Mycobacteria,* and *Salmonella* infect primarily phagocytic cells, creating disease reservoirs, often inaccessible to drug therapy. β-Lactam antibiotics, formulated as a solution, have a low uptake by phagocytic cells. By contrast, ampicillin-loaded polyisohexylcyanoacrylate nanoparticles incubated with *Salmonella typhimurium*–loaded murine macrophages were observed to colocalize within the same phagolysosomes, leading to bactericidal effects. However, the frequency with which this occurred is unclear. The low value found might reflect the inhibition of phagosome–lysosome fusion by *S. typhimurium*, which could prevent contact between the phagosomal localized bacteria and the nanoparticles localized primarily in secondary lysosomes (27). In an in vivo study, clofazimine nanosuspension targeted organs of the MPS, the liver, spleen, and lungs, effectively controlling a mycobacterium infection, primarily infecting macrophages (28).

Diverse Cells, Tissues, and Organs

Targeting biological addresses located successively higher in the organelle-cell-tissue-organ system hierarchy has been reported. Altered intracellular distribution, favoring deposition of the cytotoxic adriamycin in the Golgi apparatus, was accomplished for nanoparticles labeled with monensin, a carboxylic ionophore. The greatly enhanced resulting efficacy may be attributed to facilitating transport of the toxin molecules to ribosomal ribonucleic acid (RNA) (29). Greater specificity of targeting macrophages was accomplished by using mannan-coated nanoparticles, which selectively were taken up by macrophages expressing a mannose receptor, for greater specificity of antigen delivery (30). Delivery of nanoparticles to the intrahepatic liver parenchyma, rather than the Kupffer and endothelial cells, was accomplished by labeling iron oxide nanoparticles with the HIV tat peptide, which has membrane-translocating properties. While the organ biodistribution was determined by the properties of the nanoparticulate, whether labeled or not, the intra-organ distribution was determined by tat peptide labeling (31). Enhanced liver uptake with significant—54%—parenchymal retention was demonstrated for a low-density lipoprotein (LDL) mimicking coating of albumin nanospheres, consisting of a fatty acid and phospholipid (32). Selective targeting to a specific organ of the MPS, the bone marrow, was accomplished with poloxamer-407-coated beads of 150 nm in diameter (33). For treating experimental Chagas' disease, involving the parasite *Trypanosoma cruzi* that infects muscles, gastrointestinal tract, as well as the MPS, stealth PEG–PLA nanospheres of the bis-triazole D0870 were employed. A significant cure rate was observed as measured by hemoculture, xenodiagnosis and antibody detection (34). Targeting more diverse organs was accomplished by labeling semiconductor inorganic nanocrystal quantum dots, called qdots, with specific recognition peptides obtained from screening phage libraries for homing to specific sites. Qdot targeting separately to the endothelium of the lung, tumor vasculature and tumor lymphatic vessels was demonstrated (35).

Lymphatic

Targeting of the lymphatics has drawn attention because of the opportunity to target lymphocytes with immunomodulators, resident HIV virus with antiviral agents, and disseminated tumor metastases. While molecularly dissolved agents cannot utilize this system efficiently, nanoparticulates are ideally designed for targeting the lymphatic circulation. Subsequent to their administration to the interstitial space, their clearance through blood capillary endothelium is limited, but entry through the intercellular clefts of the lymphatic capillaries proceeds readily. However, there is an optimal size of about 10–100 nm. The particle must be large enough to drain preferentially through the lymphatics, but small enough to diffuse through the interstitial space away from the injection site (36). However, biodegradable nanoparticles will generate smaller particulate fragments, which will be able to migrate into the lymphatic capillaries (37). Hydrophilic particles clear more readily after interstitial injection than do hydrophobic ones. This happens because of favorable partitioning into the aqueous channels rather than the mucopolysaccharide ground substance and collagenous fibers of the interstitial space. Utilization of poloxamine-908 with relatively long PEO chains to coat 60 nm polystyrene nanospheres, permitted them to leave the injection site, transit through the lymph nodes, and rapidly appear in blood. Usage of poloxamer-904 with shorter PEO chains effected drainage from the injection site, but resulted in localization in the regional lymph nodes, due to phagocytosis by the macrophages contained there (38). Negatively charged entities are less retained by the negatively charged ground substance than are positively charged ones (39). Limitations to the technique arise because of metastatic invasion of the regional lymphatics in end-stage cancer, preventing an intact route to these tumor cell sanctuaries (40). Additionally, while access to the closest regional lymph nodes are achievable following subcutaneous injection, treating the entire lymphatic system is not assured (41).

Intramuscular

Subcutaneous and intramuscular injection may be used both to achieve systemic drug levels without the inconvenience of i.v. delivery, as well as to bypass poor absorption associated with oral dosage forms. The small size of nanoparticles provides for a faster dissolution than is seen with microsuspensions. This is significant because dissolution is often the rate-controlling step for systemic blood uptake following administration of the depot (42,43). Therefore, t_{max} may be decreased and C_{max} may be increased by use of nanoparticles. Nevertheless, in many cases, there is less toxicity for a depot of nanoparticulate relative to a solution dosage form, because the dissolution rate of the nanosuspension is somewhat slower (44). Pure drug nanosuspensions will provide for a higher loading in less volume within this restricted compartment, at the expense of more prolonged delivery available with sustained release polymeric nanoparticles.

Brain-Epidural

Nanoparticulates provide a means of administering a relatively large load of drug by direct injection to the central nervous system with decreased systemic side effects (45). This is especially noteworthy if the drug is poorly water soluble, because attaining therapeutic levels of such drugs in any compartment poses a challenge. Favorable efficacy studies were seen with particulate busulfan, administered by intrathecal (i.t.) administration to mice (46). This led to a clinical trial where the pharmacokinetics were determined (47). Earlier, a 10% butamben suspension, consisting of phospholipid-coated drug crystals, was found to be well tolerated in dogs as well as humans when administered epidurally (48,49).

TYPES OF CARRIERS

Injectable nanoparticulate dosage forms can be classified into three main categories: (i) crystalline drug nanosuspensions, wherein the drug is available in a stable crystalline form; (ii) polymeric nanoparticles, wherein the drug is encapsulated within a polymer matrix in an amorphous state; and (iii) solid lipid nanoparticles, wherein the drug is encapsulated within a lipid matrix in an amorphous state. Nanocrystalline drug suspensions have an advantage of higher loading (up to 90% of the crystalline particle is the drug). Higher loading of drug within polymeric or lipid nanoparticles on the other hand may lead to crystallization of the drug from the nanoparticle matrix, leading to an unstable system.

Crystalline Drug Nanosuspensions

Crystalline nanoparticles of drugs are typically produced either by controlled crystallization or by a high-energy particle size reduction process. Examples of the latter include wet milling and high-pressure homogenization (28,50). A third approach was reported recently, wherein crystallization and particle size reduction were combined to produce injectable nanosuspensions (51). This combined process, depicted in Figure 1, exploits the advantages of both crystallization and particle comminution. The crystallization step is designed to produce unstable particles that are fragile. These fragile particles are subsequently broken down and stabilized by high-pressure homogenization. Table 1 lists some of the techniques used to prepare crystalline nanoparticles. The

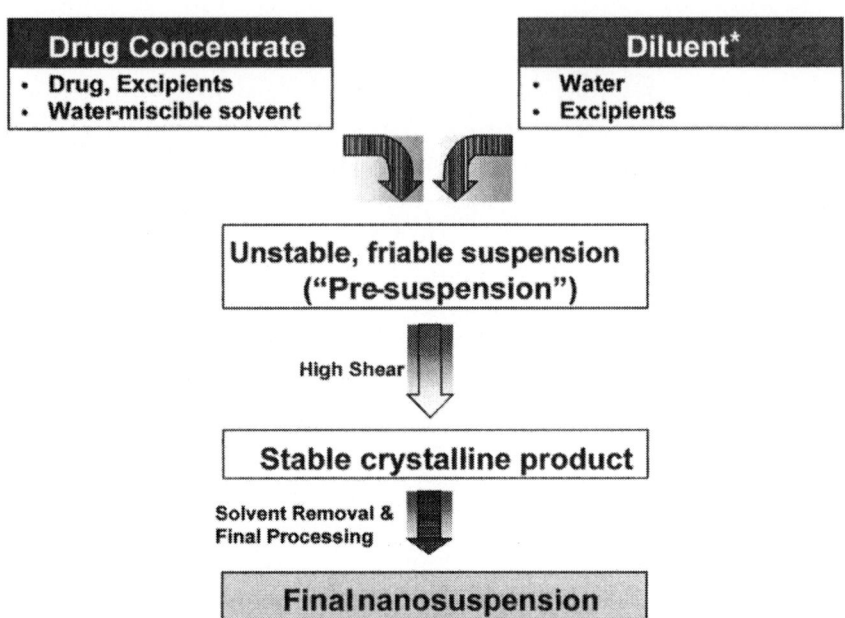

Figure 1 Combination of microprecipitation and homogenization for the production of crystalline nanoparticles.

nanocrystals thus produced are stabilized using biocompatible surfactants such as phospholipids, polysorbates, poloxamers, etc. As seen in the scanning electron micrograph (Fig. 2A), the final nanosuspension typically consists of as much as 90% drug surrounded by a layer of surfactants. Because of this feature, crystalline drug nanosuspensions can provide high drug loading. Furthermore, as very low levels of excipients are used, concerns regarding excipient-related toxicity are reduced. By choosing surfactants that are approved in injectable dosage forms, and by ensuring that 100% of the drug particles are <5 µm, crystalline drug nanosuspensions can be effectively and safely employed for i.v. administration.

Some of the important considerations in development of injectable nanosuspensions include:

1. Nanoparticles should be stable and not susceptible to phenomena such as aggregation or Ostwald ripening.

Table 1 Various Processes Used for the Preparation of Crystalline and Polymeric Nanoparticles

Process	Type of particles produced	Particle size/ comments	References
Single emulsion	Polymeric nanoparticles	Particle size depends on the size of dispersion used	52
Double emulsion	Polymeric nanoparticles	100–1000 nm	52
Spray drying	Polymeric or lipidic nanoparticles	Typically >200 nm	53
Gas-antisolvent precipitation	Polymeric nanoparticles	400–600 nm	54
Nanoprecipitation	Polymeric or crystalline nanoparticles	Down to 100 nm	55
Temperature-induced phase transition	Polymeric nanoparticles	PLGA and pluronic F-127 were comelted with the drug and cooled to cast the nanoparticles	56
High-pressure homogenization	Polymeric, lipidic, or crystalline nanoparticles	> 300 nm	28, 57
Wet milling	Crystalline nanoparticles	Down to 100 nm	50, 58
Microprecipitation– homogenization	Crystalline or polymeric nanoparticles	Down to 100 nm	51
High gravity reactive precipitation	Polymeric nanoparticles	Down to 100 nm	59

2. The nanosuspension should be free of contamination from any media used during processing.
3. The nanoparticle process should be amenable to an alternate form of sterilization, if terminal sterilization is not an option for the drug or formulation.
4. Surfactants/excipients used should be acceptable for injectable applications.

(A)   (B)

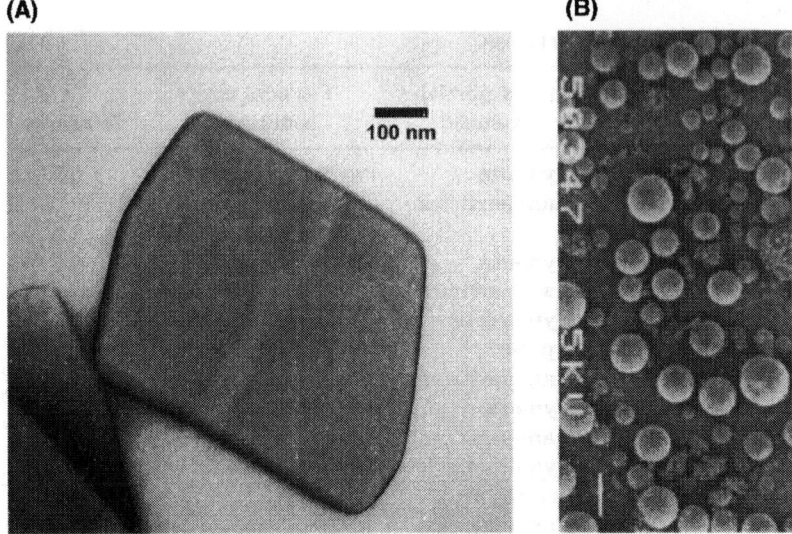

Figure 2 Scanning electron micrograph of (**A**) drug nanocrystal coated with nonphospholipid surfactants and (**B**) polymeric nanospheres incorporating drug in an amorphous form.

Nanosuspension stability can be addressed by the careful optimization of surfactants to be used in the formulation. Adsorption kinetics and affinity of the surfactant to the newly formed crystal surface play a determining factor on the final particle size and stability of the nanosuspension. A number of surfactants have been explored for the stabilization of nanocrystals. Pace et al. (60) describe the use of phospholipids for the stabilization of various drugs including carbamazepine, dantrolene, dexamethasone, indomethacin, and oxytetracycline. Up to 600-fold increase was seen in drug loading with nanocrystalline suspensions, as compared to commercial formulations. In another example, Williams et al. (61) used a mixture of lipids to coat and stabilize nanoparticles of a topoisomerase inhibitor for i.v. administration. It was shown that the lipid-coated nanoparticles effectively inhibited the degradation of the lactone ring for the drug (SN-38). Merisko-Liversidge et al (58). demonstrated the use of crystalline nanoparticles for i.v. administration of a number of anticancer drugs (piposulfan, etoposide, camptothecin, and paclitaxel). The agents were wet milled as a 2% w/v solid suspension containing 1% w/v surfactant stabilizer. Stability of the nanosuspension can also be achieved by freeze-drying, or in extreme cases, by storing the suspension in a frozen state.

Polymeric Nanoparticles

Polymeric nanoparticles consist of the drug dispersed in an amorphous form within a polymer matrix. Such particles could be prepared as nanospheres, wherein the drug is dispersed uniformly throughout the matrix of the particle (typically as a solid solution in polymer), or as nanocapsules, wherein the drug is present in the core of the particle (either as a solid solution or a solution in oil). Polymeric nanoparticles are typically prepared from biodegradable polymers to avoid accumulation of the polymer matrix on repeat dosing. Early reports of injectable polymeric nanoparticles typically involved polylactide (PLA) or its copolymer with glycolide (PLGA). Choice of these polymers in early studies was mainly based on their prior use as biomaterials for surgical sutures and related applications. Subsequent studies have reported the use of various other biodegradable polymers including polyanhydrides, polycyanoacrylates, and polyorthoesters.

Rapid uptake of polymeric (PLGA, PLA) nanoparticles by the reticuloendothelial system (RES) led to newer generation products involving the use of copolymers of polyesters (such as PLA, PLGA) and PEG. As explained elaborately in the section on 'Avoidance', optimization of the PEG length at the surface of such nanoparticles has been studied to evade the RES and extend release of the incorporated drug, making it comparable to sterically stabilized liposomes (62).

Polymeric nanoparticles are typically prepared using conventional emulsion-based processes as depicted in Figure 3. The nanoparticles produced using this process are uniformly spherical in nature as shown in the scanning electron micrograph in Figure 2(B). A number of other technologies have also been employed to manufacture polymeric nanoparticles, as listed in Table 1 .

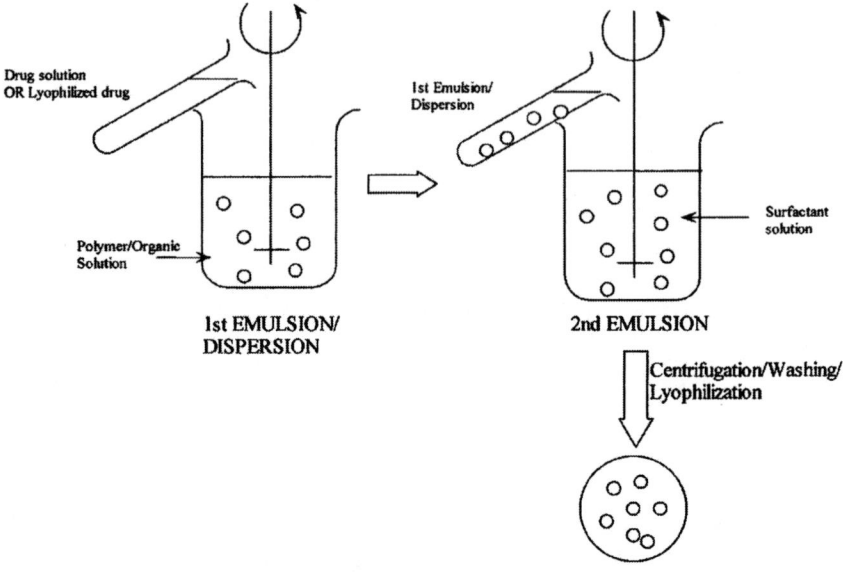

Drug solution
OR Lyophilized drug

1st Emulsion/
Dispersion

Polymer/Organic
Solution

Surfactant
solution

1st EMULSION/
DISPERSION

2nd EMULSION

Centrifugation/Washing/
Lyophilization

Drug loaded nanospheres

Figure 3   Schematic of the emulsion-based process typically used for preparation of polymeric nanoparticles.

Lipidic Nanoparticles

Lipidic nanoparticles use biocompatible lipids as carriers. The principles of preparation and stabilization of such carriers are similar to polymeric nanoparticles. Readers are referred to a recently published comprehensive review on these systems for additional information (63). An additional category of carriers is inorganic nanoparticles. Until recently these carriers were considered mainly in nontherapeutic applications. However, recent efforts in the area of quantum dots has led to research into applications of such systems as therapeutic drug carriers. Due to their early exploratory nature, further discussion on quantum dots is considered beyond the scope of this report and the readers are referred to other related publication (35). Other types of inorganic nanoparticles are those which are used in conjunction with an externally triggered system (such as magnetic nanoparticles). Such particles are discussed in a subsequent section.

COATING FUNCTIONALITY

One of the more elaborate examples of coated nanoparticles elegantly incorporates features designed to accomplish all of the drug loading, MPS avoidance, active targeting, endocytotic uptake, and endosomal escape processes. A cyclodextrin-containing polycation of imidazole was designed to electrostatically complex with a catalytic oligonucleotide, a DNAzyme, forming sub-100-nm particles termed "polyplexes." The positively charged particles can interact with the negatively charged cell surface proteoglycans for endocytotic uptake. Further, imidazole had been demonstrated to enhance endosomal escape. However, neutralization of the excess charge was required for minimizing nonspecific uptake, to enhance efficiency of active targeting. This was accomplished with addition of the anionic glutamate functionality to adamantane–PEG conjugates, which forms inclusion complexes with the exposed cyclodextrins. The exposed PEG chains confer long circulation in biological fluids. Because transferrin is often upregulated in rapidly growing cells, active targeting was considered by adding transferrin–PEG–adamantane conjugates. Biodistribution in an HT-29, high transferrin uptake, tumor xenograft mouse model was followed subsequent to different routes of administration. Intraperitoneal injection indicated high levels remaining in the peritoneum; presumably mobility was limited by their size, even at 30–50 nm. Subcutaneous injection did not result in fluorescence outside of the injection site. But i.v. delivery showed high levels in tumor, liver and kidney, all organs rich in transferrin receptor activity. Polyplexes delivered by i.v. were internalized by the tumor cells (64).

EXTERNAL ASSISTANCE IN TARGETING

An alternate approach that has been explored to provide targeting functionality to nanoparticles is via the use of an external energy source. For example, Rudge et al. (65) described a nanoparticulate system that was responsive to an external magnetic field. The particles were comprised of activated carbon to allow adsorption and loading of drug and metallic iron to provide magnetically triggered targeting of the particles. Good loading efficiency could be obtained for a number of drugs including doxorubicin, mitomycin C, methotrexate, and camptothecin. In vivo studies using magnetic doxorubicin particles showed that efficient targeting was achieved by injecting the particles using an arterial catheter, and then homing the particles to a specific tissue, by using a strong magnetic field. In another study, a much higher concentration of mitoxantrone was obtained in the tumor area, by using only 50% and 20% of the normal dose by the use of magnetic drug targeting (66). Ultrasound triggered drug delivery has also been adopted to provide targeted release of drug to tumors. Nanoparticles and micelles accumulate into the tumor as a result of passive targeting and the EPR effect. Ultrasound is then applied to trigger the release of the drug so that the entire drug load is delivered within the tumor (67).

DRUGS INCORPORATED

A number of drug classes are expected to benefit from the unique drug delivery characteristics offered by nanoparticles. Table 2 provides a brief representative list of drugs that have been delivered using nanoparticles. Passive targeting to macrophages can be exploited for antiinfective agent delivery (70). A significant amount of work has been conducted to demonstrate utility of nanoparticles to deliver chemotherapeutic agents to tumors either via active targeting or via passive targeting using the EPR effect. Brain targeting functionality of nanoparticles can also be exploited for various therapeutics including antivirals, chemotherapeutics, and antiinfective drugs. Besides the targeting functionality, drugs are also incorporated into nanoparticles for rapid dissolution and provision for an i.v. dose, without the use of toxic solubilizers. Example of this approach was demonstrated in the use of nanoparticles for prednisolone (45). Yet another need that nanoparticles provide is for sustained release via

Table 2 A Representative List of Drugs Incorporated into Nanoparticles for Targeted Drug Delivery

Drug	Class	Target organ/ cells	Technology	References
Camptothecin	Chemothera-peutic	Solid tumors	PEG–PLA nanoparticles	68
Paclitaxel	Chemothera-peutic	Arterial neointima	Albumin nanoparticles	69
Clofazimine	Antiinfective	Macrophages	Crystalline nanoparticles	70
SN-38	Chemothera-peutic	Tumor	Crystalline nanoparticles	54
Indinavir	Antiviral	Brain	Crystalline nanoparticles	71
Doxorubicin	Chemothera-peutic	Brain	Polysorbate-80-coated nanoparticles	72
Dalargin	Analgesic	Brain	Polysorbate-80-coated nanoparticles	73
Itraconazole	Antifungal	Macrophages	Crystalline nanoparticles	15
Piperine	Antileishma-nial	Liver	Lipid nanospheres	74

Abbreviation: PEG, polyethylene glycol; PLA, polylactic acid.

an extravascular site. An example of this is the sustained release of a pain medicament, tetracaine (44).

CLINICAL DEVELOPMENT

ABRAXANE^{TM}, recently approved, is the nanoparticle formulation which is furthest advanced in clinical development. It is an albumin nanoparticle formulation of paclitaxel, that does not contain the problematic excipient Cremophor EL. It is formed by dissolving paclitaxel in water immiscible methylene chloride, and adding this to a solution of human serum albumin in water with low-speed homogenization. This creates an emulsion with albumin located at the aqueous–solvent interface. Subsequent high-pressure homogenization reduces the particle size, and breaks and reforms the disulfide bonds, essentially crosslinking the albumin coating, stabilizing the particle. An evaporation step volatilizes the methylene chloride, leaving an aqueous suspension of 140–160 nm nanoparticles, consisting of an amorphous paclitaxel core surrounded by a 25 nm coating of albumin with bound paclitaxel. Because of its size this material can be sterile filtered (75).

Clinical pharmacokinetic studies for ABRAXANE indicate reduced area under the concentration-time curve (AUC) relative to Taxol® (6427 vs. 7952 ng hr/mL) and comparable $t_{1/2}$ (15 vs. 13 hours) (76,77). Ibrahim attributes the increased AUC for Taxol to vascularly retained lipophilic Cremophor® micelles, which encapsulate the drug. Despite similar circulation times, tumor levels of drug are increased with ABRAXANE (78).

In a Phase III trial involving 460 breast cancer patients, comparing the two drugs, Taxol was administered by its standard protocol, 175 mg/m^2 by three-hour infusion. Additionally, premedication with steroids and antihistamines was required to forestall Cremophor-related hypersensitivity. ABRAXANE was administered at higher doses of 260 mg/m^2 over a shorter 30-minute duration, without premedication or granulocyte-colony stimulating factor (G-CSF) support. Despite this, the toxicity of the nanosuspension was no worse: there were no hypersensitivity reactions; neutropenia was actually less; while neuropathy was somewhat higher. This is significant given the correlation that exists between the duration of plasma paclitaxel concentrations above a threshold of 0.1 µmol/L with adverse decline of absolute neutrophil count and white blood cell count (79). The shorter 30-minute duration of infusion of ABRAXANE results in higher C_{max} than that for Taxol with a three-hour infusion (6100 vs. 2170 ng/ mL) for 135-mg/m^2 doses (80,81). In this trial, ABRAXANE also produced a higher tumor response rate versus paclitaxel (31% vs. 16%) and a longer time to tumor progression (21.9 week vs. 16.1 week) (82).

The improved efficacy of the nanosuspension apparently challenges reports of the purported benefit of Cremophor EL in enhancing drug level in tumor cells by inhibiting the P-glycoprotein efflux pump (83). This probably occurs because

Table 3 Injectable Nanoparticulate-Based Formulations in Clinical Development

Drug	Indication	Status	Route	Drug delivery company	Pharma company
ABRAXANE (paclitaxel)	Anticancer	Marketed	i.v.	American BioScience	American Pharmaceutical Partners
Undisclosed multiple	Antiinfective	Phase II	i.v.	Baxter NANOEDGE	Undisclosed
Undisclosed	Anticancer	Phase I	i.v.	Baxter NANOEDGE	Undisclosed
Diagnostic agent	Imaging agent	Phase I/II	i.v.	Elan Nanosystems	Photogen
Thymectacin	Anticancer	Phase I/II	i.v.	Elan Nanosystems	NewBiotics/Ilex Oncology
Busulfan	Anticancer	Phase I	i.t./i.v.	SkyePharma	Supergen

Cremophor is retained in the central vascular compartment with a low volume of distribution of 3.70 L/m², and therefore does not enter the tumor tissue (84). The fact that corticosteroids do not have to be taken as premedication opens the possibility for combining paclitaxel with IL-2 or interferon for treatment of metastatic melanoma, renal cell carcinoma, etc. Because steroids lyse lymphokine-activated killer (LAK) cells, thus mitigating the benefits of the cytokines, the current cremophor containing formulation cannot be used (85).

Other injectable nanoparticulate formulations in development are shown in Table 3.

CONCLUSIONS

Drug delivery systems have developed in response to needs of drugs in the area of increased solubility, reduced toxicity, and increased efficacy. Evolving medical needs, coupled with the shortcomings of liposomal and emulsion-based systems, spurred the development of the more stable and flexible nanoparticulate platform. For injectables, as the complexity of the biological barriers of the manonuclear phagocyte system (MPS), the disease process and specific target organs became appreciated, multiple requirements were imposed upon the formulation. These specifications for size, MPS avoidance, active targeting, high drug payload, controlled drug release, and stability would often have resulted in mutually contradictory stipulations for the earlier platforms. The solid phase of a nanosuspension, however, offers the prospect of simultaneously optimizing multiple parameters.

Numerous, even a bewildering array of, choices for the carrier are available, which suggests an overemphasis toward the perspective of polymer design and formulation. This occurs of course because academic progress is often driven by relative expense of the programs undertaken, and biological-based studies are more expensive. The area would benefit from more extensive interdisciplinary collaboration, emphasizing biological assessment of the advantages and disadvantages of the major classes of carriers, to indicate where future effort should be expended. Undoubtedly this will occur because of the approval and marketing of an intravenous nanosuspension. This demonstrates that the platform has migrated from a laboratory curiosity to a serious contender for formulation of enhanced generics as well as for NCE pharmaceuticals. As such, industry will be more willing to provide leadership in the area, to further advance applications of the technology.

REFERENCES

1. Masson E, Zamboni WC. Pharmacokinetic optimisation of cancer chemotherapy. Clin Pharmacol 1997; 32:324–343.

2. Utoguchi N, Mizuguchi H, Dantakean A, et al. Effect of tumor cell-conditioned medium on endothelial macromolecular permeability and its correlation with collagen. Br J Cancer 1996; 73:24–28.

3. Yuan F, Dellian M, Fukumura D, et al. Vascular permeability in a human tumor xenograft: molecular size dependence and cutoff size. Cancer Res 1995; 55:3752–3756.

4. Hobbs SK, Monsky WL, Yuan F, et al. Regulation of transport pathways in tumor vessels: role of tumor type and microenvironment. Proc Natl Acad Sci USA 1998; 95:4607–4612.

5. Maeda H, Wu J, Sawa T, Matsumura Y, Hori K. Tumor vascular permeability and EPR effect in macromolecular therapeutics: a review. J Control Release 2000; 65:271–284.

6. Maeda H, Wu J, Okamoto T, Maaruo K, Akaike T. Kallikrein-kinin in infection and cancer. Immunopharmacology 1999; 43:115–128.

7. Schreiber H, Rowley DA. Inflammation and cancer. In: Gallin JI, Snyderman R, eds. Fearon DT, Haynes, BF, Nathan C, assoc. eds. Inflammation: Basic Principles and Clinical Correlates. 3rd ed. New York: Lippincott Williams & Wilkins, 1999:1117–1129.

8. Noguchi Y, Wu J, Duncan R, et al. Early phase tumor accumulation of macromolecules: a great difference between the tumor vs. normal tissue in their clearance rate. Jpn J Cancer Res 1998; 89:307–314.

9. Wu NZ, Da D, Rudoll TL, Needham D, Whorton AR, Dewhirst MW. Increased microvascular permeability contributes to preferential accumulation of stealth liposomes in tumor tissue. Cancer Res 1993; 53:3765–3770.

10. Mukherjee S, Ghosh RN, Maxfield FR. Endocytosis. Physiol Rev 1997; 77:759–803.

11. Rupper A, Cardelli J. Regulation of phagocytosis and endophagosomal trafficking pathways in *Dictyostelium discoideum*. Biochim Biophys Acta 2001; 1525:205–216.

12. Pratten MK, Lloyd JB. Pinocytosis and phagocytosis: the effect of size of a particulate substrate on its mode of capture by rat peritoneal macrophages cultured in vitro. Biochim Biophys Acta 1986; 881:307–313.

13. Steinman RM, Mellman IS, Muller WA, Cohen ZA. Endocytosis and the recycling of plasma membrane. J Cell Biol 1983; 96:1–27.

14. Gibaud S, Demoy M, Andreux JP, Weingarten C, Gouritin B, Couvreur P. Cells involved in the capture of nanoparticles in hematopoietic organs. J Pharm Sci 1996; 85:944–950.

15. Rabinow BE, White RD, Glosson J, Sun C-S, Papadopoulos P. Enhanced efficacy of NANOEDGE itraconazole nanosuspension in an immunosuppressed rat model infected with an itraconazole-resistant *C. albicans* strain (Abstract #R6184). Poster Presentation at the American Association of Pharmaceutical Scientists (AAPS)—2003 Annual Meeting and Exposition, Salt Lake City, Utah, October 26–30, 2003.

16. Christie RJ, Grainger DW. Design strategies to improve soluble macromolecular delivery constructs. Adv Drug Del Rev 2003; 55:421–437.

17. Panyam J, Zhou WZ, Prabha S, Sahoo SK, Labhasetwar V. Rapid endo-lysosomal escape of poly (d,l-lactide-co-glycolide) nanoparticles: implications for drug and gene delivery. FASEB J 2002; 16:1217–1226.

18. Farassen S, Voros J, Csucs G, Textor M, Merkle HP, Walter E. Ligand-specific targeting of microspheres to phagocytes by surface modification with poly(l-lysine)-grafted poly(ethylene glycol)conjugate. Pharm Res 2003; 20:237–246.

19. Gref R, Minamitake Y, Peracchia MT, Trubetskoy V, Torchilin V, Langer R. Biodegradable long-circulating polymeric nanospheres. Science 1994; 263:1600–1603.

20. Cabral DA, Loh BA, Speert DP. Mucoid *Pseudomonas aeruginosa* resist nonopsonic phagocytosis by human neutrophils and macrophages. Pediatr Res 1987; 22:429–431.

21. Troster SD, Muller U, Kreuter J. Modification of the body distribution of poly(methyl methacrylate) nanoparticles in rats by coating with surfactants. Int J Pharm 1990; 61:85–100.

22. Moghimi SM. Mechanisms regulating body distribution of nanospheres conditioned with pluronic and tetronic block copolymers. Adv Drug Del Rev 1995; 16:183–193.

23. Mosqueira VCF, Legrand P, Morgat J-L, et al. Biodistribution of long-circulating PEG-grafted nanocapsules in mice: effects of PEG chain length and density. Pharm Res 2001; 18: 1411–1419

24. De Sousa Delgado A, Leonard M, Dellacherie E. Surface properties of polystyrene nanoparticles coated with dextrans and dextran-PEO copolymers. Effect of polymer architecture on protein adsorption. Langmuir 2001; 17:4386–4391.

25. Li YP, Pei Y-Y, Zhang X-Y, et al. PEGylated PLGA nanoparticles as protein carriers: synthesis, preparation and biodistribution in rats. J Control Release 2001; 71:203–211.

26. Ameller T, Marsaud V, Legrand P, Gref R, Barratt G, Renoir J-M. Polyester-poly (ethylene glycol) nanoparticles loaded with the pure antiestrogen RU 58668: physico-chemical and opsonization properties. Pharm Res 2003; 20: 1063–1070.

27. Pinto-Alaphandary H, Balland O, Laurent M, Andremont A, Puisieux F, Couvreur P. Intracellular visualization of ampicillin-loaded nanoparticles in peritoneal macrophages infected in vitro with *Salmonella typhimurium*. Pharm Res 1994; 11:38–46.

28. Muller RH, Peters K. Nanosuspensions for the formulation of poorly water soluble drugs. I. Preparation by a size reduction technique. Int J Pharm 1998; 160:229–237.

29. Shaik MS, Idediobi O, Turnage VD, McSween J, Kanikkannan N, Singh M. Long-circulating monensin nanoparticles for the potentiation of immunotoxin and anticancer drugs. J Pharm Pharmacol 2001; 53:617–627.

30. Cui A, Hsu C-H, Mumper RJ. Physical characterization and macrophage cell uptake of mannan-coated nanoparticles. Drug Dev Ind Pharm 2003; 29:689–700.

31. Wunderbaldinger P, Josephson L, Weissleder R. Tat peptide directs enhanced clearance and hepatic permeability of magnetic nanoparticles. Bioconjug Chem 2002; 13:264–268.

32. Jaitely V, Vyas SP. Development and characterization of surface modified albumin nanospheres for liver targeting. Proc 26th Intl Symp Control Rel Bioact Mater 1999; Abstract #5241.

33. Porter CJH, Moghimi SM, Illum L, Davis SS. The polyoxyethylene/polyoxypropylene block co-polymer poloxamer-407 selectively redirects intravenously injected microspheres to sinusoidal endothelial cells of rabbit bone marrow. FEBS Lett 1992; 305:62–67.

34. Molina J, Urbina J, Gref R, Brener Z, Junior JMR. Cure of experimental Chagas' disease by the bis-triazole D0870 incorporated into 'stealth' polyethyleneglycol-polylactide nanospheres. J Antimicrob Chemo 2001; 47:101–104.

35. Akerman ME, Chan WCW, Laakkonen P, Bhatia SN, Ruoslahti E. Nanocrystal targeting in vivo. Proc Natl Acad Sci USA 2002; 99:12617–12621.

36. Jackson AJ. Intramuscular absorption and regional lymphatic uptake of liposome entrapped insulin. Drug Metab Dispos 1981; 9:535–540.

37. Ohsawa T, Matsukawa Y, Takakura Y, Hashida M, Sezaki H. Fate of lipid and encapsulated drugs after intramuscular administration of liposomes prepared by the freeze thawing method in rats. Chem Pharm Bull 1985; 33:5013–5018.

38. Moghimi SM, Hawley AE, Christy NM, Gray T, Illum L, Davis SS. Surface engineered nanospheres with enhanced drainage into lymphatics and uptake by macrophages of the regional lymph nodes. FEBS Lett 1994; 344:25–33.

39. Zakakura Y, Kitajima M, Matsumototo S, Hashida M, Sezaki H. Development of a novel polymeric prodrug of mitomycin C-dextran conjugate with anionic charge. I. Physicochemical characteristics and in vivo and in vitro antitumour activities. Int J Pharm 1987; 37:135s–141s.

40. Porter CJH. Drug delivery to the lymphatic system. Crit Rev Ther Drug Carrier Syst 1997; 14:333–393.

41. Perez-Soler R, Lopez-Berestein G, Jahns M, Wright K, Kasi LP. Distribution of radiolabeled multilammelar liposomes injected intralymphatically and subcutaneously. Int J Nucl Med Biol 1985; 12:261–266.

42. Zuidema J, Pieters FA, Duchateau G. Release and absorption rate aspects of intramuscularly injected pharmaceuticals. Int J Pharm 1988; 47:1–12.

43. Stout PJM, et al. Dissolution performance related to particle size distribution for commercially available prednisolone acetate suspensions. Drug Dev Ind Pharm 1992; 18:395–408.

44. Boedeker BH, Logeski E, Kline M, Haynes D. Ultra-long duration local anesthesia produced by injection of lecithin-coated tetracaine microcrystals. J Clin Pharmacol 1994; 34: 699–702.

45. Grossman SA, Krabak MJ. Leptomeningeal carcinomatosis. Cancer Treat Rev 1999; 25:103–119.

46. Archer GE, Sampson JH, McLendon RD, et al. Intrathecal busulfan treatment of human neoplastic meningitis in athymic nude rats. J Neurooncol 1999; 44:233–241.

47. Quinn JA, Glantz M, Petros W et al. Phase I trial of intrathecal Spartaject busulfan for patients with neoplastic meningitis. (Abstract No. 318) 8th ASCO Annual Meeting, Orlando, FL, May 18–21, 2002.

48. Shulman M, Hoseph NJ, Haller CA. Effect of epidural and subarachnoid injections of a 10% butamben suspension. Reg Anesth 1990; 15:142–146.

49. Shulman M. Treatment of Cancer pain with epidural butylamino benzoate suspension. Reg Anesth Jan–March 1987:1–4.

50. Merisko-Liversidge E, Liversidge GG, Cooper ER. Nanosizing: a formulation approach for poorly-water-soluble compounds. Eur J Pharm Sci 2003; 18(2):113–120.

51. Chaubal M, Doty M, Kipp J, et al. Novel process for the preparation of injectable nanosuspensions to address insolubility challenges. Proc. Control Release Soc Meeting 2003; 30.

52. Zambaux MF, Bonneaux F, Gref R, et al. Influence of experimental parameters on the characteristics of poly(lactic acid) nanoparticles prepared by a double emulsion method. J Control Release 1998; 50(1–3):31–40.

53. Raffin Pohlmann A, Weiss V, Mertins O, Pesce da Silveira N, Staniscuaski Guterres S. Spray-dried indomethacin-loaded polyester nanocapsules and nanospheres: development, stability evaluation and nanostructure models. Eur J Pharm Sci 2002; 16(4–5):305–312.

54. Elvassore N, Bertucco A, Caliceti P. Production of insulin-loaded poly(ethylene glycol)/poly(l-lactide) (PEG/PLA) nanoparticles by gas antisolvent techniques. J Pharm Sci 2001; 90(10):1628–1636.

55. Chorny M, Fishbein I, Danenberg HD, Golomb G. Lipophilic drug loaded nanospheres prepared by nanoprecipitation: effect of formulation variables on size, drug recovery and release kinetics. J Control Release 2002; 83(3):389–400.

56. Lee KE, Kim BK, Yuk SH. Biodegradable polymeric nanospheres formed by temperature-induced phase transition in a mixture of poly(lactide-co-glycolide) and poly(ethylene oxide)-poly(propylene oxide)-poly(ethylene oxide) triblock copolymer. Biomacromolecules 2002; 3(5):1115–1119.

57. Calvor A, Muller BW. Production of microparticles by high-pressure homogenization. Pharm Dev Technol 1998; 3(3): 297–305.

58. Merisko-Liversidge E, Sarpotdar P, Bruno J, et al. Formulation and antitumor activity evaluation of nanocrystalline suspensions of poorly soluble anticancer drugs. Pharm Res 1996; 13(2):272–278.

59. Chen JF, Zhou MY, Shao L, et al. Feasibility of preparing nanodrugs by high-gravity reactive precipitation. Int J Pharm 2004; 269(1):267–274.

60. Pace S, Pace G, Parikh I, Mishra A. Novel injectable formulations of insoluble drugs. Pharm Tech 1999; 116–133.

61. Williams J, Lansdown R, Sweitzer R, et al. Nanoparticle drug delivery system for intravenous delivery of topoisomerase inhibitors. J Control Release 2003; 91:1–2:167–172.

62. Fawaz F, Bonini F, Guyot M, Lagueny AM, Fessi H, Devissaguet JP. Influence of poly(DL-lactide) nanocapsules on the biliary clearance and enterohepatic circulation of indomethacin in the rabbit. Pharm Res 1993; 10(5):750–756.

63. Wissing SA, Kayser O, Muller RH. Solid lipid nanoparticles for parenteral drug delivery. Adv Drug Deliv Rev 2004; 56(9):1257–1272.

64. Pun S, Bellocq N, Cheng J, et al. Targeted delivery of RNA-Cleaving DNA Enzyme (DNAzyme) to tumor tissue by transferrin-modified, cyclodextrin-based particles. Cancer Biol Ther 2004; 3:e31–e40.

65. Rudge S, Peterson C, Vessely C, Koda J, Stevens S, Catterall K. Adsorption and desorption of chemotherapeutic drugs from a magnetically targeted carrier (MTC). J Control Release 2001; 74:335–340.

66. Alexiou C, Jurgons R, Schmid RJ, et al. Magnetic drug targeting—biodistribution of the magnetic carrier and the chemotherapeutic agent mitoxantrone after locoregional cancer treatment. J Drug Target 2003; 11(3):139–149.

67. Rapoport NY, Christensen DA, Fain HD, Barrows L, Gao Z. Ultrasound-triggered drug targeting of tumors in vitro and in vivo. Ultrasonics 2004; 42(1–9):943–950.

68. Miura H, Onishi H, Sasatsu M, Machida Y. Antitumor characteristics of methoxypolyethylene glycol-poly(dl-lactic acid) nanoparticles containing camptothecin. J Control Release 2004; 97(1):101–113.

69. Kolodgie FD, John M, Khurana C, et al. Sustained reduction of in-stent neointimal growth with the use of a novel systemic nanoparticle paclitaxel. Circulation 2002; 106(10):1195–1198.

70. Peters K, Leitzke S, Diederichs JE, et al. Preparation of a clofazimine nanosuspension for intravenous use and evaluation of its therapeutic efficacy in murine *Mycobacterium avium* infection. J Antimicrob Chemother 2000; 45(1):77–83.

71. Rabinow B, Chaubal M, Werling J, et al. Indinavir nanosuspensions for increased CNS delivery through macrophage targeting. Controlled Release Society Annual Meeting, Abstract #604, 2004.

72. Gulyaev AE, Gelperina SE, Skidan IN, Antropov AS, Kivman GY, Kreuter J. Significant transport of doxorubicin into the brain with polysorbate 80-coated nanoparticles. Pharm Res 1999; 16(10):1564–1569.

73. Schroder U, Sabel BA. Nanoparticles, a drug carrier system to pass the blood–brain barrier, permit central analgesic effects of i.v. dalargin injections. Brain Res 1996; 710(1–2): 121–124.

74. Veerareddy PR, Vobalaboina V, Nahid A. Formulation and evaluation of oil-in-water emulsions of piperine in visceral leishmaniasis. Pharmazie 2004; 59(3):194–197.

75. Ibrahim NK, Desai N, Legha S, et al. Phase I and pharmacokinetic study of ABI-007, a cremophor-free, protein-stabilized, nanoparticle formulation of paclitaxel. Clin Cancer Res 2002; 8:1038–1044.

76. Taxol® (paclitaxel) Injection. Direction Insert. Princeton, NJ: Bristol-Myers Squibb Oncology.

77. Sparreboom A, Van Tellingen O, Nooijen WJ, Beijnen JH. Nonlinear pharmacokinetics of paclitaxel in mice results from the pharmaceutical vehicle cremophor EL. Cancer Res 1996; 56:2112–2115.

78. Desai N, Yao Z, Trieu V, Soon-Shiong P, Dykes D, Noker P. Evidence of greater tumor and red cell partitioning and superior antitumor activity of cremophor free nanoparticle paclitaxel (ABI-007) compared to Taxol. Breast Cancer Res Treat 2003; 82(suppl 1):S82–S83.

79. Huizing MT, Keung ACF, Rosing H, et al. Pharmacokinetics of paclitaxel and metabolites in a randomized comparative study

in platinum-pretreated ovarian cancer patients. J Clin Oncol 1993; 11:2127–2135.

80. Ibrahim NK, Desai N, Legha S, et al. Phase I and pharmacokinectic study of ABI-007, a cremophor-free, protein-stabilized, nanoparticle formulation of paclitaxel. Clin Cancer Res 2002; 8:1038–1044 and Taxol® (paclitaxel) Injection. Direction insert. Bristol-Myers Squibb Oncology, Princeton NJ.

81. O'Shaughnessy J, Tjulandin S, Davidson N, Shaw H, Desai N, Hawkins M. ABI-007 (Abraxane), a nanoparticle albumin-bound (nab) paclitaxel demonstrates superior efficacy vs. Taxol in MBC: a phase III trial. 26th Ann San Antonio Breast Cancer Symp 2003; Abstr 44.

82. Woodcock DM, Jefferson S, Linsenmeyer ME, et al. Reversal of the multidrug resistance phenotype with cremophor EL, a common vehicle for water-insoluble vitamins and drugs. Cancer Res 1990; 50:4199–4203.

83. Sparreboom A, Verweij J, van der Burg ME, et al. Disposition of Cremophor EL in humans limits the potential for modulation of the multidrug resistance phenotype in vivo. Clinical Cancer Res 1998; 4:1937–1942.

84. Desai N, Yao Z, Trieu V, Soon-Shiong P, Dykes D, Noker P. Evidence of greater tumor and red cell partitioning and superior antitumor activity of cremophor free nanoparticle paclitaxel (ABI-007) compared to Taxol. Breast Cancer Res Treatment 2003; 82(suppl 1):S82–S83.

9

Polymeric Nanoparticles for Oral Drug Delivery

**VIVEKANAND BHARDWAJ and MAJETI NAGA
VENKATA RAVI KUMAR**

Department of Pharmaceutics, National Institute
of Pharmaceutical Education and Research,
Punjab, India

INTRODUCTION

The enteric system has been specifically designed for the uptake of foreign substances to maintain homeostasis in the body. Despite the extensive research and success stories with other routes for drug delivery, the oral route is still the most preferred route because of its basic functionality and the advantages that ensue. Nanoparticles (NPs) also have been extensively studied for peroral drug delivery, for systemic effect following uptake from the enteron, or to act locally in the gastrointestinal tract (GIT). NPs are expected to address some

231

specific issues for drug delivery like low mucosal permeability, absorption windows, low solubility of the drug and gut metabolism, and first pass effect. The potential advantages of NPs as oral drug carriers are enhancement of bioavailability, delivery of vaccine antigens to the gut-associated lymphoid tissues (GALT), controlled release, and reduction of the gastrointestinal irritation caused by drugs (1).

The utility of NPs for oral drug delivery arises out of the particulate uptake mechanisms that exist in the GIT, especially the transcellular absorptive pathways involving vesicular transport through M cells of Peyer's patches (PP). From the surface of M cells, NPs are taken up and transported to lymphocytes in the form of vesicles. The lymphatic absorption of a drug via the GALT prevents presystemic metabolism in the liver because it bypasses the portal blood circulation. This mechanism provides a chance to target cancers of the lymphatics, rapidly achieve mucosal immunity, and stain the lymph nodes before surgery (2).

After oral administration, colloidal drug carriers have the ability to increase bioavailability by protecting the drug from denaturation in the gastrointestinal lumen or by prolonging the exposure of the mucous membrane to elevated drug concentration (3).

PHYSIOLOGY OF GIT WITH RELEVANCE TO PARTICULATE UPTAKE

Gastrointestinal Tract Physiology

The GIT serves to carry out the digestion of food and the absorption of water, nutrients, and electrolytes, and provides a selective barrier between the environment and the systemic circulation. It provides a variety of physiological and morphological barriers, such as proteolytic enzymes, in the gut lumen and at the brush border membrane: the mucous layer, the bacterial gut flora, and the epithelial cell lining itself. Although the GIT is designed to prevent uptake of particulate matter (potentially toxic materials and pathogens) from the environment, it is not a completely prohibitive barrier. Specialized mechanisms exist that allow internalization of macromolecules and particles. The mucus, built up from mucin molecules, covering the absorptive enterocytes in the intestines, acts as a barrier for oral absorption of foreign matter, including NPs. The mucus is a translucent viscoelastic hydrogel and is mainly composed of glycoproteins that have both acidic and basic chemical groups (4–6). The chemical composition of mucus provides an opportunity for both acidic and basic compounds to interact with it, thereby increasing the residence time of both drugs and NPs in close proximity to the absorptive surfaces.

The primary function of the GIT is to selectively take up interest substances from the ingested bulk. To prevent harmful material from getting in, various protective mechanisms like pH variation, degrading enzymes, mucus, and nonpathogenic microflora exist. Also, the immunological load ingested makes the mucosa an ideal site for the identification and resistance of antigenic challenges. The local immune system is composed of GALT, composed of lymphoid tissues, called PPs in the small intestine, which are characterized by a monolayer of specialized epithelium containing M cells and absorptive enterocytes, i.e., follicle associated epithelium (FAE). PPs are present all along the intestine with the maximum concentration found in the ileum. FAE, adapted for endocytosis/transcytosis of antigens and microorganisms to the organized lymphoid tissue within the mucosa and M cell basement membrane, appears to play an important role in facilitating antigen-to-cell and cell-to-cell interactions during an immune response. M cells lack fully developed microvilli in comparison to the neighboring absorptive cells and deliver the particles taken up to the lymphatics from where they, in a size-dependent manner, are then released into the bloodstream (Fig. 1) (7). This specialized physiological transport mechanism is thus being widely explored for oral drug delivery through colloidal carriers, such as NPs (8). This absorption mechanism was demonstrated by LeFevre's in 1978 (9), who reported a large accumulation of 2-μm latex particles in PPs after chronic feeding.

Channels of Uptake

To deliver their drug content in the blood, lymph, or target organs, NPs have to cross the gastrointestinal barrier either

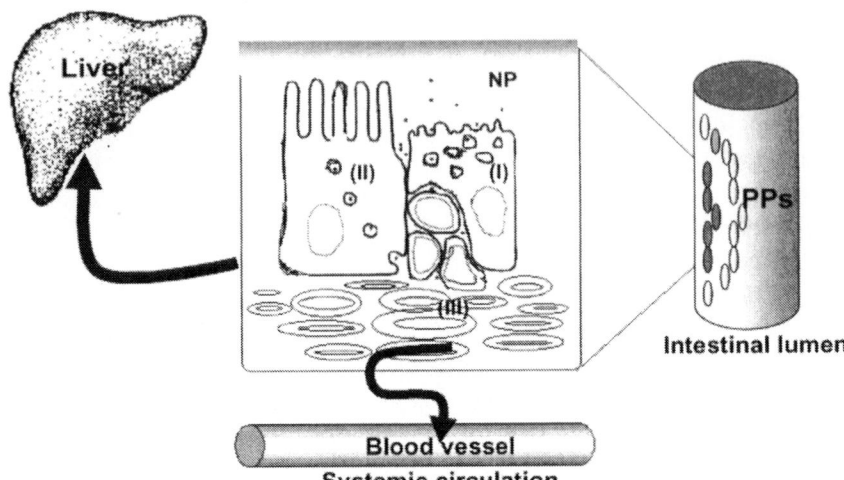

Figure 1 Mechanism of uptake of orally administered NPs: (I) M cells of the PP, (II) enterocytes, and (III) GALT. Schematic representation of the mechanism of uptake of NPs on oral administration. The direct uptake of NPs through the lymph into the systemic circulation bypassing the liver reduces the first pass metabolism, thus improving bioavailability. *Abbreviations*: NP, nanoparticles; PPs, Peyers patches.

by passive diffusion via transcellular or paracellular pathways or by active processes mediated by membrane-bound carriers or membrane-derived vesicles (10,11).

Another possible mechanism for the transport of NP across intestinal cells is paracellular uptake via aqueous channels. In humans, the equivalent pore diameter has been estimated to be between 4 and 8 Å and about 10–15 Å in rat and rabbit. The mucosal epithelium in the small intestine consists of polarized cells, connected by tight intercellular junctions, which account for < 1% of the surface area of the intestine (8). The uptake of particulate matter from between the absorptive cells is inversely proportional to the structural integrity of the tight junction barrier. The epithelial transport of larger molecules or particles can be increased by reversibly increasing the permeability of the tissue by opening the tight junctions under the influence of some mucoadhesive polymers and penetration enhancers (10).

Chitosans (CS) have been shown to transiently affect the gating function of tight junction in Caco-2 cell lines by interacting with anionic epithelial glycoproteins, such as sialic acid, to increase the transport of drugs. The intestinal permeability enhancement was dependent on the pH, which determines the degree of ionization of chitosan (12). Carbomer (a mucoadhesive polymer) also showed similar partial opening of tight junctions in Caco-2 cells as visualized by confocal laser scanning microscopy (CLSM) (13).

Lymphatic Uptake

Translocation of particulates via a transcellular mode of transport particularly across M cells was first shown in 1961 by Sanders and Ashworth (14) who, using electron microscopy, reported the endocytosis of 200-nm latex particles in ordinary enterocytes that were transported to the liver. The transcellular particle uptake can be divided into four distinct processes that are influenced by the size, surface charge, and surface characteristics of the NPs: diffusion of particles through the mucus lining the surface of the absorptive cells, initial surface contact and interaction with the enterocyte, cellular vesicular transport from the mucosal to serosal side, and, finally, the interaction with postenterocyte cells (11). M cells can carry out fluid phase endocytosis, adsorptive endocytosis, and phagocytosis, each of which results in the transport of the particles packed into endosomes and large multivesicular bodies followed by exocytosis across the basolateral membrane (8).

Initially, only a small fraction of the total particle dose appears in the blood after oral administration, which may be due to entrapment of particles in M cell pockets filled with lymphocytes and macrophages, nonfenestrated capillary endothelium in PP and trapping in lymph nodes, which impedes direct access of particulates to circulation (8). This is supported by the fact that most of the in vivo studies carried out with NPs report maximum concentration of NPs in the systemic circulation 48–72 hours after administration. The size and surface characteristics influence the fate of the endocytosed particle to a great extent and different mechanisms may exist for the same.

Another very important consideration is the difference in the expression of PPs in various species, in terms of number, type, and regional occurrence. As PPs are postulated to be the major uptake portal of NPs, overseeing these differences will result in inaccurate correlations and predictions of particle uptake (11). Animals with a higher number of PPs are expected to show a higher rate of uptake and this may not relate to human physiology to effect a meaningful extrapolation (15).

PARTICLE SIZE AND SURFACE CHARGE: CRITICAL FACTORS IN PARTICLE ABSORPTION

Mathiowitz et al. (16) have shown that particles in the size range of 40–120 nm were translocated both transcellularly and paracellularly. Size is a determining factor for both uptake and biological fate of particulate systems. In mice, Poly(lactide-*co*-glycolide) microspheres larger than 5 µm in diameter were not taken up into PPs, whereas those between 2 and 5 µm remained in the PPs, and those below 2 µm migrated from PPs to mesenteric lymph nodes (8). A size-dependent phenomenon exists in the gastrointestinal absorption of particles. The amount of 100 nm particles taken up was 2.5 and 6 times more than 1 and 10 µm particles, respectively, on a weight basis as studied in Caco-2 cell line (16). The uptake efficiencies from a 100 µg/mL dose were 41%, 15%, and 6% for 100 nm, 1 µm, and 10 µm particles, respectively.

Eldridge et al. (17) have also shown that the nature and the surface characteristics of the particles affect particle uptake as well. Hydrophobic particles are absorbed more readily than hydrophilic ones. The number of particles present in the PPs following oral administration correlate well with the relative hydrophobicities of the polymers used to make the particles. Increasing the hydrophobicity of polystyrene particles enhanced their permeability through mucus but decreased translocation through and across the absorptive cells (18). Thus, an efficient uptake process shall require, like for a drug, an optimum hydrophilic–lipophilic balance as an essential feature of the particle matrix former.

Addition of surfactants, especially polysorbate 80, increased the NP uptake significantly while hydrophobic vehicle, peanut oil, had no significant effect as compared to saline (19). Addition of oleic acid to polymethyl methacrylate NPs increased the oral absorption in Wistar rats by 50%, whereas incorporation of 5% of polysorbate 80 in a saline suspension increased the absorption by 200–300% (19). Adsorption of poloxamer 188 and 407 surfactants onto polystyrene particles of diameter 60 nm completely inhibited their uptake from the small intestine of Sprague Dawley rats (20).

BIOADHESION

As mentioned earlier, before the particles come into contact with the mucosal surface, they have to come in direct contact with mucus and may develop interactions with it. When microparticles or NPs are orally administered in the form of a suspension, they diffuse into the liquid medium and encounter the mucosal surface rapidly during the time course of their transit in the GIT, thus remaining away from the absorbing surfaces for most of the time. The particles can be immobilized at the intestinal surface by an adhesion mechanism, referred to as bioadhesion (21). When adhesion is restricted to the mucous layer lining the mucosal surface, the term mucoadhesion is employed (Fig. 2). Nanoparticulates generally display variable oral absorption, but actual figures depend on the size. This renders most encapsulated drugs ineffective after oral administration and therefore remains as the main obstacle to their practical application as oral drug delivery vehicles. Strategies, such as using mucoadhesive polymers and targeted delivery systems, have been explored in animals to improve particle absorption efficiency (7). This can increase the transit time of the particles in the GIT, allowing the particles to be present at the surface of absorptive cells for a longer duration, and thereby increases and maintains local concentration gradient at the absorption site. For example, on oral administration in rats, the bioadhesive potential of poly(methylvinylether-*co*-maleic anhydride) (PVM/MA) was

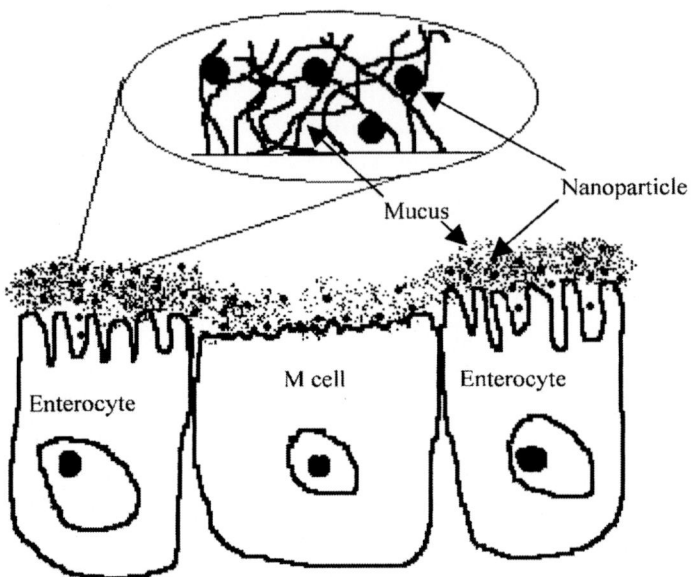

Figure 2   The particles interact with mucus (mucoadhesion) before coming into contact with the absorptive cells. The distribution of mucus is lesser over the PP. Positively charged particles thus stand more chance of uptake as they can associate to the negatively charged functional groups in the mucus. *Abbreviation*: PP, Payer's patches.

found to be 2.3 times higher when formulated as NPs than in the solubilized form in water, thus signifying that the particulate form is capable of imparting certain properties to the matrix polymer (22).

Bioadhesion can be achieved by either nonspecific or specific interactions with the mucosal surface. Nonspecific interactions are driven by the physicochemical properties and surface of the particles. Specific interactions depend on the presence of a ligand attached to the particle used for the recognition and attachment to a specific site at the mucosal surface. The process involved in the formation of such bioadhesive bonds has been described in three steps: (i) wetting and swelling of the polymer to permit intimate contact with biological tissue, (ii) interpenetration of bioadhesive polymer chains and entanglement of polymer and mucin chains, and (iii) formation of weak chemical bonds (23).

The mucoadhesives work by the "adhesion by hydration" phenomenon, where after initial contact, hydrophilic material starts attracting water by adsorption, swelling, or capillary forces. Certain mucoadhesives are known to enhance the mucosal permeability explained by the adhesion–dehydration theory, where the hydrophilic polymer absorbs water from mucosal tissue in such a way that the mucosal cells are dehydrated and shrunk until the normally tight intercellular junctions between the cells become physically separated. However, this is not a universal phenomenon, as seen with chitosans that enhance the mucosal permeability when applied as aqueous solution or gel (5). Even the best mucoadhesives are severely limited by a physiological limitation that the adhesive effect lasts only as long as the mucus itself remains firmly attached to tissue. The mucus turnover rate thus becomes a limiting factor in mucoadhesion-facilitated absorption. Besides, because of the high viscosity of the mucus, the particles entrapped inside this detached mucus have an even lesser chance of subsequent uptake.

Significant research has been done in an attempt to find the ideal ligands for mucoadhesion that will confer specificity and at the same time should be safe. In 1989, Pappo generated a monoclonal antibody that recognized M cells of rabbit's PP (24). This monoclonal antibody was adsorbed onto fluorescent polystyrene microspheres, and the particles localized specifically in the PP epithelium after administration. Antibody-coated particles accumulated in thrice the amount in PPs as compared to uncoated ones.

The concept of site specificity for mucoadhesion based on the affinity between sugars and lectins has also been intensively investigated. Either of the two approaches can be tried: the use of a sugar targeted to a lectin from the intestinal tract, or the use of a lectin targeted to a sugar from the intestinal mucus glycoproteins. In the latter case, the lectins employed are mostly from tomato, mycoplasma, or asparagus. Tomato lectin conjugates were found to be more specific for the mucous gel layer and, therefore, for intestinal regions without PPs, while the mycoplasma and asparagus lectin conjugates were more specific for the PPs (21). Lectin adsorption on the surface of NPs sometimes requires a linking agent. For example, dextran was coated onto polycaprolactone NPs (200 nm) to adsorb lectins as mucoadhesive ligands onto the surface (25). The specificity and haemagglutination properties of the adsorbed lectins were maintained. The absorption of polystyrene NPs was increased 10-fold by conjugation with tomato lectin (26).

Wheat germ agglutinin (WGA) from *Triticum vulgare* specifically binds to *N-acetyl*-D-glucosamine and sialic acid, both of which are constituents of mucus. In addition to binding to the surface of Caco-2 cells and human enterocytes, WGA is also taken up into the cells by receptor-mediated endocytosis involving the epidermal growth factor (EGF) receptor, which is present in a significant amount even on enterocytes. Poly(lactide-co-glycide) (PLGA) NPs conjugated with WGA as a carbohydrate binding ligand showed improved cytoassociation as compared to the unmodified ones (27). The mucin–lectin interaction is characterized by temperature dependence, specificity (*Dolichos biflorus* lectin), pH-dependence, and reversibility (6,10,28).

A number of issues have to be considered for the applicability of the proposed advantages. Formulations grafted with these ligands should show their specificity independent of food. As the toxicity of different plant lectins can vary significantly, their safety should be established in terms of toxicity, immunogenicity, and allergenicity. Tomato lectin can provoke high local and systemic immune responses but WGA, red kidney bean lectin, and *U. europaeus* isoagglutinin I have elicited low or no specific immune response (6). Tomato lectin ($\approx$70 kDa) is resistant to digestion and binds to rat intestinal villi without inducing any deleterious effect, but suffers from the disadvantage of strong cross-reactivity with mucus glycoproteins (5).

Besides lectins, other ligands used have been those that are substrates for receptor-mediated transporters. The endocytosis of folic acid conjugates is being tried as a promising strategy to target tumors that overexpress the folate receptor. Another concept relying on substrate recognition–mediated absorption is the utilization of the vitamin B_{12} receptor for endocytosis of conjugated drugs. Transferrin receptor involved in iron absorption and transport is being explored to enhance the absorption of DNA and proteins like insulin (6).

TRACER TECHNIQUES

Polymeric NPs are being extensively investigated for different therapeutic applications, such as sustained and targeted drug release, vaccine, and gene delivery. It is therefore important to assess the mechanism and kinetics of the cellular and the tissue uptake for these applications. However, the data generated from these techniques should be carefully interpreted because the NP concentration cannot always be assumed to reflect the drug concentration (which is why it is important to correlate these results with the polymer degradation and in vitro drug release). To have credibility, these data should be supplemented with stability studies which establish that (i) the NPs remain stable in the GIT for the time till the uptake occurs, and (ii) the drug is not released in significant amounts from the NPs before the uptake. The second point is, however, not an absolute prerequisite as the drug release data can be integrated with the transit data to predict the in vivo kinetics and the fate of NPs and the incorporated drug.

Quantification of NP uptake after oral administration in animal models is difficult and different analytical techniques are employed, including light, confocal, and electron microscopy, and fluorometry (29).

Microscopy

Quantification using microscopic techniques is cumbersome as NPs have to be counted on microscopic slides after tissue preparation from different sites along the GIT. Therefore, only a semiquantitative picture assessment can be made, that too on the assumption that the sampling is representative (29). The different studies done on these particles are complementary rather than supplementary.

The localization of NPs in a particular cell, tissue, or specific organ and the effect of various formulation parameters on uptake and distribution can be visualized by either

confocal or fluorescence microscopy (30,31). Confocal microscopy is capable of providing a three-dimensional view of the
samples and has been used to mark NP localization in histological sections of tissues exhibiting inflammation (15,32). Scanning electron microscopy (SEM) can provide magnification
from 10× to 300,000× and a resolution of 10 nm (33). SEM
allows one to differentiate M cells from enterocytes (15). Transmission electron microscopy can locate particles within the cell
in the cytoplasmic space (30). However, the particles may be
difficult to observe and some intracellular organelles could be
mistaken for nano- or microspheres. To tackle this problem,
electron-dense (ferritin-entrapped) particles are used (15).

Unlike electron microscopes, AFMs can image samples in
air and under liquids. AFM operates by measuring attractive
or repulsive forces between a tip and the sample and do not
use lenses, the resolution being limited by the refraction
angle (34). Like CLSM, AFM also is a three-dimensional imaging technique (35). It provides a high resolution (better than
1 nm) and the surfaces of Caco-2 cell monolayers have been
visualized to identify individual microvilli. The major advantage of AFM is that it does not require elaborate sample preparative techniques like fixation, gold sputtering or high
vacuum, and can even be performed on living cells in a physiological buffer and at physiological temperatures (10).

Spectroscopy

Fluorometry is rapid and nonradioactive. The marker used is
nonleaching, and it allows the simultaneous detection of multiple fluorophores such that two or more different fluorescent
types of particles can be detected in the same sample. This
may enable the study of site of uptake of differently sized particles (30). As the tissue components can also fluoresce to
interfere with the detection (especially low doses) of NPs, it
is advisable that fluorescent dyes be used in NPs having emission over 500 nm. The uptake of particles can then be studied
by carrying out spectrofluorometric measurement of the samples (36). This technique has provided evidence of internalization of NPs by Caco-2 cells, showing that surface modification

of PLGA NPs with vitamin E d-alpha-tocopheryl polyethylene glycol 1000 succinate (TPGS) improves the cellular uptake (31).

Adsorption isotherm studies using turbidimetry and FTIR–ATR analysis of polystyrene latex to rat intestinal mucosa have been carried out and it was shown that these isotherms depend on the size and surface properties of NPs (37–39).

Radionuclides

The foremost advantage of radioactivity is sensitivity; however, instability of the radioisotope can generate false results. Radiolabeled polymers have been used for making NPs and then radioactivity was measured to follow the uptake and distribution (40). Alternatively, the drug can be labeled and incorporated into polymeric NPs, and disposition studied in various tissues. After oral administration of NPs containing ^{14}C-labeled zidovudine, higher radioactivity was measured in blood, liver, reticuloendothelial system (RES) organs, and brain (41). Similarly, ^{125}I-radiolabeled tetanus toxoid (an antigen) was quantified in blood and lymphatics following oral administration of NPs to see the effect of coating of polyethylene glycol (PEG) (42). γ-Scintigraphy allows noninvasive visualization of the dosage form under normal physiological conditions and is being increasingly used to monitor the GIT transit of novel drug delivery systems by capturing images at different time points (43). The amount of radionuclide required for γ-scintigraphy is very small ($\approx$1 MBq in each dosage form). For this purpose, a very small concentration of radionuclide samarium oxide (^{153}Sm) is incorporated into the dosage form (44). One has to be careful though in interpretation of these results as the radionuclide used can possibly affect the stability of NPs and vice versa and in either case, an error in quantification can occur.

IN VITRO AND IN VIVO MODELS

In vitro and ex vivo models are being increasingly explored to study the influence of particle characteristics including size, zeta potential, nature of the polymer, hydrophobicity, and coating or complexation with mucoadhesive or other ligands. In vitro models, such as cell cultures, are used extensively to study the interactions between cells and polymeric particles. Caco-2 cell line derived from human colon adenocarcinoma is the most widely used permeability screen to study transepithelial transport (15). It has been shown that the Caco-2 cell line can be converted to M cells by coculture with PP lymphocytes (45). Establishment of an in vitro system reproducing the main characteristics of M cells relevant to particle uptake can help in designing strategies to translocate particulates (8). Ex vivo experiments using ligated intestinal segments are frequently used to determine the permeability of drugs across mammalian intestinal tissue. The intestinal tissue to be used for examining the uptake and transport of NPs must be obtained from freshly sacrificed animals as the epithelial cell layer undergoes rapid lysis and exfoliation, characterized by a complete loss of villi, after death (46).

NANOPARTICLE FORMULATION

Materials for Preparing NP Matrix

A number of polymers have been evaluated for the development of oral vaccines, including naturally occurring polymers (e.g., starch, alginates, and gelatin) and synthetic polymers (e.g., poly(lactide-*co*-glycolide) (PLGA), polyanhydrides, and phthalates). Toxicity, irritancy, and allergenicity are the factors of primary concern and hence there is a need for a biodegradable or soluble coating. The advantages of using natural polymers include their low cost, biocompatibility, and aqueous solubility. However, the natural polymers may also be limited in their use because of the presence of impurities, batch-to-batch variability, and generally low hydrophobicity. In comparison, synthetic polymers are more reproducible and can be prepared with the desired degradation rates, molecular weights, and copolymer compositions. But they may be disadvantageous because of their limited solubility;

they are often soluble only in organic solvents and, consequently, may not release the drug or may denature susceptible ones (47).

Natural Polymers and Derivatives

The use of colloidal carriers made of hydrophilic polysaccharides like CS is increasing as a promising alternative for improving the transport of drugs and macromolecules, such as peptides, proteins, oligonucleotides, and plasmids across biological membranes. CS {(1 → 4)-2-amino-2-deoxy-β-D-glucan} is a deacetylated chitin that has gained considerable interest for oral drug delivery. CS has been shown to increase the paracellular permeability of [^{14}C] mannitol (a marker for paracellular routes) across Caco-2 intestinal epithelia (48). These findings attributed the property of transmucosal absorption enhancement. CS is soluble only in solutions at pH values below 6.5, and only protonated chitosan (i.e., in its uncoiled configuration) can open the tight junctions, thereby facilitating the paracellular transport of hydrophilic compounds. The problem of CS ineffectiveness at neutral pH values can be tackled by derivatization at the amine group that renders the polymer soluble and effective for the purpose (49). However, pH above 6.5 is encountered only at the distal ends of the enteron and is expected to be of concern only when NPs are targeted to these portions of the gastrointestinal tract (GIT), such as the colon.

CS can enhance insulin absorption across human intestinal epithelial (Caco-2) cells without injuring them. CS NPs were more effective than the aqueous solution of CS in increasing the intestinal absorption of insulin (50). Because of its low production costs, biocompatibility, and very low toxicity, CS is a very interesting excipient for vaccine delivery research also. As chitosan easily forms nano- and microparticles with high loading capacities for various antigens, it is a promising candidate for designing carrier systems for oral vaccine delivery. An important advantage of CS nano- and microparticles is that, often, the use of organic solvents, which may alter the immunogenicity of antigens, is avoided during preparation and loading (51). Kumar and others have extensively reviewed the chemistry and applications of CS, especially in the pharmaceutical field (52,53).

Use of cyclodextrins in nanoparticulate delivery systems has been studied by Duchene et al. (54) in two ways: (i) using cyclodextrin NPs and (ii) incorporating cyclodextrins in polymeric NPs. Esterification of primary hydroxyl groups by hydrocarbon chains varying from C6 to C14 resulted in amphiphilic skirt-shaped cyclodextrins, which were capable of spontaneously forming both nanocapsules and nanospheres. The drug in the amphiphilic cyclodextrin NPs is dispersed at the molecular level and can be rapidly released. In the second method, natural or hydroxypropyl cyclodextrins were loaded onto poly(isobutylcyanoacrylate) NPs. The apparent solubility of saquinavir was increased 400 times by incorporation of its complex with hydroxypropyl-β-cyclodextrin into polyalkylcyanoacrylate NPs (55).

Synthetic Polymers

The foremost area of concern for these polymers is their biocompatibility and biodegradability. Polyesters and polyanhydrides are the most important class of polymers for drug delivery applications. Poly(lactic acid) (PLA) has been widely used for the preparation of NPs (56). McClean reported that PLA particles had an affinity for, and were absorbed by, both PP and non-PP tissue (57). Particle uptake was dependent on size, but was not exclusive to PP tissue. Poly(glycolic acid) (PGA), PLA, and especially their copolymers, PLGA, are the most commonly used family of biodegradable polymers. The PLGA copolymer is degraded in the body by hydrolytic cleavage of ester linkage to lactic and glycolic acid, which are formed at a very slow rate and easily metabolized in the body (58,59). Bala et al. (60) have reviewed the use of PLGA in polymeric NPs. PLGA is the most extensively studied and preferred polymer for drug delivery through NPs because of its ease of preparation, commercial availability, versatility, biocompatibility, and hydrolytic degradation into harmless products. The popularity of PLGA is further supported by the United States Food and Drug Administration's (USFDA) approval for a number of clinical applications (61).

Polycaprolactone (PCL), a polyanhydride, is also recognized as a biodegradable and nontoxic material. PCL exists in amorphous form and exhibits high permeability to low-molecular species at body temperature. These properties combined with documented biocompatibility make PCL a promising candidate for controlled release application. PCL hydrolyzes at a rate lower than the PLA and PLGA, and hence is more suitable for long-term drug delivery. Another positive aspect of PCL is its remarkable compatibility with numerous other polymers, allowing the formation of copolymers, which allows control over the drug release behavior (61).

In addition to the above polymers, a number of other polymers like methacrylic acid derivatives, Eudragit®, poly(N-isopropylacrylamide) (PNIPAAm), polyisobutylcyanoacrylate (PBCA) have also been explored for preparation of orally deliverable NPs.

Miscellaneous

Stearic acid and Gliadins, hydroxypropyl methyl cellulose phthalate, have also been used for preparation of NPs. Apart from "drug only" and polymeric forms, NPs are also prepared by utilizing simple organic or inorganic compounds (62–64). The choice of biodegradable polymers on offer is limited. Tyner et al. (65) converted the drug camptothecin into micelles with the help of a negatively charged surfactant and these micelles were then encapsulated into NPs of magnesium–aluminum layered double hydroxides by an ion exchange process. Stacked sheet-like nanostructures of 500 nm in two dimensions and 10 nm in the third dimension were obtained.

Stabilizers for NPs

Stabilizers are used to prevent the aggregation of particles by conferring a surface charge. Normally, the higher the surface charge the greater is the stability. Surface coating with the surfactant also increases the mean particle size (20). It is now becoming evident that the surfactants also modulate the particle uptake and release behavior of the incorporated drug from the NP matrix.

In 0.1 N HCl, nonionic surfactants protected indomethacin-incorporated nanosuspensions of ethyl cellulose, poly (methyl methacrylate), and cellulose acetate butyrate intended for oral use against flocculation, while anionic and macromolecular stabilizers were not effective (66).

Polyvinyl alcohol (PVA) is the most widely used stabilizer for NPs. One difficulty faced with PVA is that a fraction of it remains associated with the NPs, as explained in chapter 6, despite repeated washings because PVA forms an interconnected network with the polymer at the interface. This residual PVA on PLGA NPs can be controlled by altering the PVA (read stabilizer) concentration or the type of organic solvent employed in the emulsion. Residual PVA can influence the physical properties of the particles (like size, zeta potential, polydispersity index, and surface hydrophobicity), drug loading, cellular uptake (lower values are associated with increased hydrophilicity imparted by PVA), and release. Hence, residual PVA and the factors influencing it can be used as formulation parameters to alter the properties or application of NPs (67). A study on the influence of grade of PVA used for PLGA NPs showed that PVA with a low degree of hydrolysis gives a higher yield, uniform size distribution, and excellent redispersibility. Particle characteristics depend more on the degree of hydrolysis than on the degree of polymerization, and should be an important parameter to finalize in the initial stages itself for developing a nanoparticulate formulation (68).

Cellular uptake of vitamin E d-alpha tocopheryl polyethylene glycol 1000 succinate (TPGS) coated PLGA NPs was shown to be 1.4-fold higher than that of PVA-coated PLGA NPs and 4–6-fold higher than that of uncoated polystyrene NPs, highlighting the role of stabilizers in particulate uptake (31).

Polypeptides and macromolecular drugs often undergo molecular denaturation on surface adsorption. Kossovsky et al. (69) have described surface modification of carbon ceramic NPs and self-assembled calcium–phosphate dihydrate particles by carbohydrate adsorption, which serves the dual purpose of maintaining the dynamic freedom of peptide drugs and cryopreservation. CS, didodecyldimethyl ammonium bromide and gelatin are used as positive charge imparting stabilizers and the resulting NPs are expected to interact more strongly with mucus than the negatively charged particles.

Freeze-Drying

Freeze-drying is done for ensuring stability, ease in storage and handling, and formulation into solid-dosage forms. The presence of water accelerates degradation of various types of polymers used in NPs (70). Schaffazick et al. (71) have shown that even nanocapsules and nanospheres can be freeze-dried without fear of leakage of drug or disturbing the structural integrity of the capsule wall. They mixed colloidal silicon dioxide, a standard glidant used in oral-dosage forms, before freeze-drying of diclofenac nanocapsules to prevent aggregation. An increase in size of the NPs was seen following freeze-drying with the aid of cryoprotectants like sucrose, glucose, trehalose, and gelatin (72). PLGA and PCL NPs of cyclosporine A (CyA) became 1.5 times their original size (100 nm) after freeze-drying (160 nm), and this change in size can significantly change the uptake and the fate of the NPs, and hence the pharmacokinetics of the incorporated drug(s). An important consideration in freeze-drying of polymeric nanodispersions for oral delivery is their redispersibility as the particles have to be present in the segregated state to allow the uptake processes. To ensure a readily dispersible powder, lyo- and cryoprotectants like sugars are used. Chacon et al. (70) showed improved stability of PLGA NPs with freeze-drying. Ahlin reported greater redispersibility of PLGA NPs by using trehalose as cryoprotectant (73).

Drug Release

Drug release from some of the nanoparticulate formulations is seen to be biphasic—an initial burst is followed by a rather slow (and controlled) release (59,66,74,75). This phenomenon has been explained for NPs prepared by emulsification solvent evaporation method (76). For single emulsions, the solvent elimination concentrates the incorporated substance toward the surface and for multiple emulsions, it makes holes in the polymeric walls near the surface resulting in initial burst release. The rest of the incorporated drug is released under the dual influence of diffusion and polymer degradation.

After oral administration, the enzymes present in the lumen can influence the release of drug from the drug delivery system. Therefore, the dissolution or release medium should be incorporated with enzymes for in vitro release studies. Enzymatic degradation of the NPs depends on the polymer type and molecular weight. The presence of proteolytic enzyme trypsin in the release medium resulted in increased drug release from doxorubicin–gelatin NPs conjugate (77).

Dose

The dose to be incorporated into a nanoparticulate system depends on the extent of particle uptake. This in turn depends on the particle size (smaller particles are taken up more readily and in greater proportions than the larger ones), surface hydrophobicity/hydrophilicity (an optimum balance is required, although hydrophobicity shows a higher correlation), zeta potential, presence of other excipients (which can modulate particle uptake), and bioadhesivity of the system. Additionally, the molecular weight of the drug, its interaction with the NP system, and method of incorporation will decide the maximum drug loading. Also, a major proportion of the NPs administered orally can be excreted without absorption, depending on the particle size and surface characteristics. Based on the above discussion, it can be inferred that high-dose drugs cannot be administered in the form of NPs unless linked with carriers or when particle uptake is not the major mechanism of drug absorption; as only a fraction of NPs administered are absorbed, if the absorption at low levels fluctuates, the percentage error in dose absorbed can be significant.

The second point has profound implications in the development and adoption of nanoparticulates for oral delivery. For a given dose, an amount of particles taken up in excess of the expected figure will result in toxicity and the lower absorption will end up in failed therapy as a consequence of the subtherapeutic drug levels attained. However, this should not deter the pharmaceutical scientist from exploring this technology for oral use because there are solutions available, like bioadhesion and ligand coating, that can significantly improve the uptake by providing prolonged residence at the site of uptake or active transport.

Dosage Form

NPs are normally given orally to experimental animals in the form of suspension. An oral multiple-unit dosage form, which overcame many of the problems commonly observed during the compression of microparticles into tablets, was developed by Bodmeier et al. (78). Microparticles and NPs were entrapped in beads formed by ionotropic gelation of the charged polysaccharide, chitosan, or sodium alginate and in solutions of the counterion, tripolyphosphate, or calcium chloride, respectively.

Polymeric NPs (Eudragit® RL 30D, L 30D, NE 30D, or Aquacoat®) were incorporated into various solid-dosage forms (granules, tablets, and pellets) by Schmidt and Bodmeier (79). They were evaluated for compatibility studies with excipients commonly used with solid oral-dosage forms. Ideally, the NPs should be released from the solid-dosage forms with their original properties. Hence, the necessity of the dosage forms to disintegrate back to the constitutive NPs was stressed, identifying their wettability as a critical parameter. The addition of polymeric binders (e.g., polyvinylpyrrolidone, sodium carboxymethylcellulose, or hydroxypropyl methylcellulose) to the aqueous NP dispersions before wet granulation resulted in phase separation for many NP/binder systems. Two quality control parameters for the complete redispersibility of the NPs are: (i) a high minimum-film formation temperature of the polymer dispersion, and (ii) a good wettability of the dried polymeric NPs. Contact angle measurements are good indicators of wettability.

Murakami et al. (80) prepared long-acting matrix tablets by direct compression of the drug with PLGA NPs. The drug showed a biphasic release pattern, which was altered by variation in tablet weight and size, but the amount released per unit surface area remained constant. The release pattern of such a preparation would be based only on the swelling properties of the NPs and should be independent of the drug. For local release in GIT, the drug release should be programmed for 24 hours or lesser.

NPs have also been used as an enteric coat. Hydroxypropyl methylcellulose phthalate and ammonium hydroxide were used to prepare NPs by neutralization emulsification technique. These NPs were used to provide an enteric coating to tablets and their drug release and swelling were studied (81).

Formulation Evaluation

Degradation Studies

Degradation in NPs is indicative of their stability and the possible time period and kinetics of release of incorporated drug. The dose of the drug to be incorporated can be calculated by correlating the in vivo detectable levels of NPs with the degradation kinetics over a period of time. Thus, the effective delivery period of the drug from the NPs becomes dependent on the combined effect of polymer degradation and natural scavenging mechanisms of the body. The design of these in vitro studies should be based on the actual physiological environment to which the particles are going to be exposed. Particle size plays a significant role in determining the rate of degradation. As the particle size is reduced, more surface area is available for entry of water into NPs resulting in faster degradation and release of therapeutic agent. Polymer degradation was demonstrated to be biphasic in PLGA NPs, with an initial rapid degradation during the first 20–30 days followed by a much slower phase. It was suggested that the surface-associated PVA rather than the particle size plays a dominant role in controlling the degradation of NPs (59).

Storage

Depending on its chemistry and morphology, a polymer will absorb some water on storage in a humid atmosphere. Absorbed moisture can initiate degradation and a change in physicochemical properties, which can in turn affect the performance in vivo. Storage conditions may thus be critical to the shelf life of a polymeric NP formulation. The incorporation of the drug may also affect the storage stability of a polymer matrix. The relative strength of water polymer bonds and the degree of crystallization of polymer matrix are other important factors. To maintain the absolute physicochemical integrity of degradable polymeric drug delivery devices, storage in an inert atmosphere is recommended (61).

Commercialization of nanoparticulate systems has not been taken up because of the problems in maintaining the stability of suspensions for an acceptable shelf life (72). The colloidal suspension, in general, does not tend to separate just after preparation because submicronic particles sediment slowly and aggregation effect is counteracted by mixing tendencies of diffusion and convection. However, after several months of storage, aggregation can occur. Additionally, microbiological growth, hydrolysis of the polymer, drug leakage, and/or other component degradation in aqueous environment is possible. Freeze-drying is a good method to dry nanospheres to increase the stability of these colloidal systems. However, because of their vesicular nature, nanocapsules are not easily lyophilized, as they tend to collapse releasing the oil core.

APPLICATIONS

The potential applications of orally administered NPs are depicted in Figure 3 and described in this section.

Enhanced Oral Bioavailability

The physicochemical and biological properties of protein and peptides are different from those of conventional drugs, such as molecular size, biological half-life, conformational stability, physicochemical stability, solubility, oral bioavailability, dose requirement, and administration (82). NPs can be efficient drug carriers for achieving oral peptide delivery. Because of their special uptake mechanisms, NPs can be regarded as

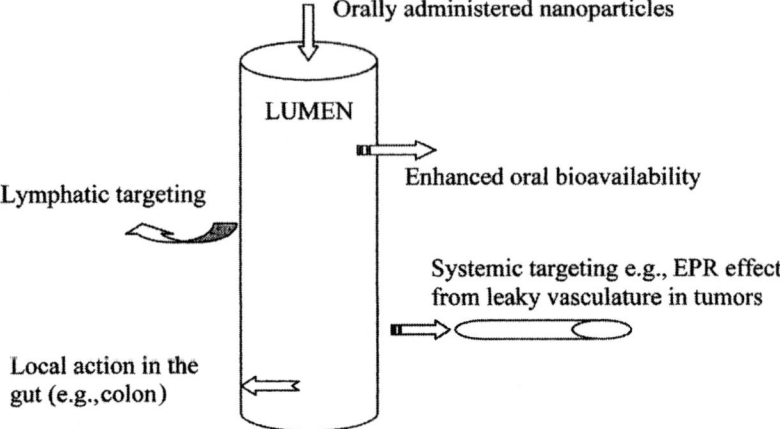

Figure 3 Potential applications of orally delivered NPs. *Abbreviation*: EPR, Enhanced Permeation and Retention; NP, nanoparticles.

interesting devices to increase the oral bioavailability of drugs. With the presently available efficiencies of particle uptake, it might not be possible to increase the bioavailability of drugs to 100%, but this technology has potential applications in achieving therapeutic levels of drugs that are considered undeliverable or have very low bioavailabilities through this route by regular means.

Damgé et al. (83) found that insulin encapsulated in PBCA nanocapsules reduced glycemia by 50–60%, although free insulin did not affect glycemia when administered orally to diabetic rats. A specific formulation, 1.6% zinc insulin in PLGA with fumaric anhydride oligomer and iron oxide additives (Fe_3O_4) has been shown to be active orally and is able to control plasma glucose levels when artificially challenged with glucose (84). A better control was hypothesized with a higher homogeneity in size.

The bioavailability of salmon calcitonin (sCT) was improved significantly by PNIPAAm [poly(N-isopropylacrylamide)] NPs composed of graft copolymers, as shown by the increased hypoglycemic effect. The absorption enhancement was explained on the basis of the dual effect of mucoadhesion of NPs and an increase in the stability of sCT against degradation by digestive enzymes in the GIT (85).

Zhang et al. (62) observed an increase in oral bioavailability and sustained release of CyA from stearic acid NPs in Wistar rats. The NPs showed reduced maximum plasma concentration and a relative bioavailability of 80% in comparison to market formulation Sandimmun Neoral® (Sandoz). The influence of charge-inducing agents on CyA NP absorption was evaluated using chitosan HCl, gelatin-A, or sodium glycocholate and the results compared against a commercially available microemulsion preparation (Neoral®). The relative bioavailability of CyA from chitosan NPs (positively charged) was increased by about 73%, and by about 18% from gelatin NPs (positively charged), whereas it was decreased by about 36% from soluble guanylyl cyclase (SGC) NPs (negatively charged) (86). Dai et al. (87) also compared the oral bioavailability of CyA from NPs of methacrylate polymers (Eudragits) with Neoral in Sprague-Dawley rats. The relative bioavailability of CyA increased for Eudragits S100, L100–55, and CyA-L100 NPs, while it decreased for E100 NPs when compared with the Neoral microemulsion.

Arbos et al. (88) evaluated the potential of certain bioadhesive NPs to increase the oral bioavailability of drugs degraded in GIT using 5-fluorouridine as a model drug. From the urine data, poly(methylvinylether-co-maleic anhydride) NPs and those coated with albumin showed higher bioavailability over the control oral solution.

Incorporation of danazol into NPs (170 nm) significantly improved its oral bioavailability over an aqueous suspension of conventional danazol particles (10 μm) in fasted male beagle dogs, and was comparable to that from a danazol-hydroxypropyl-β-cyclodextrin complex (89). Maincent et al. (90) reported improved oral bioavailability of vincamine in rabbits when administered using polyhexylcyanoacrylate NPs over the aqueous solution.

Vitamin B_{12} offers many advantages as a carrier for oral drug delivery, like immunocompliance, low cost, and can be readily modified to provide suitable functional groups for conjugation with drugs, especially proteins and peptides. This uptake system can potentially increase the oral uptake of molecules such as luteinizing hormone releasing hormone analogs, α-interferon, erythropoietin, and granulocyte colony-stimulating factor (G-CSF), which have been covalently linked to the vitamin B_{12} molecule (91). Wu et al. (92) reported that the NanoCrystal$^{\circledR}$ dispersion (a commercialized nanodrug delivery technology platform) eliminated the food effect on oral absorption in the dog at a dose of 2 mg/kg. The drug MK-0869 (aprepitant) exhibits regional specific absorption with higher amounts absorbed from upper GIT. A large increase in surface area in the drug NPs could overcome the narrow absorption window and lead to rapid in vivo dissolution, fast absorption, and increased bioavailability.

NPs may overcome immune surveillance by surface modification (e.g., PEGylation); however, it is very likely that these same surface modifications result in reduced cell uptake and thus oral absorption via M cells. However, it may happen that the stealth nature may be of a different degree for GIT uptake and RES scavenging and if the latter is more, there is a net overall benefit. PEG coating on PLA NPs was shown to increase their blood levels as compared to the noncoated ones (42). It can be implied from this study that the particles were not only absorbed but also evaded the RES system to result in higher levels of stealth NPs.

Local Delivery—Colon Specific Targeting

Numerous drugs are inactivated in the GIT, because of the stomach pH, the presence of proteolytic enzymes, and the hepatic first-pass effect. For drugs presenting a narrow intestinal absorption window, bioadhesive solid-dosage forms offer an interesting approach to prolong the residence time at or before this absorption window. Targeting a drug directly to the colon offers many advantages (21). Colon, as a site, offers a near-neutral pH (slightly alkaline), longer transit time, reduced digestive enzymatic activity, and greater response to absorption enhancers.

Polymeric nanoparticulate carrier systems can target the inflamed tissue in intestinal bowel diseases. As no sedimentation occurs with colloidal drug carriers, they might be affected to a lesser extent because of higher diffusion rates. With the anti-inflammatory drug rolipram, the oral administration of drug loaded NP formulations proved to be as effective as solubilized drug in relieving experimental colitis. When drug administration was discontinued colitis reappeared in animals treated with the drug solution, whereas animals treated with the NPs maintained reduced the level of inflammation. An important advantage of this strategy is direct contact of the carriers with the inflammation site, which in principle can provide higher drug concentration. Moreover, NPs were found not only to accumulate in the ulcer but also to adhere to nonulcerated inflamed tissue as mucus production is significantly increased in inflamed tissue (32,93).

Lymphatic Targeting

The problems inherent with the oral route of delivery, including low pH, gastric enzymes, and rapid transit away from the absorption sites and poor absorption of large molecules, pose significant challenges to antigen delivery. Thus, an effective delivery system shall protect the antigen in the gut, target the antigen to the GALT, or increase the residence time of the antigen in the intestine through bioadhesion.

PPs are the main targets for orally delivered vaccines. M cells play a determining role in the sampling and transport of luminal antigens into lymphoid tissues for immunological surveillance and initiation of appropriate immunological response. By incorporating the vaccine into the nanoparticulate drug delivery system, the vaccine is protected against degradation on its way to the mucosal tissue and efficiently targeted to and taken up by the M cells (51). Oral administration of vaccines might result in improvements in efficacy, as oral immunization would stimulate mucosal immunity at the sites at which many pathogens initially infect the host (47).

Lectins can act as transport molecules to cotransport the haptens or proteins into and across the intestine. The binding ability of the targeting molecule, as well as the immunogenicity of the antigen to be delivered must be preserved during the preparation of bioconjugates. Russell-Jones et al. (94) incubated Caco-2 and opossum kidney (OK) cells with Leucotriene B, Concanavalin A, and WGA-coated NPs and examined the essential criteria for binding, uptake, and transport of various sized commercial Fluorescent yellow-green (YG) NPs from Polysciences. Their studies proved that a wide range of targeting molecules like these three can enhance the uptake of nanoparticulates in a range of sizes 50–500 nm.

Systemic Targeting

After their uptake from the intestine, NPs can appear in the systemic circulation and can thus be used to deliver drugs in virtually any organ perfused by blood. Drug targeting essentially involves exploiting the altered/unique physiology at the target site (e.g., a tumor). The functionalities used to enable NPs this way pose processing and cost challenges, besides the primary requirement of specificity to the target tissue, which may have implications on the particle uptake also.

The biggest challenge in antineoplastic chemotherapy is to achieve selective localization of the drug at the tumor site for the desired period without causing cytotoxic effects on other organs. The tumor vasculature is hyperpermeable and selectively retains macromolecules and colloidal carriers of diameter up to 600 nm (95). WGA has affinity for the EGF receptor that is overexpressed in tumors, including those of the liver, breast, lung, and bladder. Thus, prodrugs or drug delivery systems containing WGA are expected to be appropriate for targeting of anticancer drugs (6). Brannon-Peppas and Blanchette (96) have reviewed the application of NPs for cancer chemotherapy by utilizing the concepts of enhanced permeation and retention effect, gene delivery, and avoiding RES.

A study was carried out to explore lectin-functionalized PLGA NPs as bioadhesive drug carriers against tuberculosis (TB) to reduce the drug dosage frequency of antitubercular drugs and thus improve patient compliance in TB chemotherapy. On administration of lectin-coated NPs in the size range of 350–400 nm, through the oral/aerosol route, the presence of drugs in plasma was observed for 6–7 days for rifampicin and

13–14 days for isoniazid and pyrazinamide. However, upon oral or aerosolized administration of uncoated particles, rifampicin was detectable in plasma for 4–6 days, whereas isoniazid and pyrazinamide were detectable for 8–9 days. All three drugs were present in lungs, liver, and spleen for 15 days. Administration of WGA-coated NPs caused a significant increase in the relative bioavailability of antitubercular drugs (97).

The cationic polymers bind to the negatively charged DNA to deliver the payload directly inside the cell (98). Hexylcyanoacrylate NPs were used as drug carriers for azidothymidine (AZT) to investigate specific drug targeting of antiretroviral drugs to reticuloendothelial cells by the oral route. An increase in bioavailability and the longer duration of action was observed at sites containing abundant macrophages, that is, in blood, brain, and organs of RES. This may allow a reduction in dosage, and hence a decrease in systemic toxicity. This type of delivery system can be applied to other drugs also which do not cross the blood–brain barrier (41). A new chemical entity CGP 70726, which is a very poorly water-soluble HIV-1 protease inhibitor, was incorporated into NPs made of poly(methacrylic acid–*co*–ethylacrylate) copolymer Eudragit L100–55 and oral administration of aqueous dispersions of these NPs to beagle dogs provided pH sensitive drug delivery (99).

Reduction in particle size from 20–30 μm to 270 nm reduced the gastric irritation induced after oral administration of naproxen, a nonsteroidal anti-inflammatory drug. The size reduction also resulted in a fourfold increase in the rate of absorption because of increase in surface area available for dissolution (89).

Lipidic peptide dendrimers of 2.5 nm (NPs by definition) were orally administered to female Sprague Dawley rats to assess oral absorption (100). Although the dendrimers were taken up in the intestine, their absorption was lower than that documented with 50–3000 nm polystyrene NPs suggesting that size is not the only and the most important determinant in particle uptake. Table 1 summarizes a few examples of NPs explored for oral delivery of NPs.

Table 1 A Few Examples of NPs Explored for Oral Delivery

NP matrix	Drug	References
Polystyrene	Trinitrobenzenesulfonic acid	32
Chitosan	Insulin	50
Fumaric acid and sebacic acid copolymer	Dicumarol	101
Stearic acid	Cyclosporin A	62
β-cyclodextrin, Poly (alkylcyanoacrylate)	Saquinavir	55
Polyvinylpyrrolidone	Danazol	102
Polystyrene	Fluorescein (marker)	20
PLGA and polystyrene	Uptake studies in excised intestine	46
N-Isopropylacrylamide, *tert*-butyl methacrylate	sCT	85
Polylysine dendrimer	None	100
Poly(methacrylic acid-*co*-ethacrylate) (Eudragit L100–55)	CGP 70726	99
n-hexylcyanoacrylate	AZT	41
Commercial YG (Polysciences)	Vitamin B_{12} derivatized (Caco-2 uptake studies)	91
PLGA, polymethylmethacrylate	Enalaprilat	73
Polymethylmethacrylate	None (radioactively labeled)	19
PLA, PLGA, and poly(fumaric-*co*-sebacic) anhydride	Insulin	84

Abbreviations: NP, nanoparticles; PLGA, poly(lactide-*co*-glycocide); PLA, poly(lactic acid).

FUTURE DIRECTIONS

Drug products for human use should be safe, efficacious, and of an acceptable quality. The safety of NPs has to be established not only for the drug but also for the matrix used for preparing the NPs. Safety is a relative term and is generally defined in terms of an upper maximum limit up to which a substance can be used in the FDA's generally recognized as safe list (GRAS), based on the toxicological data of use in other preparations given by the same route. An additional concern is whether the requirement of sterility can and should be imposed on the NPs for oral use designed for uptake in the GIT, because there is a possibility of microorganisms and their spores gaining access to the systemic circulation along with these carriers.

The critical parameters of a nanoparticulate formulation to set and monitor quality standards have to be based on simplicity (for routine analysis), reliability, and correlation to the in vivo performance. These can include particle size, zeta potential, pH of the suspension, (absence of) visible aggregation, redispersibility (contact angle measurement), assay of the incorporated drug, maximum allowable limit of solvents, residual stabilizer, and degradation products (oligomers/monomers) for ensuring quality assurance.

Dissolution tests can be developed for nanoparticulate formulations of only the drug or polymer entrapped drug with or without surfactant. Similarly, if the NPs are formulated into a solid-dosage form–like tablet, then a disintegration test has to be developed that will ensure total recovery of constitutive particles in the original nanosize range and with the same physicochemical properties. As the mode of absorption of the drug can be from either a (faster and locally generated) solution or direct uptake through the PPs, the drug release from the NPs within the expected time of residence of the particles in the GIT has to be accounted for in both qualitative and quantitative terms. This is especially important in the light of the fact that NPs can give an initial quicker release for the drug at or near the surface where polymer degradation and dissolution are not controlling the drug release. The drug release before the particles are absorbed (and when uptake is the only mechanism of drug absorption) is not going to contribute to the overall bioavailability of the drug, and thus the drug release has to be seen in the background of the mechanism of drug absorption.

ACKNOWLEDGMENTS

V. B. is grateful to NIPER for providing Ph.D. fellowship. Start-up fund from NIPER and research grants from Royal

Society of Chemistry, London, Third World Academy of Science, Italy, Department of Science and Technology, India, to M. N. V. R. K. are gratefully acknowledged.

REFERENCES

1. Sakuma S, Hayashi M, Akashi M. Design of nanoparticles composed of graft copolymers for oral peptide delivery. Adv Drug Deliv Rev 2001; 47:21–37.

2. Hans ML, Lowman AM. Biodegradable nanoparticles for drug delivery and targeting. Curr Opin Solid State Mater Sci 2002; 6:319–327.

3. Ponchel G, Montisci M, Dembri A, Durrer C, Duchene D. Mucoadhesion of colloidal particulate systems in the gastrointestinal tract. Eur J Pharm Biopharm 1997; 44:25–31.

4. Leo E, Angela VM, Cameroni R, Forni F. Doxorubicin-loaded gelatin nanoparticles stabilized by glutaraldehyde: involvement of the drug in the cross-linking process. Int J Pharm 1997; 155:75–82.

5. Lehr CM. From sticky stuff to sweet receptors-achievements, limits, and novel approaches to bioadhesion. Eur J Drug Metab Pharmacokin 1996; 21:139–148.

6. Gabor F, Bogner E, Weissenboeck A, Wirth M. The lectin-cell interaction and its implications to intestinal lectin-mediated drug delivery. Adv Drug Deliv Rev 2004; 56:459–480.

7. Chen H, Langer R. Oral particulate delivery: status and future trends. Adv Drug Deliv Rev 1998; 34:339–350.

8. Yeh P, Ellens H, Smith PL. Physiological considerations in the design of particulate dosage forms for oral vaccine delivery. Adv Drug Deliv Rev 1998; 34:123–133.

9. LeFevre ME, Warren JB, Joel DD. Particles and macrophages in murine Peyer's patches. Exp Cell Biol 1985; 53:121–129.

10. Lehr CM. Lectin-mediated drug delivery: the second generation of bioadhesives. J Control Release 2000; 65:19–29.

11. Hussain N, Jaitley V, Florence AT. Recent advances in the understanding of uptake of microparticulates across the gastrointestinal lymphatics. Adv Drug Deliv Rev 2001; 50:107–142.

12. Lemarchand C, Gref R, Couvreur P. Polysaccharide-decorated nanoparticles. Eur J Pharm Biopharm 2004; 58:327–341.

13. Borchard G, Lueben HL, de Boer AG, Verhoef JC, Lehr CM, Junginger HE. The potential of mucoadhesive polymers in enhancing intestinal peptide drug absorption. III: effects of chitosan–glutamate and carbomer on epithelial tight junctions in vitro. J Control Release 1996; 39:131–138.

14. Sanders E, Ashworth CT. A study of particulate intestinal absorption and hepatocellular uptake. Use of polystyrene latex particles. Exp Cell Res 1961; 22:37–45.

15. Delie F. Evaluation of nano- and microparticle uptake by the gastrointestinal tract. Adv Drug Deliv Rev 1998; 34:221–233.

16. Mathioweitz E, Jacob JS, Jong YS, et al. Biologically erodable microspheres as potential oral drug delivery systems. Nature 1997; 386:410–414.

17. Eldridge JH, Hammond CJ, Meulbroek JA, Staas JK, Gilley RM, Tice TR. Controlled vaccine release in the gut-associated lymphoid tissues. I. Orally administrated biodegradable microspheres target the Peyer's patches. J Control Release 1990; 11:205–214.

18. Norris DA, Puri N, Sinko PJ. The effect of physical barriers and properties on the oral absorption of particulates. Adv Drug Deliv Rev 1998; 34:135–154.

19. Araujo L, Sheppard M, Lobenberg R, Kreuter J. Uptake of PMMA nanoparticles from the gastrointestinal tract after oral administration to rats: modification of the body distribution after suspension in surfactant solutions and in oil vehicles. Int J Pharm 1999; 176:209–224.

20. Hillery AM, Florence AT. The effect of adsorbed poloxamer 188 and 407 surfactants on the intestinal uptake of 60-nm polystyrene particles after oral administration in the rat. Int J Pharm 1996; 132:123–130.

21. Duchene D, Ponchel G. Bioadhesion of solid oral dosage forms, why and how? Eur J Pharm Biopharm 1997; 44:15–23.

22. Arbos P, Campanero MA, Arangoa MA, Renedo MJ, Irache JM. Influence of the surface characteristics of PVM/MA nano-

particles on their bioadhesive properties. J Control Release 2003; 89:19–30.

23. Jaeghere FD, Doelker E, Gurny R. Encyclopedia of controlled drug delivery. In: Mathiowitz E, ed. Nanoparticles. New York: John Wiley & Sons, 1999:641–663.

24. Pappo J. Generation and characterization of monoclonal antibodies recognizing follicle epithelial M cells in rabbit gut-associated lymphoid tissues. Cell Immunol 1989; 120:31–41.

25. Rodrigues JS, Santos-Magalhaes NS, Coelho LCBB, Couvreur P, Ponchel G, Gref R. Novel core(polyester)-shell(polysaccharide) nanoparticles: protein loading and surface modification with lectins. J Control Release 2003; 92:103–112.

26. Florence AT, Hillery AM, Hussain N, Jani PU. Nanoparticles as carriers for oral peptide absorption: studies on particle uptake and fate. J Control Release 1995; 36:39–46.

27. Weissenbock A, Wirth M, Gabor F. WGA-grafted PLGA-nanospheres: preparation and association with Caco-2 single cells. J Control Release 2004; 99:383–392.

28. Arangoa MA, Ponchel G, Orecchioni AM, Renedo MJ, Duchene D, Irache JM. Bioadhesive potential of gliadin nanoparticulate systems. Eur J Pharm Sci 2000; 11:333–341.

29. Jung T, Kamm W, Breitenbach A, Kaiserling E, Xiao JX, Kissel T. Biodegradable nanoparticles for oral delivery of peptides: is there a role for polymers to affect mucosal uptake? Eur J Pharm Biopharm 2000; 50:147–160.

30. Panyam J, Sahoo SK, Prabha S, Bargar T, Labhasetwar V. Fluorescence and electron microscopy probes for cellular and tissue uptake of poly(D,L-lactide-co-glycolide) nanoparticles. Int J Pharm 2003; 262:1–11.

31. Win KY, Feng S. Effects of particle size and surface coating on cellular uptake of polymeric nanoparticles for oral delivery of anticancer drugs. Biomaterials 2005; 26:2713–2722.

32. Lamprecht A, Schafer U, Lehr CM. Size-dependent bioadhesion of micro- and nanoparticulate carriers to the inflamed colonic mucosa. Pharm Res 2001; 18:788–793.

33. www.unl.edu (accessed Oct 20, 2004).

34. www.stm2.nrl.navy.mil (accessed Oct 13 2004).

35. www.chembio.uoguelph.ca (accessed Oct 13 2004).

36. Arbos P, Arangoa MA, Campanero MA, Irache JM. Quantification of the bioadhesive properties of protein-coated PVM/MA nanoparticles. Int J Pharm 2002; 242:129–136.

37. Durrer C, Irache JM, Puisieux F, Duchene D, Ponchel G. Mucoadhesion of latexes: I. Analytical methods and kinetic studies. Pharm Res 1994; 11:674–679.

38. Durrer C, Irache JM, Puisieux F, Duchene D, Ponchel G. Mucoadhesion of latexes: II. Adsorption isotherms and desorption studies. Pharm Res 1994; 11:680–683.

39. Takeuchi H, Yamamoto H, Kawashima Y. Mucoadhesive nanoparticulate systems for peptide drug delivery. Adv Drug Deliv Rev 2001; 47:39–54.

40. Kreuter J, Muller U, Munz K. Quantitative and microautoradiographic study on mouse intestinal distribution of polycyanoacrylate nanoparticles. Int J Pharm 1989; 55:39–45.

41. Lobenberg R, Araujo L, Kreuter J. Body distribution of azidothymidine bound to nanoparticles after oral administration. Eur J Pharm Biopharm 1997; 44:127–132.

42. Tobio M, Sanchez A, Vila A, et al. The role of PEG on the stability in digestive fluids and in vivo fate of PEG–PLA nanoparticles following oral administration. Colloids Surf B 2000; 18:315–323.

43. Yang L, Chu JS, Fix JA. Colon-specific drug delivery: new approaches and in vitro/in vivo evaluation. Int J Pharm 2002; 235:1–15.

44. Newman SP, Wilding IR. Imaging techniques for assessing drug delivery in man. Pharm Sci Technol Today 1999; 2:269.

45. Savidge TC, Smith MW, James PS, Aldred P. *Salmonella*-induced M-cell formation in germ-free mouse Peyer's patch tissue. Am J Pathol 1991; 139:177–184.

46. Pietzonka P, Walter E, Duda-Johner S, Langguth P, Merkle HP. Compromised integrity of excised porcine intestinal epithelium obtained from the abattoir affects the outcome of

in vitro particle uptake studies. Eur J Pharm Sci 2002; 15:39–47.

47. Singh M, O'Hagan D. The preparation and characterization of polymeric antigen delivery systems for oral administration. Adv Drug Deliv Rev 1998; 34:285–304.

48. Artursson P, Lindmark T, Davis SS, Illum L. Effect of chitosan on the permeability of monolayers of intestinal epithelial cells (Caco-2). Pharm Res 1994; 11:1358–1361.

49. Thanou M, Verhoef JC, Junginger HE. Oral drug absorption enhancement by chitosan and its derivatives. Adv Drug Deliv Rev 2001; 52:117–126.

50. Pan Y, Li Y, Zhao H, et al. Bioadhesive polysaccharide in protein delivery system: chitosan nanoparticles improve the intestinal absorption of insulin in vivo. Int J Pharm 2002; 249:139–147.

51. van der Lubben IM, Verhoef JC, Borchard G, Junginger HE. Chitosan for mucosal vaccination. Adv Drug Deliv Rev 2001; 52:139–144.

52. Kumar MNVR, Muzzarelli RAA, Muzzarelli C, Sashiwa H, Domb AJ. Chitosan chemistry and pharmaceutical perspectives. Chem Rev 2004; 104:6017–6084.

53. Kumar MNVR. A review of chitin and chitosan applications. React Funct Polym 2000; 46:1–27.

54. Duchene D, Ponchel G, Wouessidjewe D. Cyclodextrins in targeting: application to nanoparticles. Adv Drug Deliv Rev 1999; 36:29–40.

55. Boudad H, Legrand P, Lebas G, Cheron M, Duchene D, Ponchel G. Combined hydroxypropyl-[beta]-cyclodextrin and poly(alkylcyanoacrylate) nanoparticles intended for oral administration of saquinavir. Int J Pharm 2001; 218:113–124.

56. Paul M, Fessi H, Laatiris A, et al. Pentamidine-loaded poly(D,L,-lactide) nanoparticles: physicochemical properties and stability work. Int J Pharm 1997; 159:223–232.

57. McClean S, Prosser E, Meehan E, et al. Binding and uptake of biodegradable poly-lactide micro- and nanoparticles in intestinal epithelia. Eur J Pharm Sci 1998; 6:153–163.

58. Anderson JM, Shive MS. Biodegradation and biocompatibility of PLA and PLGA microspheres. Adv Drug Deliv Rev 1997; 28:5–24.

59. Panyam J, Dali MM, Sahoo SK, et al. Polymer degradation and in vitro release of a model protein from poly(D,L-lactide-co-glycolide) nano- and microparticles. J Control Release 2003; 92:173–187.

60. Bala I, Hariharan S, Kumar MNVR. PLGA nanoparticles in drug delivery: the state of the art. Crit Rev Ther Drug Carrier Syst 2004; 21:387–422.

61. Edlund U, Albertsson AC. Degradable polymer microspheres for controlled drug delivery. Adv Polym Sci 2002; 157:67–112.

62. Zhang Q, Yie G, Li Y, Yang Q, Nagai T. Studies on the cyclosporin A loaded stearic acid nanoparticles. Int J Pharm 2000; 200:153–159.

63. Duclairoir C, Orecchioni AM, Depraetere P, Osterstock F, Nakache E. Evaluation of gliadins nanoparticles as drug delivery systems: a study of three different drugs. Int J Pharm 2003; 253:133–144.

64. Wang X, Dai J, Chen Z, et al. Bioavailability and pharmacokinetics of cyclosporine A-loaded pH-sensitive nanoparticles for oral administration. J Control Release 2004; 97:421–429.

65. Tyner KM, Schiffman SR, Giannelis EP. Nanobiohybrids as delivery vehicles for camptothecin. J Control Release 2004; 95:501–514.

66. Bodmeier R, Huagang C. Indomethacin polymeric nanosuspensions prepared by microfluidization. J Control Release 1990; 12:223–233.

67. Sahoo SK, Panyam J, Prabha S, Labhasetwar V. Residual polyvinyl alcohol associated with poly (D,L-lactide-co-glycolide) nanoparticles affects their physical properties and cellular uptake. J Control Release 2002; 82:105–114.

68. Murakami H, Kawashima Y, Niwa T, Hino T, Takeuchi H, Kobayashi M. Influence of the degrees of hydrolyzation and polymerization of poly(vinylalcohol) on the preparation and properties of poly(D,L-lactide-co-glycolide) nanoparticle. Int J Pharm 1997; 149:43–49.

69. Kossovsky N, Gelman A, Rajguru S, et al. Control of molecular polymorphisms by a structured carbohydrate/ceramic delivery vehicle—aquasomes. J Control Release 1996; 39:383–388.

70. Chacon M, Molpeceres J, Berges L, Guzman M, Aberturas MR. Stability and freeze-drying of cyclosporine loaded poly(D,L- lactide-glycolide) carriers. Eur J Pharm Sci 1999; 8:99–107.

71. Schaffazick SR, Pohlmann AR, Dalla-Costa T, Guterres SS. Freeze-drying polymeric colloidal suspensions: nanocapsules, nanospheres, and nanodispersion. A comparative study. Eur J Pharm Biopharm 2003; 56:501–505.

72. Saez A, Guzman M, Molpeceres J, Aberturas MR. Freeze-drying of polycaprolactone and poly(D,L-lactic-glycolic) nanoparticles induce minor particle size changes affecting the oral pharmacokinetics of loaded drugs. Eur J Pharm Biopharm 2000; 50: 379–387.

73. Ahlin P, Kristl J, Kristl A, Vrecer F. Investigation of polymeric nanoparticles as carriers of enalaprilat for oral administration. Int J Pharm 2002; 239:113–120.

74. Mu L, Feng SS. A novel controlled release formulation for the anticancer drug paclitaxel (Taxol®): PLGA nanoparticles containing vitamin E TPGS. J Control Release 2003; 86:33–48.

75. Lamprecht A, Ubrich N, Hombreiro Perez M, Lehr CM, Hoffman M, Maincent P. Influences of process parameters on nanoparticle preparation performed by a double emulsion pressure homogenization technique. Int J Pharm 2000; 196:177–182.

76. Rosca ID, Watari F, Uo M. Microparticle formation and its mechanism in single and double emulsion solvent evaporation. J Control Release 2004; 99:271–280.

77. Leo E, Cameroni R, Forni F. Dynamic dialysis for the drug release evaluation from doxorubicin–gelatin nanoparticle conjugates. Int J Pharm 1999; 180:23–30.

78. Bodmeier R, Chen HG, Paeratakul O. A novel approach to the oral delivery of micro- or nanoparticles. Pharm Res 1989; 6:413–417.

79. Schmidt C, Bodmeier R. Incorporation of polymeric nanoparticles into solid dosage forms. J Control Release 1999; 57: 115–125.

80. Murakami H, Kobayashi M, Takeuchi H, Kawashima Y. Utilization of poly(D,L-lactide-co-glycolide) nanoparticles for preparation of mini-depot tablets by direct compression. J Control Release 2000; 67:29–36.

81. Kim IH, Park JH, Cheong IW, Kim JH. Swelling and drug release behavior of tablets coated with aqueous hydroxypropyl methylcellulose phthalate (HPMCP) nanoparticles. J Control Release 2003; 89:225–233.

82. Li Y, Pei Y, Zhou Z, et al. PEGylated polycyanoacrylate nanoparticles as tumor necrosis factor-[alpha] carriers. J Control Release 2001; 71:287–296.

83. Damgé C, Michel C, Aprahamian M, Couvreur P. New approach for oral administration of insulin with polyalkylcyanoacrylate nanocapsules as drug carriers. Diabetes 1988; 37: 246–251.

84. Carino GP, Jacob JS, Mathiowitz E. Nanosphere based oral insulin delivery. J Control Release 2000; 65:261–269.

85. Sakuma S, Suzuki N, Kikuchi H, et al. Oral peptide delivery using nanoparticles composed of novel graft copolymers having hydrophobic backbone and hydrophilic branches. Int J Pharm 1997; 149:93–106.

86. El-Shabouri MH. Positively charged nanoparticles for improving the oral bioavailability of cyclosporin-A. Int J Pharm 2002; 249:101–108.

87. Dai J, Nagai T, Wang X, Zhang T, Meng M, Zhang Q. pH-sensitive nanoparticles for improving the oral bioavailability of cyclosporine A. Int J Pharm 2004; 280:229–240.

88. Arbos P, Campanero MA, Arangoa MA, Irache JM. Nanoparticles with specific bioadhesive properties to circumvent the presystemic degradation of fluorinated pyrimidines. J Control Release 2004; 96:55–65.

89. Liversidge GG, Conzentino P. Drug particle size reduction for decreasing gastric irritancy and enhancing absorption of naproxen in rats. Int J Pharm 1995; 125:309–313.

90. Maincent P, Verge RL, Sado P, Couvreur P, Devissaguet JP. Disposition kinetics and oral bioavailability of vincamine-loaded polyalkyl cyanoacrylate nanoparticles. J Pharm Sci 1986; 75:955–958.

91. Russell-Jones GJ, Arthur L, Walker H. Vitamin B12-mediated transport of nanoparticles across Caco-2 cells. Int J Pharm 1999; 179:247–255.

92. Wu Y, Loper A, Landis E, et al. The role of biopharmaceutics in the development of a clinical nanoparticle formulation of MK-0869: a Beagle dog model predicts improved bioavailability and diminished food effect on absorption in human. Int J Pharm 2004; 285:135–146.

93. Lamprecht A, Ubrich N, Yamamoto H. Biodegradable nanoparticles for targeted drug delivery in treatment of inflammatory bowel disese. J Pharmacol Exp Ther 2001; 299:775–781.

94. Russell-Jones GJ, Veitch H, Arthur L. Lectin-mediated transport of nanoparticles across Caco-2 and OK cells. Int J Pharm 1999; 190:165–174.

95. Chawla JS, Amiji MM. Biodegradable poly([epsiv]-caprolactone) nanoparticles for tumor-targeted delivery of tamoxifen. Int J Pharm 2002; 249:127–138.

96. Brannon-Peppas L, Blanchette JO. Nanoparticle and targeted systems for cancer therapy. Adv Drug Deliv Rev 2004; 56:1649–1659.

97. Sharma A, Sharma S, Khuller GK. Lectin-functionalized poly (lactide-*co*-glycolide) nanoparticles as oral/aerosolized antitubercular drug carriers for treatment of tuberculosis. J Antimicrob Chemother 2004; 54:761–766.

98. Azzam T, Domb AJ. Current developments in gene transfection agents. Curr Drug Deliv 2004; 1:165–193.

99. Jaeghere FD, Allemann E, Kubel F, et al. Oral bioavailability of a poorly water soluble HIV-1 protease inhibitor incorporated into pH sensitive particles: effect of particle size and nutritional state. J Control Release 2000; 68:291–298.

100. Florence AT, Sakthivel T, Toth I. Oral uptake and translocation of a polylysine dendrimer with a lipid surface. J Control Release 2000; 65:253–259.

101. Desai MP, Labhasetwar V, Walter E, Levy RJ, Amidon GL. The mechanism of uptake of biodegradable microparticles in Caco-2 cells is size dependent. Pharm Res 1997; 14:1568–1573.

102. Liversidge GG, Cundy KC. Particle size reduction for improvement of oral bioavailability of hydrophobic drugs: I. Absolute oral bioavailability of nanocrystalline danazol in beagle dogs. Int J Pharm 1995; 125:91–97.

10

Brain Delivery by Nanoparticles

SVETLANA GELPERINA

Institute of Molecular Medicine,
Moscow Sechenov Medical Academy,
Moscow, Russia

INTRODUCTION

The brain is probably one of the least accessible organs for the delivery of drugs due to the presence of the blood–brain barrier (BBB) that controls the transport of endogenous and exogenous compounds, thus providing the neuroprotective function. The structural BBB is formed by the cerebral capillary endothelial cells that, in contrast to endothelial cells in capillary blood vessels in most other tissues, are closely joined to each other by tight junctions produced by the interaction of several transmembrane proteins (Fig. 1). Moreover, these endothelial cells demonstrate very little fenestration and display only low pinocytic activity. This physical barrier effectively abolishes any aqueous paracellular diffusional pathways

273

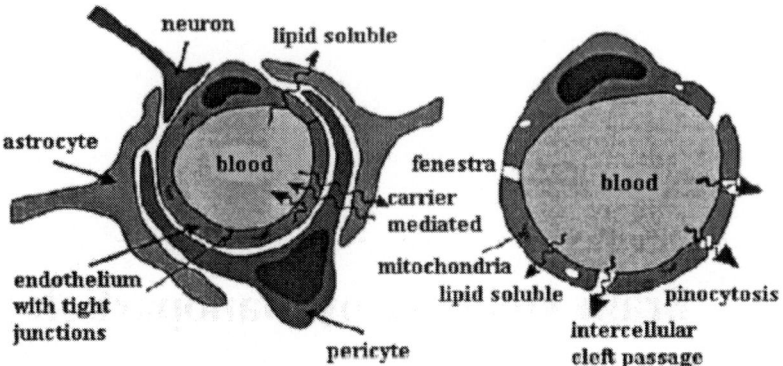

Figure 1 Schematic comparison between brain (*left*) and general (*right*) capillaries. *Source*: Adapted from Ref. 1.

between the extracellular fluid in the blood and brain. The endothelial cells forming the BBB also exhibit a number of bidirectional and unidirectional transporters. Essential compounds such as amino acids, hexoses, neuropeptides, and proteins employ these transporters or specific carriers to permeate the brain (1,2). Lipophilic solutes are able to diffuse across the BBB by direct permeation through the cell membrane if their molecular weight is not more than 500 Da (3). However, many of these lipophilic molecules will be actively removed from the cerebral compartment by the adenosine triphosphate binding cassette (ABC) efflux transporters, such as P-glycoprotein (P-gp) or multidrug resistance proteins (MRP) (4). Thus, many potential drugs with activity at a particular site or receptor in the brain have failed in the treatment of central nervous system (CNS) disorders. These drugs simply do not enter the CNS in sufficient quantities to be effective, which, consequently, diminishes their therapeutic value.

Generally, drug targeting to the brain could be achieved by going either "through" or "behind" the BBB. Several strategies employ craniotomy-based drug delivery, including either intraventricular drug diffusion or local intracerebral implants. Although these procedures can significantly increase drug levels in the brain, all of them are highly invasive.

In addition, craniotomy-based drug delivery relies on diffusion from the local depot sites; since diffusion decreases with the square of the diffusion distance, the effective treatment volume is often limited (5). Some of the approaches attempt to increase delivery of systemically administered drug by disruption of the BBB by infusion of hyperosmotic solutions or vasoactive agents, such as bradykinin or, recently, by circumvention of the drug efflux mechanisms. These approaches are probably most suited to short-term treatments, where a single or infrequent exposure to a drug is required. Another approach is the chemical modification of the drug for increasing its lipophilicity; however, lipophilic compounds are also potential substrates for the ABC efflux transporters. The conjugation of the drug with the BBB specific transport vector takes advantage of the normal endogenous transport pathways within the brain capillary endothelium; the disadvantage is a low carrying capacity of the vector molecules that is generally limited by 1:1 stoichiometry of a carrier to a drug (5). The liposomes have a much higher capacity and could enable drug transport to the brain; however, liposomal formulation of the antitumor antibiotic doxorubicin (DOX) displayed only moderate activity in glioblastoma patients (6). Receptor-mediated brain targeting of another potent anthracycline, daunomycin, was achieved using immunoliposomes (7). Yet, the efficacy of this delivery system for chemotherapy of brain tumors has not been demonstrated so far.

Therefore, despite the certain progress in this area, development of drug-targeting technology that enables safe and noninvasive access to the brain remains a challenge, resulting in the emergence of novel strategies.

Thus, it was shown recently that the drugs normally unable to cross the BBB could be delivered to the brain after binding to the surface-modified poly(butylcyanoacrylate) (PBCA) nanoparticles (NP) (8–10). Further investigations provided evidence that the NP-based drug delivery systems possess a significant potential for brain targeting.

This chapter addresses the various aspects of systemic drug delivery to the brain by means of polymeric NP. The use of NP for imaging is beyond the scope of the review.

BIODISTRIBUTION STUDIES

The ability of various NP to deliver drugs to the brain has been evidenced by a number of biodistribution studies (typical physicochemical parameters of the carriers are shown in Table 1).

The pharmacokinetic rule states that the mass of drug delivered to the brain is equally proportional to the BBB permeability coefficient and the area under curve (AUC) plasma concentration versus time (5).

One way to increase the drug circulation time is the application of long-circulating NP that avoid rapid clearance by the tissues of the mononuclear phagocyte system (MPS). Evasion of particle uptake by macrophages could be achieved to a certain extent by interference with protein adsorption and opsonization, thus preventing complement activation and recognition of NP (steric stabilization of the particles or the so-called "stealth" effect). The approaches for design and engineering of long-circulating vehicles have been described elsewhere (24,25). Generally, steric stabilization of nanocarriers in the bloodstream can be performed by physical adsorption of nonionic surfactants or amphiphilic block copolymers, such as poloxamers or poloxamines [block copolymers of poly(oxyethylene) and poly(oxypropylene)], or by their incorporation during the production of NP. Alternatively, particles can be formed from an amphiphilic copolymer, in which the hydrophobic block is able to form a solid phase, whereas the hydrophilic part provides protection of the surface. All of these technological approaches have proved to be successful for extension of the circulation time of NP and, in accordance with the pharmacokinetic rule, afforded enhancement of the brain delivery of NP and bound drugs.

Binding of camptothecin (Ca) and two anthracyclines, DOX and idarubicin, to solid lipid nanoparticles (SLN) afforded a very considerable increase of plasma and brain AUC (11,26–28). This effect was especially pronounced for Ca, a very hydrophobic antitumor drug. After intravenous (i.v.) administration, SLN produced an ∼10-fold increase of brain AUC, whereas plasma AUC was increased only five-fold

Table 1 Physicochemical Characteristics of the NP

NP type	Size (nm)	Surface charge (mV)	Drug	References
SLN	196.8 ± 21.3	−45.2 ± 5.2	Ca	11
Nonstealth SLN	58 (polydispersity 0.15)	−33.50	DOX	12
Stealth SLN (0.15% stealth agent)	70 (polydispersity 0.19)	−23.88	DOX	12
Stealth SLN (0.30% stealth agent)	86 (polydispersity 0.21)	−16.50	DOX	12
Stealth SLN (0.45% stealth agent)	95.5 (polydispersity 0.22)	−11.00	DOX	12
PMMA NP	130 ± 30	N/I	Unloaded	13
Nanogel™	237 ± 3	+5	Unloaded	14
Nanogel complexes with ODN				
N/P ratio 4	76 ± 1	−0.2	ODN	14
N/P ratio 8	94 ± 2	+2.3	ODN	14
PBCA NP	270	N/I	DOX	15
	228 (polydispersity 0.05)	N/I	Unloaded	16
	251 (polydispersity 0.053)	N/I	Probenecid	16
	221 (polydispersity 0.349)	N/I	MRZ 2/576	16
PHDCA NP	135–161 (polydispersity < 0.2)	−45	Unloaded	17
PEG–PHDCA NP	100–140 (polydispersity < 0.2)	−24.4 ± 1	Unloaded	18
PEG–PHDCA NP (organic formulation)	180–205 (polydispersity < 0.2)	+15 ± 2	DOX	19

(Continued)

Table 1 Physicochemical Characteristics of the NP (*Continued*)

NP type	Size (nm)	Surface charge (mV)	Drug	References
Poly(isobutylcyanoacrylate) NP	260 ± 80	N/I	Cyclosporine A	19
	139 ± 46	N/I	DOX	
FITC-labeled albumin NP	289 ± 38	−24.2 ± 2.1	Unloaded	20
Biotin-labeled albumin NP	304 ± 47	−25.1 ± 1.7		
SLN (E. Wax/Brij 78)	58 ± 0.8	N/I	[³H]-Cetyl-	21
SLN (Brij 72/Ps 80)	98 ± 1.8		alcohol	
Ps 80–coated poly(lactide) NP	202.6	−10.17	FITC dextran	22
Polysaccharide NP	60 ± 15	N/I	Fluorescein	23

Abbreviations: NP, nanoparticles; SLN, solid lipid nanoparticles; Ca, camptothecin; DOX, doxorubicin; PMMA, poly(methylmethacrylate); N/I, not indicated; ODN, oligonucleotide; N/P ratio, nitrogen (nanogel) to phosphate (ODN) ratio; PBCA, poly(butylcyanoacrylate); PHDCA, poly(hexadecylcyanoacrylate); PEG, poly(ethylene)glycol; FITC, fluorescein isothiocyanate; Ps 80, polysorbate 80.

as compared to the drug solution (Ca-sol). Moreover, brain AUC of Ca bound to SLN (Ca-SLN) was ~6 times higher than plasma AUC of this formulation; the ratio of brain AUC of Ca-SLN to brain AUC of Ca-sol was the highest among the tested organs. The SLN formulation acted as a sustained release system: a low but detectable concentration of Ca-SLN (4.7 ng/g) could be found in the brain 72 hr after i.v. injection of a dose of 1.3 mg/kg (11).

The brain delivery of DOX with SLN was less efficient, even though plasma AUC was considerably increased; however, the delivery was significantly improved by steric stabilization of the particles (27). The stealth effect was achieved using poly(ethylene)glycol (PEG) derivative (PEG 2000—stearic acid) (12,29). The influence of the increasing concentrations of the stealth agent (0.15–0.45%) on the distribution of DOX bound to SLN was demonstrated in rabbits (12). After i.v. administration, plasma and brain concentrations of DOX were increasing in parallel with the increasing amount of the stealth agent in the SLN (Fig. 2A and B). Compared to the nonstealth SLN, the SLN containing 0.45% of the stealth agent produced a nine-fold increase of DOX concentration in the brain that was achieved 30 minutes after injection. The concentrations were gradually decreasing with time and after six hours only the SLN containing 0.45% of the stealth agent provided a detectable DOX concentration in the brain. It is noteworthy that all SLN preparations decreased the heart concentrations of DOX, which suggests that these carriers may reduce the cardiotoxicity of this drug.

The same tendency was observed for drug-free [^{14}C]-labeled poly(methylmethacrylate) (PMMA) NP coated with block copolymer poloxamine 908 or the nonionic surfactant polysorbate 80 (Tween® 80, Ps 80) (13). Coating was performed by incubation of the particles in a surfactant solution before injection; the concentration range of the coating agents was 0.001–5%. Measurements were made 30 minutes after i.v. administration to rats. It was shown that at concentrations below 0.1% the NP behaved like uncoated particles, whereas concentrations above 0.1% for poloxamine 908 and 0.5% for Ps 80 significantly influenced the body distribution.

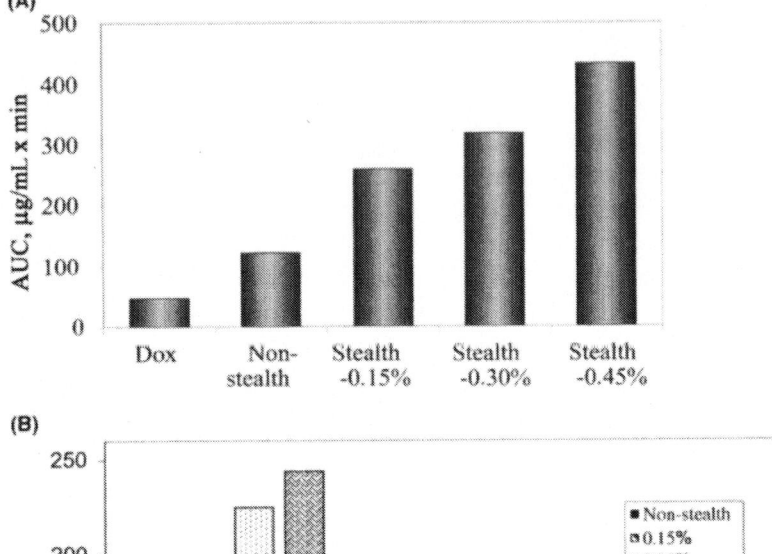

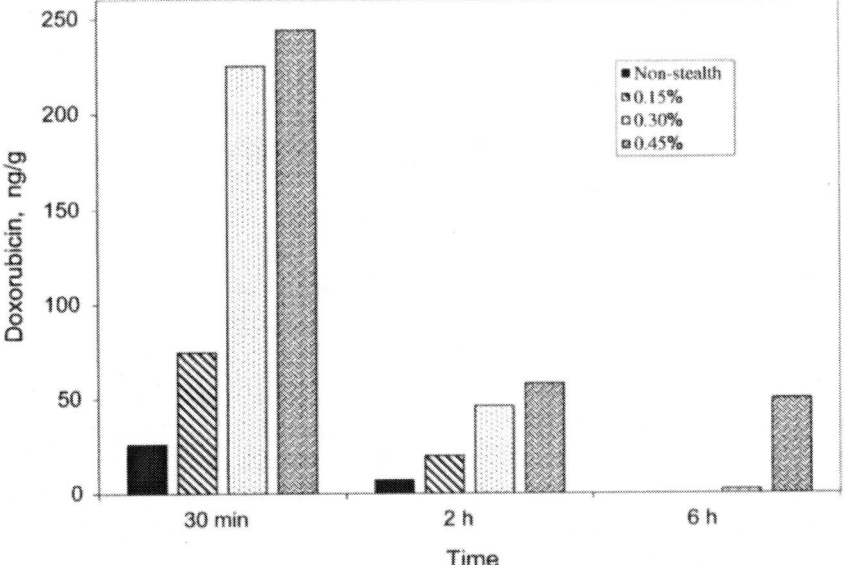

Figure 2 (**A**) Plasma AUC of DOX bound to various types of SLN after i.v. administration to rabbits in a dose of 1 mg/kg. (**B**) Brain distribution of DOX bound to various types of SLN after i.v. administration to rabbits in a dose of 1 mg/kg. *Abbreviations*: AUC, area under curve; SLN, solid lipid nanoparticles; i.v., intravenous; DOX, doxorubicin. *Source*: Adapted from Ref. 12.

Again, brain concentrations of the particles were increasing in a parallel manner to the concentration of the stealth agent. The maximal brain concentration was achieved after injection of PMMA NP in 1% solution of poloxamine 908 and reached 0.55% of the injected dose (~9 µg/g). A similar tendency was observed for Ps 80, although this surfactant was less effective in increasing the blood and brain concentration.

Similar results were obtained earlier by Tröster et al. (30). These authors extensively studied the influence of different surfactants on the biodistribution of PMMA NP in rats. The preparations were administered i.v. after incubation of the NP in 1% surfactant solutions. Again, poloxamine 908 was the most effective among other surfactants for increasing plasma concentration of the NP (100-fold increase 30 minutes after injection), whereas Ps 80 produced only a five-fold increase, yet showed a similar brain uptake.

The efficacy of Ps 80 coating for brain targeting, however, was clearly demonstrated by Gulyaev et al. (15). Ps 80–coated PBCA NP only moderately increased the plasma AUC of DOX (by ~70%) (Fig. 3A) but enabled a very efficient delivery of DOX to the rat brain (Fig. 3B).

For surfactant coating, 1% Ps 80 was added to the NP preparation, and the suspension was incubated for one hour prior to injection. The preparations were administered i.v. in a dose of 5 mg/kg. After administration of DOX loaded in PBCA NP coated with Ps 80 (DOX-NP + Ps 80), the concentration of the drug in the brain reached very high levels of 6 µg/g, which were maintained between two and four hours after administration. Brain AUC of this formulation was approximately 10 times higher than plasma AUC. The three other preparations used as controls [DOX solution in saline (DOX), DOX solution in 1% Ps 80 (DOX + Ps 80), and DOX bound to uncoated PBCA NP (DOX-NP)] were inefficient; the brain concentrations were below the detection limit of 0.1 µg/g. Both nanoparticulate preparations considerably decreased the heart concentrations of DOX and yielded levels below the detection limit after two hours. This phenomenon was observed earlier by Couvreur et al. (31) and suggests that cardiotoxicity of DOX could be decreased by means of PBCA NP.

(A)

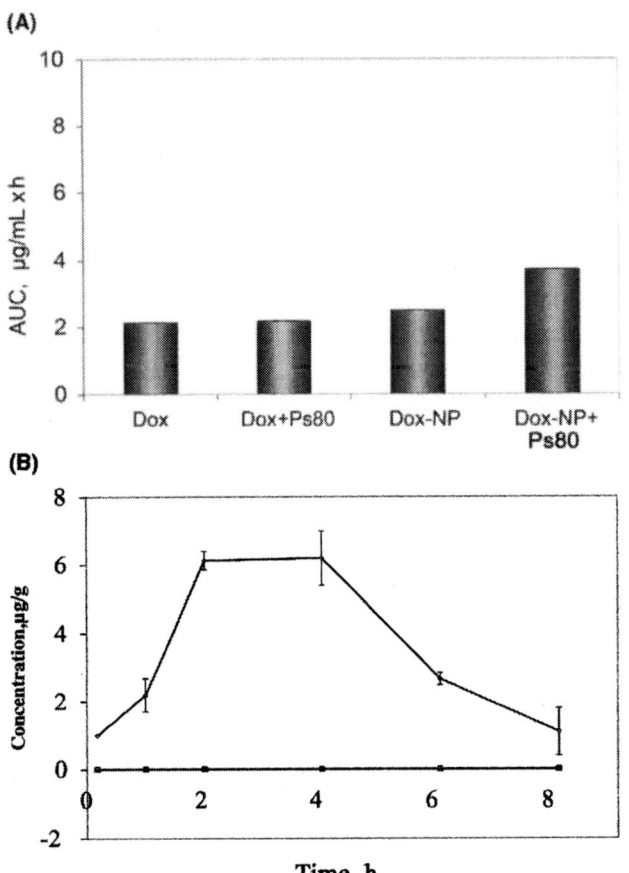

Figure 3 (**A**) Plasma AUC of DOX in solutions and bound to uncoated and Ps 80–coated PBCA NP after i.v. administration to rats in a dose of 5 mg/kg. (**B**) Brain distribution of DOX in solutions and bound to Ps 80–coated PBCA NP after i.v. administration to rats in a dose of 5 mg/kg: ●, DOX; ■, DOX + Ps 80;   , DOX-NP; ◇, DOX-NP + Ps 80. *Abbreviations*: AUC, area under curve; PBCA, poly(butylcyanoacrylate); NP, nanoparticles; i.v., intravenous; Ps 80, polysorbate 80; DOX, doxorubicin. *Source*: Adapted from Ref. 15.

Calvo et al. (32) compared the biodistribution of different types of [^{14}C]-labeled poly(hexadecylcyanoacrylate) (PHDCA) NP: unmodified PHDCA NP, long-circulating PHDCA NP modified by simple adsorption of poloxamine 908 or Ps 80, and PEG–PHDCA NP made from an amphiphilic copolymer of poly (methoxy-PEG-cyanoacrylate-co-hexadecylcyanoacrylate). The study was conducted in rats and mice. Again, the highest plasma concentration was achieved by coating the particles with poloxamine 908 (84.5% of the injected dose in rats and 43% in mice after one hour), whereas the plasma concentration of PEG–PHDCA NP was lower (50% dose in rats, 29% dose in mice) (Fig. 4A). However, in this case, PEG-modified NP produced the highest brain concentrations in both animal species (Fig. 4B and C). This effect may be attributed to the specific interaction of the PEGylated NP with the BBB. The brain uptake of PHDCA NP coated with Ps 80 again depended on the concentration of this surfactant.

Interestingly, the brain concentrations of all nanoparticulate formulations were much higher in mice than in rats (Fig. 4B and C). In contrast, the plasma concentrations were higher in rats; this difference was especially pronounced for Ps 80–coated particles (10-fold). As mentioned by the authors, this difference could be explained by distinctly different mechanisms of blood clearance in these animal species, in particular, mechanisms of liver and spleen capture of the surfactant-coated particles (33,34).

Nanogel[TM] is a new nanoscale carrier system that has been recently proposed for brain delivery of macromolecules, such as antisense oligodeoxynucleotide (ODN) (14,35,36). This system represents a nanoscale size polymer network of cross-linked ionic poly(ethyleneimine) (PEI) and nonionic PEG chains (PEG-*cl*-PEI). In solutions, PEG-*cl*-PEI forms dispersed swollen cross-linked NP that can absorb spontaneously, through ionic interactions, a variety of biomacromolecules, including negatively charged ODN. Upon binding of ODN through electrostatic interaction of this drug with the PEI chain, collapse of Nanogel occurs resulting in decreased size of the particles. Due to the effect of PEG chains, the collapsed ODN-loaded Nanogel forms a stable dispersion with the particle size of ca. 80 nm. The charge of this delivery system depends on its composition, which is usually described in terms of N/P ratio, i.e., the ratio of nitrogen (nanogel) to phosphate (ODN) concentrations in the final nanogel and ODN

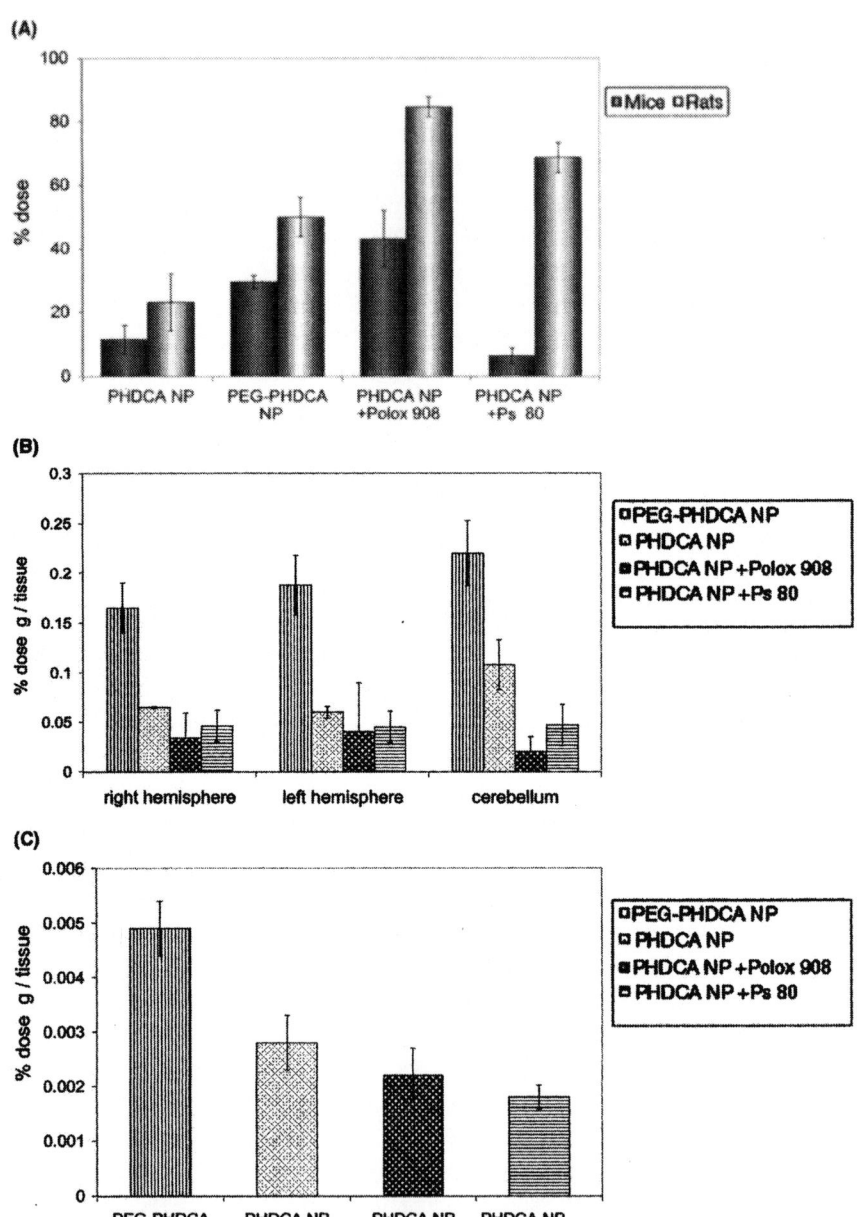

Figure 4 (*caption on facing page*)

mixture. At N/P = 8, particles exhibited a small positive charge (zeta potential ca. +2), whereas at N/P = 4, particles were electroneutral.

In vivo biodistribution study of ODN incorporated in Nanogel particles in mice demonstrated that this delivery system enabled enhanced ODN delivery to the brain, as compared to the free phosphorothioate ODN (35). The study employed two types of radiolabeled formulations: Nanogel-[^{3}H]-ODN and [^{3}H]-Nanogel–ODN. Both Nanogel formulations produced significantly higher levels of radioactivity in the brain one hour after i.v. injection, as compared to the free drug (5.34% and 2.67% of the dose, respectively, vs. 0.18%). This fact suggests that a significant portion of ODN in the brain remained associated with the carrier. The brain/plasma ratio for the Nanogel-ODN formulation increased by one order of magnitude compared to the free ODN. It also appears from this study that the cationic Nanogel formulation was more efficacious in brain delivery of ODN than the electroneutral formulation. Moreover, the accumulation of radioactivity in liver and spleen was significantly decreased, whereas plasma and lungs displayed relatively fewer changes.

Pathological Conditions of the CNS

Whereas under normal conditions the BBB limits the passage of solutes from the blood to the CNS, its function can be considerably compromised during various CNS diseases. Thus,

---

Figure 4 *(Facing Page)* (**A**) Concentration of radioactivity (% dose) in blood after i.v. administration of various types of [^{14}C]-labeled PHDCA NP in rats and mice at one hour postinjection: PEGylated, poloxamine 908 coated, Ps 80 coated, and uncoated. Values are means and SD, $n = 4$. (**B, C**) Concentration of radioactivity (% dose) in brain after i.v. administration of various types of [^{14}C]-labeled PHDCA NP at one hour postinjection: PEGylated, poloxamine 908 coated, Ps 80 coated, and uncoated. (**B**), mice; (**C**), rats. Values are means and SD, $n = 4$. *Abbreviations*: i.v., intravenous; PHDCA, poly(hexadecylcyanoacrylate); NP, nanoparticles; PEG, poly(ethylene)glycol; Ps 80, polysorbate 80. *Source*: Adapted from Ref. 32.

the development of highly malignant brain tumors is characterized by both neovascularization and vascular hyperpermeability. In contrast to normal cerebral capillaries, vessels in gliomas are tortuous and sinusoidal; the size of interendothelial gaps may reach $0.3 \times 3\,\mu m$ (37). Another common feature of glioma vasculature is an increased vessel wall thickness (endothelial thickness in glioma $\sim 0.50\,\mu m$ vs. $0.26\,\mu m$ in cerebral vessels) that contributes to an increase in nonselective transendothelial transport (38). It has been hypothesized that, as in other tumors, the structural abnormalities of glioma vessels facilitate penetration of NP due to passive extravasation across the impaired endothelium at the tumor site [an enhanced permeability and retention (EPR) effect] (39). This effect can be further improved by steric stabilization of the particles, which prevents their rapid clearance from the circulation and uptake by the MPS, thus enhancing their extravasation in the target tissue.

This hypothesis was confirmed by the results of the comparative biodistribution study of the stealth PEG–PHDCA NP and nonstealth PHDCA NP in rats bearing intracranial 9L gliosarcoma (17). As expected, accumulation in the tumor was more than three times higher for the long-circulating (PEG–PHDCA) NP than for PHDCA NP; the latter had a very short circulation time due to a rapid and massive uptake by the MPS tissues. Nevertheless, both carriers were able to extravasate across the BBB at the tumor site and to accumulate preferentially in the tumor rather than in the peritumoral brain or the healthy contralateral hemisphere (Fig. 5A and B). In addition, the four- to eight-fold higher accumulation of the PEGylated NP was observed also in parts of the brain protected by the normal BBB, as compared to PHDCA NP. This result correlated with the results of the previously discussed study conducted in healthy animals (32).

In the study of Lode et al. (40), poloxamine 908 and poloxamer 407 coating of $[^{14}C]$-PMMA NP provided a classic biodistribution profile, increasing blood concentrations and circulation time and decreasing liver uptake; however, they failed to promote considerable extravasation of the particles in the intracerebral U-373 glioblastoma or normal brain

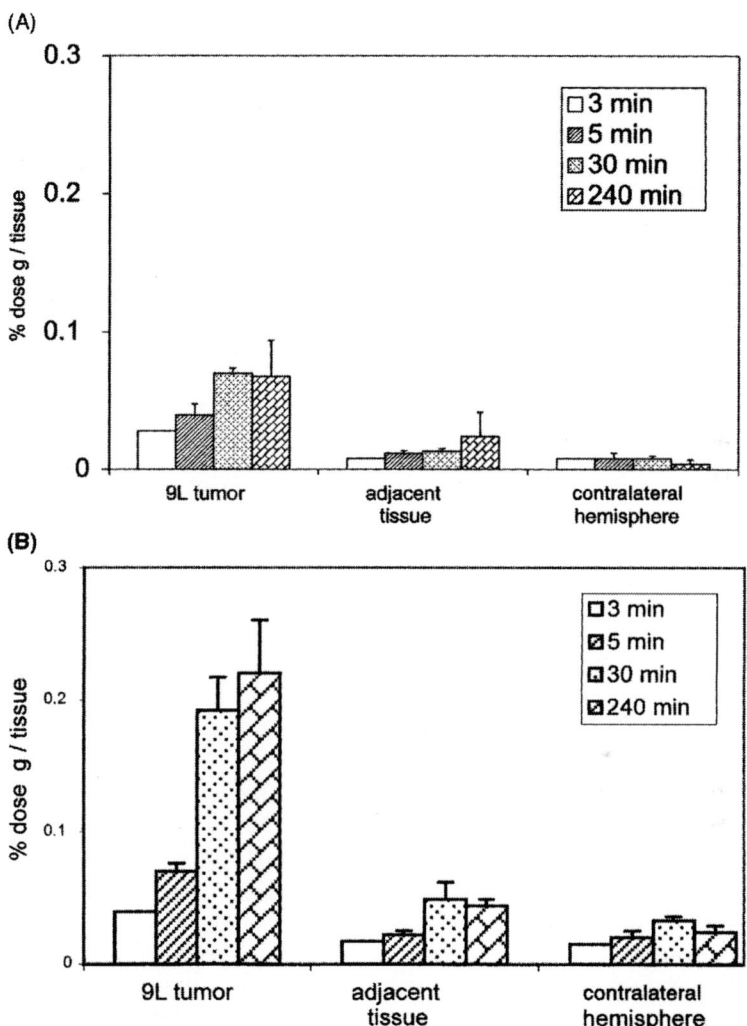

Figure 5 Brain biodistribution of the various types of [^{14}C]-labeled PHDCA NP after IV injection to rats bearing intracranial 9L gliosarcoma. (**A**), rats receiving PHDCA NP; (**B**), rats receiving PEG–PHDCA NP. ($n = 1$ at 3 min and $n = 4$ at 5, 30, and 240 min). Statistical differences between PEG–PHDCA NP and PHDCA NP are indicated by $*$ ($p < 0.05$), (nonparametric Mann–Whitney U test). *Abbreviations*: PHDCA, poly(hexadecylcyanoacrylate); NP, nanoparticles; IV, intravenous; PEG, poly(ethylene)glycol. *Source*: Adapted from Ref. 17.

tissue. At the same time, these surfactants considerably enhanced accumulation of PMMA NP in B16 melanoma implanted intramuscularly and in human breast cancer MaTu implanted subcutaneously in mice. This difference was explained by distinctive features of the tumor models. The immunohistological study revealed the correlation of the NP uptake in the tumors with the expression of the vascular endothelial growth factor (VEGF), which was interpreted as a marker of tumor-induced angiogenesis. The highest uptake was achieved in B16 melanoma, which was also characterized by the highest VEGF expression and, accordingly, by the highest growth rate, whereas a negligible NP uptake in glioblastoma paralleled a lack of VEGF expression in this tumor. As mentioned above, the abnormality of the tumor-associated vasculature is a prerequisite for the enhanced extravasation of the particles into tumor. Taken together with a parallel increase of blood concentrations and circulation time, these facts suggest that the phenomenon of the enhanced accumulation of PMMA particles in the peripheral tumors could be explained by the EPR effect. Ps 80 coating in this study was ineffective.

Another example of a CNS disorder associated with impaired BBB integrity is the experimental allergic encephalomyelitis (EAE), a well-established animal model of multiple sclerosis. Brain and spinal cord concentrations of [^{14}C]-PEG–PHDCA NP were compared with another long-circulating carrier, poloxamine 908–coated PHDCA NP, and with conventional PHDCA NP (41). The microscopic localization of fluorescent NP in the CNS was also investigated in order to further understand the mechanism by which the particles penetrate the BBB. In general, the results of this study showed the same tendency as the aforementioned study of Calvo et al. (32) conducted in healthy animals. Poloxamine 908–coated PHDCA NP again showed the smallest brain concentrations even though they produced the highest prolongation of the circulation time. The concentration of PEGylated NP in the brain, especially in white matter, was greatly increased in comparison to conventional non-PEGylated NP. As predicted, this increase was significantly higher in EAE rats than in control animals. In EAE rats, PEGylated NP were colocalized with macrophage infiltrations, suggesting the loss of BBB integrity in such lesions. The mechanism underlying particle penetration into the brain is most probably passive diffusion and macrophage uptake in inflammatory lesions.

Another possible mechanism of particle penetration across the BBB is their uptake by circulating macrophages that can cross the barrier during the EAE inflammation and target particles to the inflammatory foci. This hypothesis is supported by the immunohistochemical study of Merodio et al. (20). These authors investigated the distribution of albumin NP after intraperitoneal administration in EAE rats. The results of this study revealed that circulating macrophages (ED1) that migrate to damaged sites and resident activated microglial cells (OX42) were involved in the distribution of albumin NP.

PHARMACOLOGICAL ACTIVITY

Neuroactive Agents

The transport of neuroactive agents across the BBB by means of PBCA NP coated with Ps 80 has been extensively studied. These agents include peptides, such as the Leu-enkephalin analog dalargin (Dal) and kytorphin, the opioid loperamide, the alkaloid tubocurarine, and the N-methyl-D-aspartate (NMDA) antagonist MRZ 2/576 (8–10,42–47). All of these substances are not transported across the BBB after i.v. administration and, therefore, do not produce effects in the CNS.

The transport of the NP-bound Dal, kytorphin, and loperamide across the BBB was evidenced by the pronounced antinociceptive effects in mice demonstrated by the tail-flick test or the hot-plate test (8,42,45,47). In contrast to the drugs bound to Ps 80–coated NP, the drug solutions or uncoated nanoparticulate formulations, used as controls, did not exhibit any significant effects (Fig. 6). The antinociceptive effect of Dal bound to Ps 80–coated NP was accompanied by a pronounced Straub effect (tail erection) and was totally blocked by a prior injection of naloxone (the μ-opiate receptor antagonist),

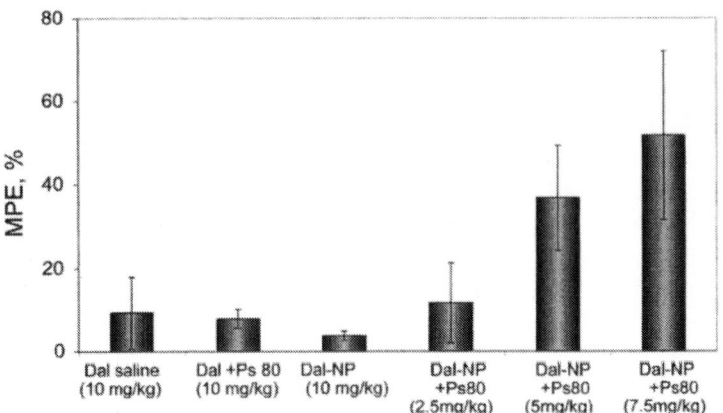

Figure 6 Analgesia in male ICR mice (% MPE) determined by the tail flick test 45 minutes after IV injection of Dal in solutions and bound to uncoated and Ps 80–coated PBCA NP ($n = 5$). Values are mean and SD. *Abbreviations*: ICR, Institute of Cancer Research; MPE, maximal possible effect; IV, intravenous; Dal, dalargin; Ps 80, polysorbate 80; PBCA, poly(butylcyanoacrylate); NP, nanoparticles. *Source*: Adapted from Ref. 45.

demonstrating the involvement of the opioid receptors. These CNS effects provided additional evidence that Dal was indeed transported across the BBB. Antinociceptive effects of Dal were obtained also if the NP were coated with polysorbates 20, 40, 60, or 85, whereas other surfactants, such as poloxamers 184, 188, 338, 407, poloxamine 908, Brij®35, Cremophor®EL, or Cremophor®RH40 were ineffective (43,46).

Transport of tubocurarine across the BBB was demonstrated using an in situ perfused rat brain technique together with a simultaneous recording of the electroencephalogram (10). Tubocurarine (a quaternary ammonium salt) does not penetrate into the brain across the normal BBB. However, direct intraventricular injection of tubocurarine provokes development of epileptiform spikes that can be registered by the encephalogram. Tubocurarine solution, tubocurarine-loaded NP without Ps 80, or a mixture of Ps 80 and tubocurarine was unable to influence the encephalogram. However, addition of tubocurarine-loaded NP coated with Ps 80 to the perfusate caused frequent severe spikes, as after intraventricular injection of the drug.

A novel noncompetitive NMDA receptor antagonist, MRZ 2/576 is a potent but rather short-acting anticonvulsant. The short effect of this drug (5–15 min) is most probably due to its rapid elimination from the CNS by efflux transporters that can be blocked by probenecid. Administration of the drug bound to PBCA NP coated with Ps 80 prolonged the duration of the anticonvulsive activity in mice up to 210 minutes and after probenecid pretreatment up to 270 min compared to 150 min with probenecid and MRZ 2/576 alone (16). The results of this study demonstrate that Ps 80–coated PBCA NP not only enhance brain delivery of drugs that are not able to freely penetrate the BBB but also can prolong the CNS availability of drugs that have a short duration of action.

Doxorubicin

The therapeutic potential of brain targeting using PBCA NP was most clearly demonstrated in experiments for the chemotherapy of intracranial glioblastoma (48). As mentioned above, malignant brain tumors are characterized by vascular hyperpermeability. However, if disruption of the BBB is evident in the tumor core, the barrier is still retained in peritumoral regions. DOX is a widely used antitumor antibiotic that has been shown to poorly cross the BBB because of the efflux transporters. Indeed, the clinical trials demonstrated that after i.v. administration DOX did not reach cytotoxic levels in glioma tissue due to delivery problems (49). At the same time, a significant increase in survival rate was achieved in patients with malignant gliomas treated with intratumoral injections of DOX (50). The enhanced brain delivery of DOX with Ps 80–coated PBCA NP suggested that this delivery system had a potential for chemotherapy of brain tumors (15).

Indeed, a high efficacy of DOX bound to Ps 80–coated PBCA was demonstrated in rats bearing 101/8 glioblastoma (48). Groups of five to eight glioblastoma-bearing rats (total $n = 151$) were subjected to 3×1.5 mg/kg or 3×2.5 mg/kg of DOX in different formulations injected i.v. on days two, five,

and eight after tumor implantation. The most prominent
result was achieved in the group treated with 3×1.5 mg/kg
of DOX bound to Ps 80–coated NP: a significant increase in
survival time (IST) was obtained (IST 84%, as compared to
the untreated control) and more than 20% animals showed
a long-term remission (Fig. 7).

These animals were sacrificed after six months and no
histological evidence of tumor was observed. Preliminary his-
tology confirmed lower tumor sizes and lower values for proliferation and apoptosis in this group. The mean survival
time was even more prolonged in the group treated with
3×2.5 mg/kg (IST 169%), indicating a dose dependence of
the antitumor effect. However, long-surviving animals in this
group died before day 180, most probably due to the higher

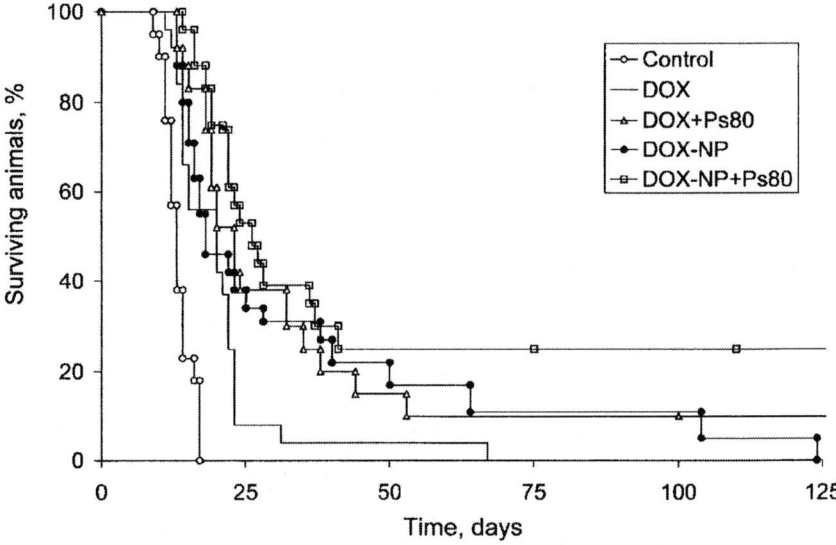

Figure 7 Percentage of surviving rats (Kaplan–Meier plot) with
intracranially transplanted 101/8 glioblastoma after IV injections
of DOX in solutions or bound to uncoated or Ps 80–coated PBCA
NP (3×1.5 mg/kg) (summarized data of three experiments,
$n = 6 \div 12$ in each run). *Abbreviations*: IV, intravenous; DOX, doxorubicin; Ps 80, polysorbate 80; PBCA, poly(butylcyanoacrylate); NP,
nanoparticles. *Source*: Adapted from Ref. 48.

toxicity of this regimen. Interestingly, the survival time was also increased in the groups treated with DOX solution in 1% Ps 80 and DOX bound to uncoated PBCA NP. This is in contrast to the previous studies conducted in healthy animals, where none of the similar controls (drug solutions or drug bound to noncoated NP) was able to produce the CNS effect. The phenomenon could most probably be explained by the EPR effect associated with a higher permeability of the BBB at the tumor site that allowed entry of other formulations into the brain (38).

It is noteworthy that clinical signs of neurotoxicity were absent throughout the study. Moreover, the histological study of the animals treated with a dose of 3×2.5 mg/kg and sacrificed on day 12 did not reveal any signs of neurotoxicity.

Brigger et al. (18) evaluated the efficacy of DOX bound to PEG–PHDCA NP against intracranial 9L gliosarcoma. As mentioned above, unloaded PEG–PHDCA NP displayed a significant accumulation in this tumor, as well as an affinity for the healthy brain tissue (17,40). Accordingly, it was assumed that the increase of DOX distribution in the bulk tumor and the adjacent tissue due to its association with the carrier would also enhance drug efficacy. However, DOX loaded in PEG–PHDCA NP failed to produce an antitumor effect against 9L gliosarcoma. Nevertheless, this study, together with the biodistribution studies of PEG–PHDCA NP discussed earlier, represents an excellent example of a thoroughly planned development of the drug delivery system designed for the treatment of brain tumors.

Two nanoparticulate formulations were prepared. DOX was incorporated in the PEG–PHDCA NP by nanoprecipitation, either by dissolution in the aqueous phase (DOX aqueous formulation), or in the organic phase (DOX organic formulation) before precipitation. These formulations displayed different patterns of the in vitro drug release: whereas the aqueous formulation released DOX with a burst effect (90% of the drug after two hours), the DOX organic formulation was characterized by better drug retention. The latter formulation displayed a two-step drug leakage, the slow release phase following zero-order kinetics that depended on the copolymer biodegradation rate [biodegradation products are polycyanoacrylic acid, MePEG, and hexadecanol (17)]. The simultaneity of these processes (polymer degradation and drug release) enables the formation of an ion-pair between DOX and polycyanoacrylic acid, necessary for reversing P-gp–dependent efflux of DOX, which could enhance drug penetration through the BBB (discussed later). Hence, the chemotherapy was conducted using DOX organic formulation.

The treatment started three days after intracerebral tumor transplantation. At that time, the permeability of the BBB at the tumor site was already increased, as shown by extravasation of a hydrophilic tracer (Evans Blue), which is a premise for the EPR-mediated drug delivery.

Then, the drug administration schedule was planned with a consideration of the individual features of the tumor model, i.e., the cell kinetics. The schedule consisted of three or five i.v. injections daily, since the 9L cells doubling time was reported to be ~20 hours (51).

The result of chemotherapy was disappointing: incorporation of DOX in PEG–PHDCA NP was unable to improve its antitumor effect against 9L gliosarcoma. The treatment effect of the DOX organic formulation (5×1.8 mg/kg) was not above 30% IST, as compared to the control group that received unloaded NP. This result did not differ significantly from that of free DOX or the aqueous formulation: 50% and 49% IST, respectively.

Additional experiments were carried out to gain insight into the question why the DOX-loaded PEG–PHDCA NP failed in the 9L gliosarcoma model. The most important findings are as follows:

1. First, it was shown that loading of the particles with DOX had a considerable impact on their biodistribution profile. The most intriguing finding was, perhaps, the 2.5-fold lower concentration of the DOX-loaded NP in the tumor and the adjacent tissue, as compared to that of unloaded NP, even though the blood concentration of the loaded particles was increased. Moreover, the loaded particles

showed a different distribution within MPS organs, with considerable accumulation in the lungs and the spleen after a single injection.

2. Serious interaction of the DOX-loaded particles with plasma proteins was observed after in vitro incubation with pure serum. Such an interaction was not observed for the unloaded PEG–PHDCA NP. This phenomenon was explained by a reversion of the surface charge of the particles due to adsorption of positively charged DOX molecules (+15.5 mV for loaded NP vs. −24.4 mV for unloaded). Furthermore, after incubation in rat plasma, the surface charge of the DOX-loaded NP was reversed again and became negative, whereas unloaded NP displayed only a slightly more negative zeta potential.

The interaction with plasma proteins was associated with an instantaneous increase of the effective size of DOX-loaded NP, which could be responsible for the increased accumulation of the particles in the lungs, since lung capillaries, as a capillary bed of the first passage, retain larger particles. Finally, it was assumed that increased accumulation of DOX-loaded PEG–PHDCA NP in lungs and spleen could divert them from non-MPS organs, thus interfering with brain delivery.

On the other hand, it may be relevant to mention here the study of Sharma et al. (52). These authors treated 9L-bearing rats with high doses of DOX encapsulated in long-circulating liposomes (cumulative dose 17 mg/kg, three weekly injections). The liposomal formulation was more effective than free DOX; however, this effect (median IST 29%) was not greater than that of DOX-loaded PEG–PHDCA NP. At the same time, the physicochemical parameters of DOX-loaded liposomes (Caelyx®) were beneficial: the surface charge was negative (−25.5 mV) and the size was smaller (~80 nm), as compared to the NP (18). Therefore, it may be speculated that 9L gliosarcoma is relatively refractory to DOX, which, among other factors, may be responsible for the limited efficacy of DOX formulations in this tumor model.

MECHANISMS OF DRUG DELIVERY TO THE
BRAIN BY MEANS OF POLYMERIC NP

A number of possibilities exist that could explain the mechanism of drug delivery across the BBB by means of NP:

1. A mechanism of general toxicity involving an opening of the BBB due to surfactant and/or NP effects characterized by an increased permeability of the endothelial cell membranes and/or an opening of the tight junctions between the endothelial cells. The NP and/or the drug could then permeate through the BBB.

2. Increased retention of NP in the brain blood capillaries due to their adhesion to the capillary wall. This could create a higher concentration gradient that would enhance the transport to the brain.

3. Endocytosis of the NP by the endothelial cells followed by the release of the drugs within these cells and delivery to the brain.

4. Transcytosis of the NP with bound drugs through the endothelial cell layer.

5. Interaction of the NP with the membrane of the endothelial cell in the brain vessels. This could induce changes in the cell membrane viscosity/fluidity thus inhibiting the efflux system, such as P-gp, and facilitating the brain uptake of P-gp-dependent drugs.

These mechanisms could be also cooperative. Among these mechanisms, mechanism 1 (opening of the BBB) is unlikely to contribute to the NP-mediated drug delivery to the brain. A number of facts provide evidence that the brain uptake of Ps 80–coated PBCA NP is not associated with an opening of the BBB due to toxic effects.

First, a number of independent studies in healthy animals employing Ps 80–coated PBCA NP involved administration of a drug in a surfactant solution as a control (8,15,45,48). These preparations were ineffective in terms of either enhancement of brain concentration or pharmacological effects. These results suggest that the doses/concentrations of Ps 80 used did not induce the BBB opening.

Second, it is unlikely that the BBB opening is induced by a nonspecific permeabilization related to the toxicity of PBCA NP coated with Ps 80, as suggested by Olivier et al. (53). If this would be the case, prior binding of drug to these NP would not be necessary, as the drug would have free diffusional access to the brain through the opened tight junctions. In order to test this hypothesis in vivo, free Dal was injected into mice 5 or 30 minutes after the injection of unloaded Ps 80–coated PBCA NP (54). The antinociceptive effect of this treatment was negligible and identical to those of a Dal solution or empty uncoated particles. In contrast, Dal bound to NP prior to their coating with Ps 80 exhibited a pronounced and statistically significant effect, indicating that the binding of Dal to NP was a prerequisite for brain delivery. These observations correlated with the results of earlier studies of Alyautdin et al. (8) and Olivier et al. (53).

Moreover, the integrity of the BBB in rats treated with Ps 80–coated PBCA NP was evaluated by the measurement of the inulin spaces (55). The increase of the spaces by 10% after 10 minutes and 99% after 45 minutes was found. This increase would suggest that the coated NP were increasing the volume available to the intravascular inulin slightly but were not significantly disrupting the BBB, as this would have required an increase by a factor of 10–20.

Inconsistent results were obtained in the in vitro studies of the modification of the BBB permeability due to Ps 80–coated PBCA NP. The permeability was evaluated by measuring the flux of the exrtracellular markers [^{14}C]-sucrose and [^{3}H]-inulin across a cell monolayer.

In the experiments of Kreuter et al. (54) no significant changes of [^{14}C]-sucrose and [^{3}H]-inulin permeability were observed in the in-vitro BBB model after coincubation with Ps 80–coated or uncoated PBCA NP. This model consisted of a coculture of bovine brain capillary endothelial cells and rat astrocytes and was shown to establish a barrier.

Olivier et al. (53) used the same BBB model and observed an over 10-fold increase in the sucrose and inulin fluxes after incubation with PBCA NP coated with Ps 80 (53). However, in this case, serum was not added to the cell medium, which could impair the integrity of the cell layer.

Steiniger et al. (56) cultivated bovine brain capillary endothelial cells (no coculture with astrocytes) originating from the gray matter onto precoated Transwell® inserts. After incubation with 10 µg/mL of the NP preparation, the [^{14}C]-sucrose flux increased twofold with uncoated and 6.5-fold with Ps 80–coated PBCA NP.

These results demonstrate that slight changes in the in-vitro models of the BBB can lead to considerable discrepancies in the experimental results.

The transport of other types of NP also was not associated with disruption of the BBB. The evaluation of the modification of the BBB permeability due to PHDCA NP or surfactants using [^{14}C]-sucrose was performed in the study by Calvo et al. (32). None of the nanoparticulate preparations modified the low passage of sucrose, which indicates that penetration of the NP was not associated with the increase of the BBB permeability. However, 1% solution of Ps 80 increased noticeably the concentration of sucrose in all brain structures.

The influence of two novel types of SLN on the permeability of the BBB was investigated by Koziara et al. (21) and Lockman et al. (57). The SLN were composed either of emulsifying wax (E. Wax/Brij 78) or Brij 72/Ps 80 and labeled with entrapped [^{3}H]-cetyl alcohol. For both SLN types, significant brain uptake was measured by an in situ brain perfusion. At the same time, these NP did not induce statistically significant changes in BBB integrity, permeability, or choline transport. The presence of the particles did not significantly influence cerebral perfusion flow, and the [^{14}C]-sucrose brain distribution space was not increased, indicating that these SLN had minimal effect on the BBB integrity. Additionally, Western blot analysis confirmed that the incubation of these NP with bovine brain microvessel endothelial cells did not alter expression of the BBB junctional proteins, such as occludin and claudin-1. The above data suggest that the brain uptake of these NP also was not associated with paracellular movement due to the opening of tight junctions.

Alternatively, the enhanced brain delivery could be explained by increased retention of NP in the brain blood capillaries due to their adsorption to the capillary wall. This could create a higher concentration gradient, thus enhancing transport across the endothelial wall to the brain (mechanism 2). Indeed, the interaction of various NP with brain microvessels was observed in several studies (see later). Tröster et al. (30) believed that PMMA particles were not engulfed by the endothelial cells lining the vasculature but rather adhered to these cells. This mechanism was also suggested, among other possibilities, for DOX-loaded stealth SLN (12). In this study, the enhanced uptake of DOX in the brain was associated with the increase of the concentration of the stealth agent in SLN. At the same time, the increasing concentration of the stealth agent (PEG derivative) also suggested a parallel increase of the surface hydrophilicity of the carriers. This increasing hydrophilicity could be expected to hinder NP interaction with the cell membrane and passage through the BBB; however, this was not the case. This observation implies involvement of an additional mechanism in DOX transport to the brain with the stealth SLN.

Another mechanism is endocytosis of the NP by brain microvessel endothelial cells (mechanism 3). Indeed, there are facts indicating that the enhanced brain delivery of the drugs loaded in PBCA NP is the result of their internalization by the endothelial cells forming the BBB.

This hypothesis is supported by the study of the interaction of PBCA NP with the BBB in vitro and in vivo (55). As mentioned earlier, an increase in inulin spaces by 10–99% was found in rats treated with Ps 80–coated PBCA NP. This slight increase could be interpreted as a result of an upfolding of the cell membrane due to endocytic events, or an increase in fluid phase endocytosis of inulin associated with the internalization of the NP. In addition, the uptake of fluorescent PBCA NP labeled with Rhodamine 6G was observed in cultured human, bovine, and murine brain microvessel endothelial cells (58). The uptake was followed by fluorescence, as well as by laser confocal microscopy. Uptake of the surfactant-coated NP was far more pronounced compared to the uncoated particles, even though in the bovine cells a slight increase in uptake of the uncoated particles was observed with increasing time of incubation. Using image analysis software, a 20-fold increase in uptake of coated with respect to uncoated NP was observed in two hours. Human cells also exhibited the enhanced uptake of the coated NP.

The uptake of fluorescent PBCA NP by rat brain endothelial cells of the RBE4 cell line also was demonstrated by Alyautdin et al. (55). The PBCA NP were labeled with fluorescein isothiocyanate (FITC) dextran 70,000. After the addition of Ps 80–coated NP, the cells showed a punctate appearance of fluorescence concentrated within the cells. In contrast, after treatment with the uncoated NP no fluorescence was observable within the cells, even after the addition of a 10-fold higher concentration of NP, while a strong fluorescence was apparent in the surrounding medium. In none of the above experiments did the addition of Ps 80–coated or uncoated NP appear to damage the RBE4 cells.

Taken together, the results of these extensive studies suggest that Ps 80–coated PBCA NP are endocytosed by the brain capillary endothelial cells. Furthermore, Kreuter et al. (59,60) suggested that endocytosis of these NP is mediated by plasma apolipoprotein B and apolipoprotein E (ApoE) adsorbed on the surface due to coating with Ps 80. These apolipoproteins interact with the low-density lipoprotein (LDL) receptors expressed in the BBB and promote uptake of the NP by the brain capillary endothelial cells via receptor-mediated endocytosis. This hypothesis was based on the following findings: Kreuter et al. (46) observed that besides Ps 80, coating of the PBCA NP with polysorbate 20, 40, or 60 also enabled an antinociceptive effect after injection of Dal-loaded PBCA NP, whereas other surfactants, such as poloxamers and poloxamines, were unable to achieve this effect. At the same time, Lück found that coating of PBCA NP with these surfactants significantly increased the amount of ApoE adsorbed on the surface of these particles after their incubation in human plasma (61). Moreover, human and bovine brain capillary endothelial cells expressed high levels of LDL receptor, as demonstrated by immunohistochemical staining (58).

These findings were corroborated by another experiment, in which the possible involvement of a number of other apolipoproteins in the transport of drugs bound to PBCA NP into the brain was investigated (60). PBCA NP loaded with Dal or loperamide were coated with the apolipoproteins AII, B, CII, E, or J without coating or after precoating with Ps 80. After i.v. injection to mice the antinociceptive threshold was measured by the tail-flick test. An antinociceptive effect was achieved only after treatment with Dal or loperamide-loaded PBCA NP coated with Ps 80 and/or with apolipoprotein B or ApoE. The effect was higher when NP were first coated with Ps 80 and then overcoated with ApoE (Fig. 8). Furthermore, the antinociceptive threshold of Ps 80–coated Dal-loaded

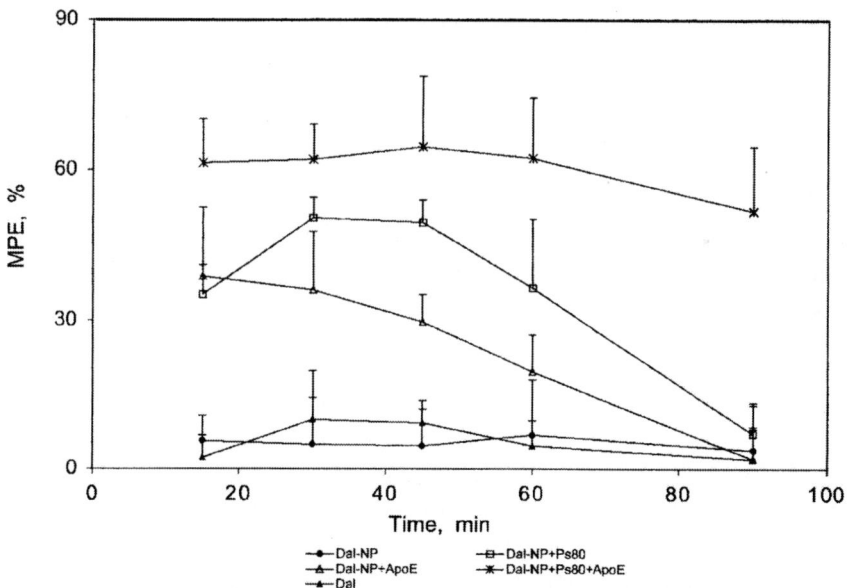

Figure 8 Analgesia in male ICR mice (% MPE) determined by tail flick test after i.v. injection of Dal bound to PBCA NP coated with Ps 80 and/or ApoE ($n = 5$) in a dose of 7.5 mg/kg. *Abbreviations*: ICR, Institute of Cancer Research; MPE, maximal possible effect; i.v., intravenous; Dal, dalargin; PBCA, poly(butylcyanoacrylate); NP, nanoparticles; Ps 80, polysorbate 80; ApoE, apolipoprotein E. Values are mean and SD. *Source*: Adapted from Ref. 60.

PBCA NP was determined in ApoE-deficient and normal mice. In the ApoE-deficient mice, the antinociceptive effect was considerably reduced in comparison to normal mice. A similar antinociceptive effect was also achieved after coating of Dal-loaded particles with apolipoprotein B.

Therefore, it was concluded that apolipoprotein B and ApoE are involved in the PBCA NP–mediated transport of drugs across the BBB. Polysorbate coating promotes adsorption of the circulating apolipoproteins, so that NP are assumed as lipoprotein particles that could be taken up by the brain capillary endothelial cells via receptor-mediated endocytosis. Bound drugs then may be transported into the brain by diffusion following release from the particles within the endothelial cells, or together with the carrier by transcytosis.

The role of Ps 80 coating of NP that facilitated their interaction with brain microvessel endothelial cells has been also demonstrated for other types of NP. Lipid drug conjugate (LDC) NP were composed of stearic acid and diminazene; Ps 80 was used as an emulsifier (62). Confocal laser scanning microscopy of the murine brain tissue showed Nile Red–labeled LDC particles adhering to the endothelium of the brain vessels and the dye diffusion into the brain tissue. The plasma protein adsorption pattern investigated by two-dimensional electrophoresis revealed strong adsorption of apolipoproteins A-I and A-IV onto LDC NP after their incubation in murine plasma; ApoE could not be identified. The authors hypothesized that the ability of Ps 80–coated NP to deliver drugs to the brain is not only mediated by adsorption of apolipoprotein B and ApoE but probably involves "team work" of other apolipoproteins that prevent the hepatic uptake of the NP, thus facilitating brain delivery.

The recent study of Sun et al. (22) employed poly(lactic) NP loaded with FITC dextran and coated with Ps 80 by 24-hours incubation. The direct observation of NP delivery to the brain was carried out using fluorescent microscopy of murine brain sections obtained after vascular perfusion fixation. The fluorescence was observed only in the animals treated with FITC dextran bound to Ps 80–coated NP, whereas treatment with other preparations, such as solution of FITC dextran in Ps 80, FITC dextran bound to uncoated NP, or in situ mixture of all components, did not produce any fluorescence of the brain sections. The fluorescence was mainly located at the wall of brain microvessels, which is indicative of the interaction between Ps 80–coated NP and brain microvessel endothelial cells.

In the study by Koziara et al. (21) the transfer rate of the SLN stabilized by Ps 80 (Brij® 72/Ps 80) NP from perfusion fluid into the brain was significantly higher than that of the NP stabilized by Brij® 78 (E. wax/Brij 78).

The hypothesis that Ps 80 mediates an interaction of the NP with the brain endothelial cells is, in a way, corroborated by the pharmacokinetic data discussed earlier. Indeed, although the coating of various NP with Ps 80 considerably enhanced drug delivery to the brain, the effect of this surfactant on the circulating characteristics of the particles was not very pronounced (13,15,30,32). In contrast, poloxamine 908 most efficiently extended the circulation time of various NP; nevertheless, its effect on brain uptake was often moderate or minimal. It could be hypothesized that poloxamine 908 created the steric barrier that not only protected the NP from opsonization, thus increasing circulation time, but also interfered with the cell membrane recognition step, which prevented transport of the NP across the BBB (32). On the other hand, and in accordance with the pharmacokinetic rule cited earlier, Ps 80 is likely to enable a specific interaction of the particles with the BBB endothelial cell.

This assumption is supported by the in vitro results of Borchardt et al. (63). These authors investigated the influence of surfactants on the uptake of [^{14}C]-PMMA NP by bovine brain microvessel endothelial cells isolated from the gray matter of the cerebral cortex. The highest and fastest uptake (>300% compared to uncoated controls after two hours) was observed after coating with Ps 80, whereas coating of the NP with poloxamine 908 yielded an insignificant uptake enhancement.

Interestingly, long-circulating PEGylated PHDCA NP not only provided higher accumulation in the brain tumor tissue but also displayed an affinity for brain regions protected by the normal BBB (17). Obviously, and in concert with other studies, the concentration of the particles in the brain was improved due to reduced clearance of the long-circulating carrier by MPS. Moreover, the analysis of the pharmacokinetic data allowed authors to conclude that if the mechanism of intratumoral accumulation was similar for PEGylated and non-PEGylated carriers, an affinity of PEGylated particles for the normal brain could not be considered as a simple diffusion/convection process. It can be speculated that, as with Ps 80, PEG coating enables a specific interaction of the particles with the BBB endothelial cells.

Above all, these results suggest that prediction of the brain uptake of the NP-bound drug on the basis of their circulation behavior is not always unequivocal and cell/particle interactions must be considered to achieve efficient brain delivery.

Adsorptive endocytosis is a likely mechanism of the enhanced transport of ODN across the BBB with the positively charged Nanogel particles (14,35). The main obstacle to effective therapy with ODN compounds is their anionic character and relatively large molecular structure, which hampers their access to the target sites localized in the cell cytoplasm and/or nucleus. On the other hand, the positively charged NP are believed to interact electrostatically with the negatively charged cell membranes, which is followed by the internalization of these particles within these cells via adsorptive endocytosis. Indeed, the positively charged Nanogel formulation allowed more effective ODN transfer across the monolayers of brain microvessel endothelial cells, as compared to the electroneutral formulation. This result is in concert with the in vivo data demonstrating the substantial brain/plasma ratio of ODN achieved after injection of ODN-loaded Nanogel particles (discussed earlier). The cationic nature of this carrier system may also influence intracellular trafficking of ODN. Thus, delivery with Nanogel particles afforded effective release of ODN and its accumulation within the nucleus, whereas free ODN molecules are mainly localized within endosomal and lysosomal compartments and their access to the nucleus is usually achieved only after addition of chloroquine. It is possible that following internalization, the cationic NP interact with the negatively charged endosomal membrane, which may cause destabilization of the membrane and facilitate the release of ODN and its access to the nucleus (Fig. 9).

The influences of the surface charge and the size of the carriers on brain delivery are not yet clear; however, certain tendencies can be noted. Thus, in contrast to positively charged DOX-loaded PEG–PHDCA NP that failed to produce a high brain/blood ratio, ODN-loaded Nanogel particles had a lower zeta potential (+2.3 vs. +15.5 mV for PEG–PHDCA NP) and were considerably smaller (90 vs. 180 nm). In the study

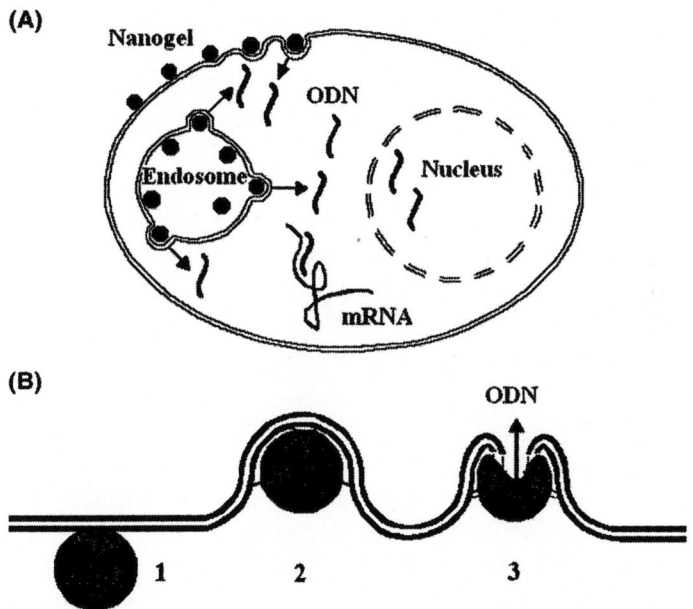

Figure 9 (**A**) Internalization of oligonucleotide (ODN)-loaded Nanogel particles by adsorptive endocytosis and putative mechanism of ODN release into the cytosol. (**B**) Destabilization of the endosomal membrane: (1) interaction of the positively charged Nanogel particles with negatively charged phospholipids; (2) Nanogel wrapping with interacting phospholipids, and (3) collapse of Nanogel–phospholipid complex and release of ODN into the cytosol. *Source*: Adapted from Ref. 14.

by Zara et al., (12) DOX accumulation in the brain increased parallel to the decrease of a negative zeta potential of the stealth SLN, which, among other factors, could contribute to successful brain targeting of these carriers (Table 1). These observations evidently imply the importance of both parameters for brain targeting.

Transcytosis (mechanism 4) of cationic polysaccharide NP coated with a lipid bilayer across the in vitro BBB model was observed by Fenart et al. (23). The BBB model consisted of a coculture of bovine brain capillary endothelial cells and rat astrocytes. Neutral, anionic, and cationic 60-nm NP were prepared from cross-linked maltodextrin derivatized or not (neutral) with anionic (phosphates) or cationic (quaternary ammonium) ligands. The particles were labeled with fluorescein and coated (or not) with a lipid bilayer. Cationic lipid-coated NP were found to be the best for permeating across the BBB, whereas coating of the neutral particles did not significantly alter their permeation characteristics. No modification of the paracellular permeability was observed during the incubation of cells with the NP, so this increase was not due to a breakdown of the barrier. The distribution of these particles throughout the cytoplasm was characteristic of transcytosis. In contrast, the perinuclear localization of uncoated polysaccharide NP showed an intracellular accumulation of these NP in a degradation compartment.

Finally, the enhanced drug delivery to the brain with Ps 80–coated PBCA NP may be associated with the inhibition of the transmembrane efflux pumps, such as MRP and P-gp (mechanism 5). As could be seen from the data above, coating with Ps 80 enables interaction of the NP with the membranes of the brain microvessel endothelial cells. Apart from other effects, this interaction could influence the membrane fluidity/viscosity, which can cause the conformational change and inhibition of the transmembrane efflux pumps. The efflux pumps are important constituents of the BBB and most of the drugs delivered to the brain by means of Ps 80–coated PBCA NP (such as loperamide, Dal, DOX, and MRZ 2/576, described earlier) are P-gp and/or MRP substrates. Although the possibility of involvement of P-gp has been mentioned in earlier publications, the role of the efflux mechanisms in the NP-mediated drug delivery to the brain has not been investigated (55,59). At the same time, this hypothesis is supported by the following facts.

Polyalkylcyanoacrylate NP display a unique ability to overcome multidrug resistance (MDR) mediated by P-gp (64–66). Thus, poly(isobutylcyanoacrylate) NP could reverse P-gp–dependent MDR to DOX and produce considerable cytotoxic effects in P388/ADR cells resistant to DOX (65–67). Intracellular accumulation of DOX and cytotoxicity clearly depended on the release of the drug from the particles. However, in contrast to what was believed, internalization of the particles in the P388/ADR cells was not required for overcoming MDR. The suggested mechanism of action was that the NP adsorb to the surface of the tumor cells and simultaneously release the encapsulated drug and NP degradation products (polycyanoacrylic acid) that form an ion pair, which could cross the membrane without being recognized by P-gp (Fig. 10). It was demonstrated that the contact of the particles with the cell membrane was essential for MDR reversion. The authors assumed that the MDR reversion was more related to the changes in the membrane permeability or fluidity than to the direct interaction with P-gp.

Furthermore, the effect of poly(alkylcyanoacrylate) NP was enhanced in the presence of inhibitors of P-gp or MRP. Thus, it was shown that the Ps 80–coated NP significantly prolonged the anticonvulsive effect of MRZ 2/576; being a substrate of MRP efflux pump, this drug is actively pumped out of the brain (66). A similar effect was achieved when MRZ 2/576 bound to uncoated NP was injected after pretreatment with probenecid, which is a known MRP inhibitor. It can be speculated that inhibition of the drug efflux by Ps 80 coating of the NP is similar to the effect of probenecid. Moreover, the inhibiting effect of probenecid was enhanced if this agent was bound to the NP. These data are in concert with the in vitro results of Soma et al. (19) who demonstrated that the cytotoxicity of DOX bound to poly(isobutylcyanoacrylate) NP against P388/ADR cell line resistant to DOX, could be enhanced by cyclosporine® A, a potent P-gp inhibitor. The

(A)   (B)

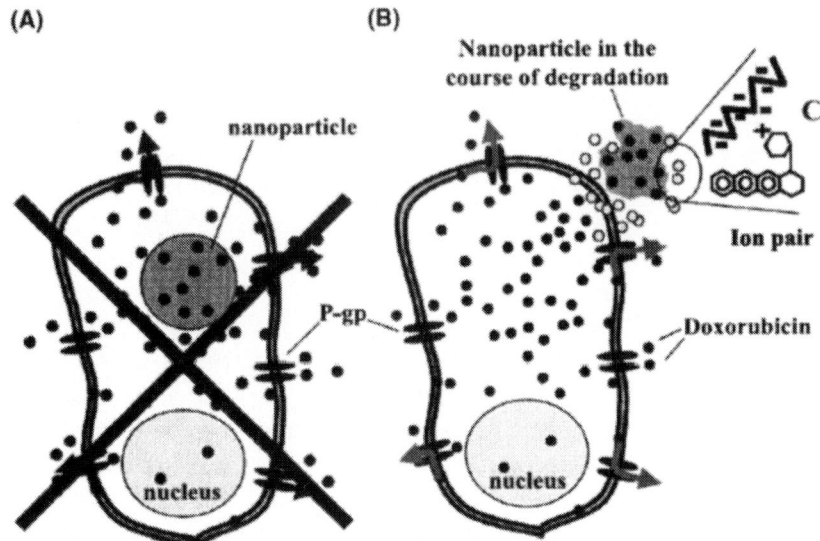

Figure 10 Hypothetic mechanism of DOX-loaded poly(alkylcya-noacrylate) NP bypassing MDR at the cellular level. DOX-loaded NP are not endocytosed by the resistant cells (**A**) but adhere to the cell surface where they simultaneously release degradation products and the drug (**B**). The degradation products and the drug form ion-pairs (C) that can penetrate the cells avoiding recognition by the P-gp. *Abbreviations*: DOX, doxorubicin; NP, nanoparticles; MDR, mul-tidrug resistance; P-gp, P-glycoprotein. *Source*: Adapted from Ref. 64.

effect of cyclosporine, in turn, could be enhanced if the drug was bound to the NP.

Although DOX loaded in uncoated PBCA NP could reverse P-gp–associated MDR in the cell culture, in vivo only DOX bound to Ps 80–coated NP could be delivered to the brain (15). It is possible that the ability of DOX-loaded PBCA NP to circumvent P-gp is assisted by Ps 80. Indeed, this sur-factant was found to reverse P-gp–associated drug resistance by increasing drug influx into the cells (68–70). In vivo, Ps 80 enhanced the adsorption of methotrexate from the mouse gas-trointestinal tract (GIT) and drug uptake into the brain in a dose-dependent manner (71). Moreover, Ps 80 could facilitate the particle interaction with the endothelial cell membrane.

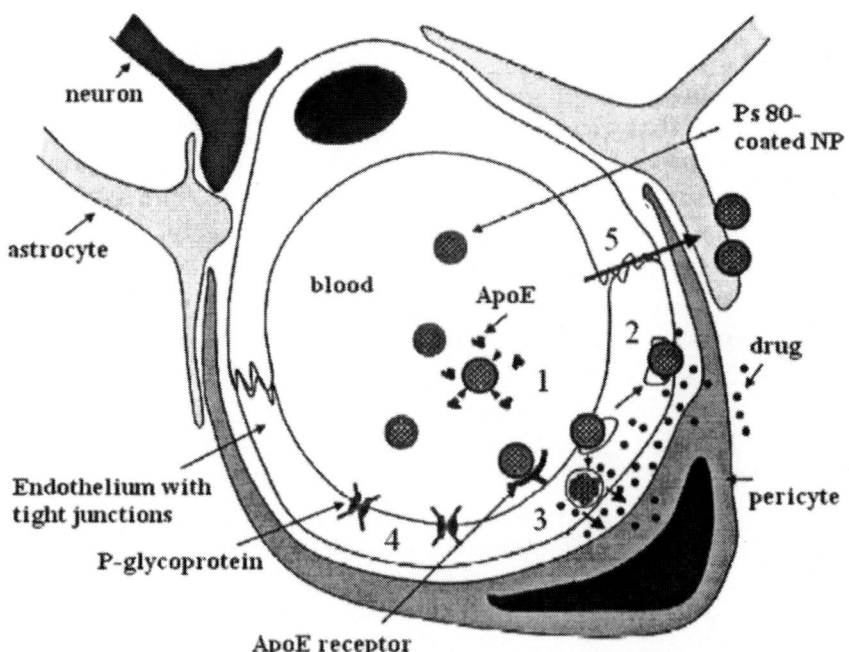

Figure 11 Hypothetic mechanism of drug delivery to the brain by means of Ps 80–coated poly(alkylcyanoacrylate) NP: (1) receptor-mediated endocytosis enabled by adsorption of ApoE onto the NP; (2) transcytosis; (3) endocytosis followed by intracellular degradation of NP, resulting in release of the drug and diffusion across the BBB; (4) inhibition of P-gp; and (5) modulation of permeability of the BBB by opening of tight junctions. *Abbreviations*: Ps 80, polysorbate 80; NP, nanoparticles; ApoE, apolipoprotein E; BBB, blood-brain barrier. *Source*: Adapted from Ref. 67.

Hence, it can be hypothesized that PBCA NP coated with Ps 80 cause indirect inhibition of P-gp through changes in the cell membrane. Possible mechanisms of DOX transport across the BBB with polyalkylcyanoacrylate NP are summarized in Figure 11.

CONCLUSIONS

NP-based drug delivery systems represent a new and interesting perspective among other strategies for drug targeting to the CNS, offering opportunities for noninvasive chemotherapy of CNS disorders.

The mechanism(s) of drug transport to the brain by means of NP is presently not fully elucidated. There are a number of processes that can be involved in this phenomenon; they may run in parallel or may be cooperative. The available data suggest that the mechanisms are likely to depend on the physicochemical parameters of the delivery systems (such as size, charge, and hydrophilicity/hydrophobicity) and the chemistry of surface-modifying agents. Hence, it is probable that different particles enable drug delivery by different pathways. The endocytic uptake of the particles by brain microvessel endothelial cells obviously plays an important role for most types of the carriers reviewed in this chapter. On the other hand, the correlation of plasma and brain concentrations observed by a number of authors suggests that the enhanced drug transport into the brain with the NP can be dependent on their pharmacokinetic parameters governing the blood–brain gradient of the drug concentration.

Obviously, an ideal drug delivery system capable of crossing the BBB must combine adequate circulation and drug release characteristics with the feasibility of specific cell/particle interactions. The diversity of the targets in the CNS will probably call to life a wide variety of colloidal carriers. It can be expected that future research will concentrate on the development of the vectorized delivery systems combining the advantages of the colloidal carriers, such as large payloads of a drug, with active targeting. In this respect, the results obtained with the NP conjugated with the surface ligands, such as thiamine, insulin, transferrin, or an anti-transferrin receptor monoclonal antibody, are encouraging (14,35,72–74).

Our understanding of the BBB has advanced considerably and it is now recognized as a highly complex and reactive interface interacting with numerous blood-borne factors, which modulate its functions. This improved knowledge is expected to move drug development into a more rational phase. Overall, the design of formulations for CNS delivery of low molecular drugs and biomacromolecules will demand an in-depth understanding and careful consideration of drug receptors and sites of action, as well as the processes trafficking the drugs to their cellular and intracellular targets, such as endocytic pathways, and other transport mechanisms governing the accumulation and elimination of the drugs in the brain. Moreover, a careful choice of adequate in vivo and in vitro models for evaluation and prediction of pharmacological activity cannot be disregarded. Finally, the success of this technology will depend on toxicological issues that have been only marginally addressed so far.

REFERENCES

1. Misra A, Ganesh A, Shahiwala A, Shah SP. Drug delivery to the central nervous system: a review. J Pharm Pharm Sci 2003; 6(2):252–273 (http://www.ualberta.ca/~csps).

2. Begley DJ, Brightman MW. Structural and functional aspects of the blood–brain barrier. In: Prokai L, Prokai-Tatrai K, eds. Progress in Drug Research. ; 61. Basel, Switzerland: Birkhauser Verlag , 200339–78.

3. Pardridge WM. CNS drug design based on principles of blood-barrier transport. J Neurochem 1998; 70:1781–1792.

4. Tsuji A, Tamai I. Carrier-mediated or specialized transport of drugs across the blood-brain barrier. Adv Drug Deliv Rev 1999; 36:277–290.

5. Pardridge WM. Non-invasive drug delivery to the human brain using endogeneous blood-brain barrier systems. PSTT 1999; 2:49–59.

6. Fabel K, Dietrich J, Hau P, et al. Long-term stabilization in patients with malignant glioma after treatment with liposomal doxorubicin. Cancer 2001; 92:1936–1942.

7. Huwyler J, Wu D, Pardridge WP. Brain drug delivery of small molecules using immunoliposomes. Proc Natl Acad Sci USA 1996; 93:14,164–14,169.

8. Alyautdin R, Gothier D, Petrov V, Kharkevich D, Kreuter J. Analgesic activity of the hexapeptide dalargin adsorbed on

the surface of Ps 80-coated poly(butyl cyanoacrylate) nanoparticles. Eur J Pharm Biopharm 1995; 41:44–48.

9. Alyautdin RN, Petrov VE, Langer K, Berthold A, Kharkevich DA,
    Kreuter J. Delivery of loperamide across the blood-brain barrier
    with Ps 80-coated polybutylcyanoacrylate nanoparticles. Pharm
    Res 1997; 14:325–328.

10. Alyautdin RN, Tezikov EB, Ramge P, Kharkevich DA, Begley
    DJ, Kreuter J. Significant entry of tubocurarine into the brain
    of rats by adsorption to Ps 80-coated polybutylcyanoacrylate
    nanoparticles: an in situ brain perfusion study. J Microencap-
    sul 1998; 15(1):67–74.

11. Yang SC, Lu LF, Cai Y, Zhu JB, Liang BW, Yang CZ. Body dis-
    tribution in mice of intravenously injected camptothecin solid
    lipid nanoparticles and targeted effect on brain. J Control
    Release 1999; 59:299–307.

12. Zara GP, Cavalli R, Bargoni A, Fundaro A, Vighetto D,
    Gasco MR. Intravenous administration to rabbits of non-
    stealth and stealth doxorubicin-loaded solid lipid nanoparticles
    at increasing concentrations of stealth agent: pharmacoki-
    netics and distribution of doxorubicin in brain and other
    tissues. J Drug Targeting 2002; 10:327–336 (http://www.
    tandf.co.uk/journals).

13. Araujo L, Loebenberg R, Kreuter J. Influence of the surfactant
    concentration on the body distribution of nanoparticles. J Drug
    Targeting 1999; 6:373–385.

14. Vinogradov SV, Bronich TK, Kabanov AV. Nanosized cationic
    gels for drug delivery: preparation, properties and interactions
    with cells. Adv Drug Deliv Rev 2002; 54:135–147.

15. Gulyaev AE, Gelperina SE, Skidan IN, Antropov AS, Kivman
    GYA, Kreuter J. Significant transport of doxorubicin into the
    brain with Ps 80-coated nanoparticles. Pharm Res 1999;
    16:1564–1569.

16. Friese A, Seiller E, Quack G, Lorenz B, Kreuter J. Increase of
    the duration of the anticonvulsive activity of a novel NMDA
    receptor antagonist using poly(butylcyanoacrylate) nanoparti-
    cles as a parenteral controlled release system. Eur J Pharm
    Biopharm 2000; 49:103–109.

17. Brigger I, Morizet J, Aubert G, et al. Poly(ethylene glycol)-coated hexadecylcyanoacrylate nanospheres display a combined effect for brain tumor targeting. J Pharmacol Exp Ther 2002; 303:928–936.

18. Brigger I, Morizet J, Laudani L, et al. Negative preclinical results with stealth® nanospheres-encapsulated doxorubicin in an orthotopic murine brain tumor model. J Control Release 2004; 100:29–40.

19. Soma CE, Dubernet C, Bentolila D, Benita S, Couvreur P. Reversion of multidrug resistance by co-encapsulation of doxorubicin and cyclosporin A in polyalkylcyanoacrylate nanoparticles. Biomaterials 2000; 21:1–7.

20. Merodio M, Irache JM, Eclancher F, Mirshahi M, Villarroya H. Distribution of albumin nanoparticles in animals induced with the experimental allergic encephalomyelitis. J Drug Targeting 2000; 8(5):289–303.

21. Koziara JM, Lockman PR, Allen DD, Mumper RJ. In situ blood-brain barrier transport of nanoparticles. Pharm Res 2003; 20:1772–1778.

22. Sun W, Xie C, Wang H, Hu Y. Specific role of Ps 80 coating on the targeting of nanoparticles to the brain. Biomaterials 2004; 25(15):3065–3071.

23. Fenart L, Casanova A, Dehouck B, et al. Evaluation of effect of charge and lipid coating on ability of 60-nm nanoparticles to cross an in vitro model of the blood-brain barrier. J Pharmacol Exp Ther 1999; 291(3):1017–22.

24. Moghimi SM, Hunter AC, Murray JC. Long-circulating and target-specific nanoparticles: theory to practice. Pharmacol Rev 2001; 53(2):283–318.

25. Moghimi SM, Szebeni J. Stealth liposomes and long circulating nanoparticles: critical issues in pharmacokinetics, opsonization and protein-binding properties. Prog Lipid Res 2003; 42(6):463–478.

26. Yang S, Zhu J, Lu Y, Yang C. Body distribution of camptothecin solid lipid nanoparticles after oral administration. Pharm Res 1999; 16(5):751–757.

27. Zara GP, Cavalli R, Fundaro A, Bargoni A, Caputo O, Gasco MR. Pharmacokinetics of doxorubicin incorporated in solid lipid nanospheres (SLN). Pharmacol Res 1999; 40(3):281–286.

28. Zara GP, Bargoni A, Cavalli R, Fundaro A, Vighetto D, Gasco MR. Pharmacokinetics and tissue distribution of idarubicin-loaded solid lipid nanoparticles after duodenal administration to rats. J Pharm Sci 2002; 91(5):1324–1333.

29. Fundaro A, Cavalli R, Bargoni A, Vighetto D, Zara GP, Gasco MR. Non-stealth and stealth solid lipid nanoparticles carrying doxorubicin: pharmacokinetics and tissue distribution after i.v. administration to rats. Pharm Res 2000; 42: 337–343.

30. Tröster SD, Muller U, Kreuter J. Modification of the body distribution of poly(methyl methacrylate) nanoparticles in rats by coating them with surfactants. Int J Pharm 1990; 61:85–100.

31. Couvreur P, Kante B, Grislain L, Roland M, Speiser P. Toxicity of polyalkylcyanoacrylate nanoparticles II: doxorubicin-loaded nanoparticles. J Pharm Sci 1982; 71(7):790–792.

32. Calvo P, Gouritin B, Chacun H, et al. Long-circulating PEGylated polycyanoacrylate nanoparticles as new drug carrier for brain delivery. Pharm Res 2001; 18:1157–1166.

33. Harashima H, Komatsu S, Kojima S, et al. Species difference in the disposition of liposomes among mice, rats and rabbits: allometric relationship and species dependent hepatic uptake mechanism. Pharm Res 1966; 13:1049–1054.

34. Demoy M, Andreux JP, Weingarten C, Gaouritin B, Guilloux V, Couvreur P. Spleen capture of nanoparticles, influence of animal species and surface characteristics. Pharm Res 1999; 16:37–41.

35. Vinogradov SV, Batrakova EV, Kabanov AV. Nanogels for oligonucleotide delivery to the brain. Bioconjug Chem 2004; 15:50–60.

36. Kabanov AV, Batrakova EV. New technologies for drug delivery across the blood brain barrier. Curr Pharm Des 2004; 10:1355–1363.

37. Schlageter KE, Molnar P, Lapin GD, Groothuis DR. Microvessel organization and structure in experimental brain tumors: microvessel populations with distinctive structural and functional properties. Microvasc Res 1999; 58:312–328.

38. Vajkoczy P, Menger MD. Vascular microenvironment in gliomas. Cancer Treat Res 2004; 117:249–262.

39. Maeda H, Matsumura Y. Tumoritropic and lymphotropic principles of macromolecular drugs. Crit Rev Ther Drug Carrier Syst 1989; 6:193–210.

40. Lode J, Fichtner I, Kreuter J, Berndt A, Diederichs JE, Reszka R. Influence of surface-modifying surfactants on the pharmacokinetic behavior of ^{14}C-poly(methylmethacrylate) nanoparticles in experimental tumor models. Pharm Res 2001; 18(11): 1613–1619.

41. Calvo P, Gouritin B, Villarroya H, et al. Quantification and localization of PEGylated polycyanoacrylate nanoparticles in brain and spinal cord during experimental allergic encephalomyelitis in the rat. Eur J Neurosci 2002; 15:1317–1326.

42. Schroeder U, Sabel BA. Nanoparticles, a drug carrier system to pass the blood-brain barrier, permit central analgesic effects of intravenous dalargin injections. Brain Res 1996; 710: 121–124.

43. Schroeder U, Sommerfeld P, Ulrich S, Sabel BA. Nanoparticles technology for delivery of drugs across the blood-brain barrier. J Pharm Sci 1998; 87(11):1305–1307.

44. Schroeder U, Schroeder H, Sabel BA. Body distribution of ^{3}H-labelled dalargin bound to poly(butyl cyanoacrylate nanoparticles) after intravenous injections to mice. Life Sci 2000; 66:495–502.

45. Kreuter J, Alyautdin RN, Kharkevich DA, Ivanov AA. Passage of peptides through the blood-brain barrier with colloidal polymer particles (nanoparticles). Brain Res 1995; 674:171–174.

46. Kreuter J, Petrov VE, Kharkevich DA, Alyautdin RN. Influence of the type of surfactant on the analgesic effects induced by the peptide dalargin after its delivery across the blood-brain barrier using surfactant-coated nanoparticles. J Control Release 1997; 49:81–87.

47. Ramge P, Kreuter J, Lemmer B. Circadian phase-dependent antinociceptive reaction in mice after intravenous injection of dalargin-loaded nanoparticles determined by the hot-plate test and the tail-flick test. Chronobiol Int 1999; 17:767–777.

48. Steiniger SCJ, Kreuter J, Khalansky AS, et al. Chemotherapy of glioblastoma in rats using doxorubicin-loaded nanoparticles. Int J Cancer 2004; 109:159–167.

49. von Holst H, Knochenhauer E, Blomgren H, et al. Uptake of adriamycin in tumor and surrounding brain tissue in patients with malignant gliomas. Acta Neurochir (Wien) 1990; 104:13–6.

50. Walter KA, Tamargo RJ, Olivi A, Burger PC, Brem H. Intratumoral chemotherapy. Neurosurgery 1995; 37:1128–1145.

51. Weizsaeker M, Deen DF, Rosenblum ML, Hoshino T, Gutin PH, Barker M. The 9L rat brain tumor: description and application of an animal model. J Neurol 1981; 224:183–192.

52. Sharma US, Sharma A, Chau RI, Straubinger RM. Liposome-mediated therapy of intracranial brain tumors in a rat model. Pharm Res 1997; 14(8):992–998.

53. Olivier J-C, Fenart L, Chauvet R, Pariat C, Cecchelli R, Couet W. Indirect evidence that drug brain targeting using Ps 80-coated polybutylcyanoacrylate nanoparticles is related to toxicity. Pharm Res 1999; 16:1836–1842.

54. Kreuter J, Ramge P, Petrov V, et al. Direct evidence that polysorbate-80-coated poly(butylcyanoacrylate) nanoparticles deliver drugs to the CNS via specific mechanisms requiring prior binding of drug to the nanoparticles. Pharm Res 2003; 20(3):409–416.

55. Alyautdin R, Reichel A, Loebenberg R, Ramge P, Kreuter J, Begley D. Interaction of poly(butylcyanoacrylate) nanoparticles with the blood-brain barrier in vivo and in vitro. J Drug Targeting 2001; 9:209–221.

56. Steiniger S, Zenker D, von Briesen H, Begley D, Kreuter J. The influence of Ps 80-coated nanoparticles on bovine brain capillary endothelial cells in vitro. Proc Int Symp Control Release Bioact Mater 2000; 26:789–790.

57. Lockman PR, Koziara J, Roder KE, et al. In vivo and in vitro assessment of baseline blood-brain barrier parameters in the presence of novel nanoparticles. Pharm Res 2003; 20(5):705–13.

58. Ramge P, Unger RE, Oltrogge JB, et al. Ps 80-coating enhances uptake of polybutylcyanoacrylate (PBCA) nanoparticles by human, bovine and murine primary brain capillary endothelial cells. Eur J Neurosci 2000; 12:1931–1940.

59. Kreuter J. Transport of drugs across the blood-brain barrier by nanoparticles. Curr Med Chem—Cent Nerv Syst Agents 2002; 2(3):241–249.

60. Kreuter J, Shamenkov D, Petrov V, et al. Apolipoprotein-mediated transport of nanoparticle-bound drugs across the blood-brain barrier. J Drug Targeting 2002; 10:317–326.

61. Lück M. Plasmaproteinadsorption als möglicher Schlüsselfaktor für eine kontrollierte Arzneistoffapplikation mit partikulären Trägern. Vols. 14–24. Ph.D. thesis, Freie Universität Berlin, 1997:137–54.

62. Gessner A, Olbrich C, Schroeder W, Kayser O, Muller RH. The role of plasma proteins in brain targeting: species dependent protein adsorption patterns on brain-specific lipid drug conjugate (LDC) nanoparticles. Int J Pharm 2001; 214:87–91.

63. Borchardt G, Audus KL, Shi F, Kreuter J. Uptake of surfactant-coated poly(methyl methacrylate)-nanoparticles by bovine brain microvessel endothelial cell monolayers. Int J Pharm 1994; 110:29–35.

64. Vauthier C, Dubernet C, Chauvierre C, Brigger I, Couvreur P. Drug delivery to resistant tumors: the potential of poly(alkyl cyanoacrylate) nanoparticles. J Control Release 2003; 93(2):151–160.

65. Hu Y-P, Jarillon S, Dubernet C, Couvreur P, Robert J. On the mechanism of action of doxorubicin encapsulation in nanospheres for the reversal of multidrug resistance. Cancer Chemother Pharmacol 1996; 37:556–650.

66. Colin de Verdiere A, Dubernet C, Nemati F, et al. Reversion of multidrug resistance with polyalkylcyanoacrylate nanoparticles: towards a mechanism of action. Br J Cancer 1997; 76:198–205.

67. Vauthier C, Dubernet C, Fattal E, Pinto-Alphandary H, Couvreur P. Poly(alkylcyanoacrylates) as biodegradable materials for biomedical applications. Adv Drug Deliv Rev 2003; 55:519–548.

68. Friche E, Jensen PB, Sehested M, Demant EJ, Nissen NN. The solvents cremophor EL and Tween 80 modulate daunorubicin resistance in the multidrug resistant Ehrlich ascites tumor. Cancer Commun 1990; 2(9):297–303.

69. Lo YL. Relationships between the hydrophilic-lipophilic balance values of pharmaceutical excipients and their multidrug resistance modulating effect in Caco-2 cells and rat intestines. J Control Release 2003; 90(1):37–48.

70. Yamazaki T, Sato Y, Hanai M, et al. Non-ionic detergent Tween 80 modulates VP-16 resistance in classical multidrug resistant K562 cells via enhancement of VP-16 influx. Cancer Lett 2000; 149:153–161.

71. Azmin MN, Stuart JFB, Florence AT. The distribution and elimination of methotrexate in mouse blood and brain after concurrent administration of polysorbate 80. Cancer Chemother Pharmacol 1985; 14:238–242.

72. Li Y, Ogris M, Wagner E, Pelisek J, Ruffer M. Nanoparticles bearing polyethyleneglycol-coupled transferrin as gene carriers: preparation and in vitro evaluation. Int J Pharm 2003; 259(1–2):93–101.

73. Olivier J-C, Huertas R, Lee HJ, Calon F, Pardridge WM. Synthesis of pegylated immunoparticles. Pharm Res 2002; 19(8):1137–1143.

74. Lockman PR, Oyewumi MO, Koziara JM, Roder KE, Mumper RJ, Allen DD. Brain uptake of thiamine-coated nanoparticles. J Control Release 2003; 93(3):271–282.

11

Nanoparticles for Ocular Drug Delivery

ANIRUDDHA C. AMRITE
Department of Pharmaceutical
Sciences, University of Nebraska
Medical Center, Omaha,
Nebraska, U.S.A.

UDAY B. KOMPELLA
Department of Pharmaceutical
Sciences, College of Pharmacy,
University of Nebraska Medical
Center, Omaha, Nebraska, U.S.A.

INTRODUCTION

Anatomically, ocular drug delivery targets either the anterior segment or the posterior segment of the eye. While the target tissues of interest in the anterior segment include the cornea, iris-ciliary body, and lens, those in the posterior segment include the choroid, retina, vitreous, and optic nerve.

Anterior segment drug delivery via topical drops is impeded by several precorneal and corneal factors including the tear flow, blinking, and the epithelial barriers with the resultant drug bioavailability to the aqueous humor being

319

< 5% (1–4). Tear flow and blinking result in short precorneal residence of the eye drops, because of the drainage of most of the dose into the nasolacrimal duct within a few minutes, and subsequently into the systemic circulation. Drugs from drops can also enter the systemic circulation via the conjunctival circulation. Frequent dosing is hence a necessity when topical ocular delivery is utilized. For drugs with long half-lives the dosing can be two to four times a day, which is manageable. However, for drugs with short half-lives dosing every one to two hours may be required, which may lead to the loss of patient compliance. Approaches to prolonged precorneal residence time are needed to increase drug bioavailability to the anterior segment following topical drop administration. Additionally, approaches are needed to rapidly lodge the drug in the corneal epithelium during the short precorneal residence times of a dosage form. Nanoparticles can be potentially designed to enhance drug delivery to the anterior segment on both these counts for some drugs. Alternative approaches to prolong drug delivery by the topical route include the use of drug loaded ointments, viscous vehicles, inserts, contact lenses, and collagen shields to prolong the precorneal residence time of the drug in the tear film.

Eye drops are ineffective for all practical purposes in delivering drugs to the posterior segment disorders, especially those afflicting the neural retina. In addition to the aforementioned factors limiting drug absorption into the eye, the drug has to cross multiple tissue and vascular barriers [conjunctiva, sclera, choroid, Bruch's membrane, and retinal pigment epithelium (RPE) or cornea, aqueous humor, lens, and vitreous] to reach the neural retina following topical administration. Because of these multiple barriers, eye drops are currently not useful for retinal drug delivery. Therefore, alternative approaches are needed to provide therapeutic concentrations of the drug in the posterior segment. Systemic route can deliver the drug to the retina; however, the delivery is limited because of the presence of the outer and the inner blood–retinal barriers. In addition, only a small fraction of the drug given systemically can actually access the ocular tissues. For instance, although the vitreous area under the concentration (AUC) versus time curve for fleroxacin following intravenous administration is ~10% that of serum, the actual dose fraction delivered to the vitreous is about 0.02% (5). This is because 0.1 mg of intravitreal fleroxacin resulted in an estimated vitreal AUC of ~70 mg/hr/L, when compared to 3.4 mg/hr/L vitreal AUC obtained following about 23–30 mg intravenous dose. Because of such low fractions delivered to the posterior segment, large systemic doses are required, which might lead to systemic toxicity. To deliver drugs to the posterior segment, alternative routes, such as periocular or intraocular routes, are being investigated. However, these routes require invasive administration and therefore, frequent administrations are unwarranted as patient safety and compliance could be compromised. Thus, even for the posterior segment, there is a need for the development of systems that can sustain drug delivery. Additionally, when intracellular targeting is desired, as is the case with gene delivery, systems capable of targeting intracellular compartments are required. Nanoparticles will likely be useful in sustaining retinal drug delivery as well as providing intracellular drug targeting in the posterior segment. Alternative delivery systems for prolongation of the posterior segment drug delivery include implants, scleral plugs, microparticles, and liposomes. Nanoparticles are not a universal solution for drug delivery in the eye. The choice of a delivery system ultimately has to be made based on the drug, the disease, and the target anatomy of interest.

An ideal ocular drug delivery system would be able to provide therapeutic concentrations of the drug at the target tissue by overcoming the blood ocular barriers, provide targeted delivery to the ocular tissues with minimal systemic effects, be safe and nonirritating to the tissues, and provide prolonged delivery, thereby reducing the dosing frequency. This chapter mainly focuses on the use of polymeric nanoparticles as drug delivery systems for ocular drug delivery. To better develop and utilize nanoparticulate systems for ocular drug delivery, it is important to understand the disposition and safety of these particulate carriers besides their ability to sustain drug delivery and enhance intracellular uptake.

Therefore, the purpose of this chapter is to describe the disposition, sustained delivery, and safety aspects of nanoparticles in the eye. Topical, intravitreal, as well as periocular routes are addressed in this chapter. Wherever the information is available, studies at the level of cells and excised tissues have been presented. Comparisons have been made between nanoparticles and other delivery systems, particularly using microparticles, wherever appropriate.

DISPOSITION OF NANOPARTICLES IN THE EYE

Topical Disposition of Nanoparticles

Topically applied nanoparticle suspensions can be eliminated in a fashion similar to other aqueous topical ophthalmic formulations. The probable disposition pathways for topically applied nanoparticles can be envisioned as outlined in Figure 1. The nanoparticles in the eyedrop formulation can enter either the cornea or the conjunctiva or drain via the nasolacrimal duct. Particulate systems in the cornea and conjunctiva might contribute to drug levels in the various eye tissues with contributions primarily to the anterior segment. Nanoparticles entering the nose can be further cleared to the gastrointestinal tract. The particles lodged in the nasal and gastrointestinal tissues might release the drug and contribute to the systemic drug levels. As a major fraction of the topical dose is drained via the nasolacrimal duct, particles entering the nasolacrimal duct might be the primary source of drug levels in the circulation following topical administration of nanoparticles. It is not unlikely that very small nanoparticles might escape the nasal and gastrointestinal epithelial barriers to enter the systemic and portal circulations, respectively.

A number of investigators employed topical nanoparticles for ocular drug delivery (6–27). One goal of topical nanoparticulate systems is to enhance the precorneal residence time of the drug. Even for nanoparticulate systems with high surface area available for adsorption, it is difficult to extend the half-time for precorneal residence time by more than a few minutes for some nanoparticles. A precorneal clearance

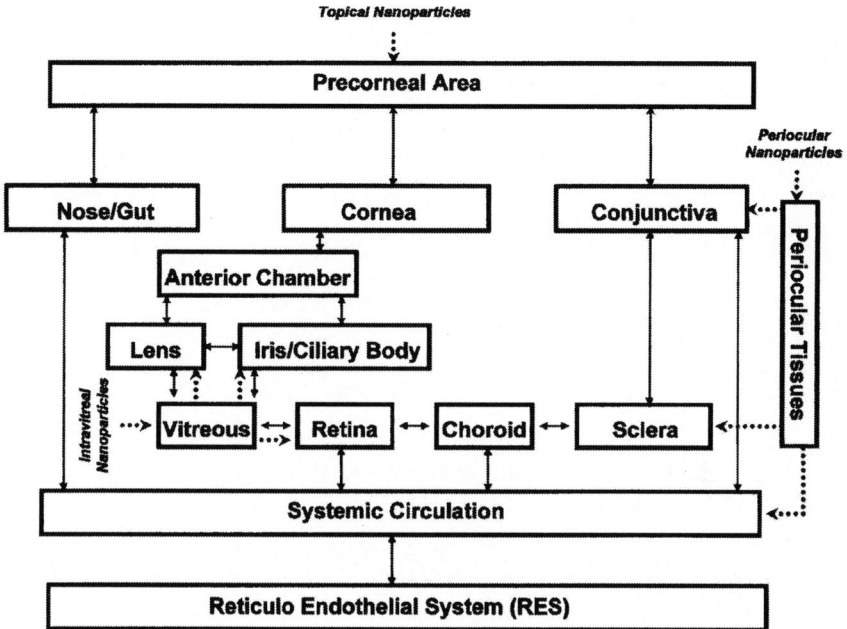

Figure 1 Probable pathways for the clearance of nanoparticles following topical, periocular, or intravitreal routes of administration. The dotted arrows indicate potential nanoparticle transport into target tissues in the immediate vicinity of the administration site. The double-sided arrows represent other possible nanoparticle disposition pathways. Disposition is expected to be driven by particle concentration gradients and fluid flow directions.

study of 100–300 nm dextran particles in a rabbit model using γ-scintigraphy indicated biphasic drainage of particles from the cornea with an initial rapid decline in 15 seconds and a later slower terminal half-life of 2.15 ± 0.09 minutes (14). However, poly(hexyl-cyanoacrylate) particles are probably better retained in the tissues and cleared by a slower rate process (24). Following topical administration in rabbits, radiolabeled poly(hexyl-cyanoacrylate) particles exhibited the highest concentrations in tears and much lower levels in the cornea and the conjunctiva. The concentrations in tears were high initially and then declined rapidly with first-order kinetics over a period of six hours. On the other hand, the levels in the cornea and the conjunctiva were fairly constant throughout the six hours (Fig. 2). The authors speculated that the nanoparticles might have adhered to the cornea and conjunctiva, with nasolacrimal drainage being the major route of precorneal elimination. The work of Calvo et al. (9) suggested that coating poly(ε-caprolactone) (PECL) nanocapsules with chitosan potentially elevates the precorneal residence time of nanoparticles.

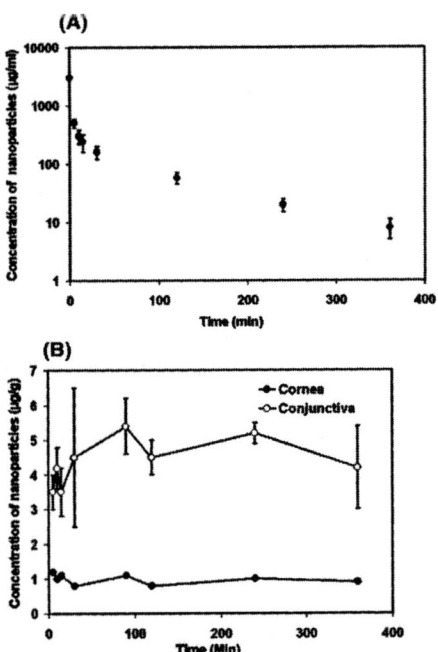

Figure 2 Concentration time profile of poly(hexyl cyanoacrylate) nanoparticles in the tear film, cornea, and the conjunctiva after topical administration of the nanoparticles. Male albino rabbits were dosed with 25 µL of 0.385% (w/v) suspension of nanoparticles by topical application to the cornea. (**A**) Concentration time profile in the tear film over a period of six hours. The tear volume was assumed to be 7 µL. (**B**) Concentration time profile of the nanoparticles in the cornea (*filled circles*) and in the conjunctiva (*open circles*) over a period of six hours. Data is expressed as mean ± SEM for $n \geq 8$. *Source:* From Ref. 24.

A second objective of topical nanoparticles is to enhance the cellular uptake of the drug. A prerequisite for such a possibility is a better uptake or accumulation of the nanoparticles compared to the drug of interest. Rabbit corneal and conjunctival uptake studies with 120 ± 20 nm poly(butyl cyanoacrylate) (PBCA) nanoparticles labeled with propidium iodide or rhodamine 6G over 30 minutes in a standard perfusion cell, indicated particle entry into these cells (25). The particles likely entered the corneal and conjunctival cells via endocytosis, as the particles were visible in intracellular vesicles. The particles stained the conjunctiva more intensely compared to the cornea. In the cornea, the penetration of the PBCA particles was limited to the superficial cell layers, with no particles observed in the corneal stroma and endothelium. In vitro studies with conjunctival cells have demonstrated uptake of the nanoparticles by an endocytic process not mediated by clathrin or caveolin containing vesicles (28–30). Nanoparticulate systems are expected to be useful in enhancing the cellular uptake of drugs with poor membrane permeability. In addition, they might improve the cellular accumulation of drugs with significant enzymatic instability in the precorneal area.

In general, nanoparticle uptake is greater in the conjunctiva when compared to cornea and inflammation can further influence the tissue uptake of topical nanoparticles with the uptake being usually higher. Evidence for this comes from a study employing [14C]poly(hexyl-cyanoacrylate) nanoparticles in rabbits with healthy or inflamed eyes (31). The inflammation was induced in the eyes by topical application of clove oil. In healthy rabbit eyes, the accumulation of poly(hexyl-cyanoacrylate) nanoparticles in the conjunctiva was four- to fivefold higher compared to the cornea (24). In inflamed eyes, the nanoparticle concentration in the conjunctiva was about half when compared to the healthy eyes for the initial time points up to 30 minutes and higher at subsequent time points for up to four hours (31). The profiles in the cornea, nictitating membrane, and the aqueous humor showed similar trends with higher concentrations of the nanoparticles in the inflamed eyes as compared to the healthy eyes at all time points up to four hours. Corneal and conjunctival concentrations of the particles increased during the first 10 minutes and declined in a biphasic manner with initial and terminal half-lives of about 30 and 240 minutes, respectively. Thus, particle uptake is in general greater under inflammatory conditions.

Thus, a major fraction of the topically applied nanoparticles remains in the tears and disappears via nasolacrimal drainage, with the kinetics of drug clearance being marginally altered by nanoparticles when compared to a solution form. Conjunctival uptake of topical nanoparticles is higher in general compared to the corneal uptake. Under inflammatory conditions, the uptake of nanoparticles by both these tissues is elevated. The nanoparticles entering the tissue are cleared slowly with the potential of contributing drug levels for prolonged periods.

Intravitreal Disposition of Nanoparticles

Following intravitreal injection, nanoparticles settle onto the inner limiting membrane of the retina within a few hours (32–34). The settling is governed by Stoke's law

$$\nu = \frac{2gr^2(\rho_1 - \rho_2)}{9\mu} \tag{1}$$

where ν is the settling velocity of the particle suspended in the vitreous (cm/sec), g is acceleration due to gravity (cm/sec^2), r is the equivalent radius of the particle (cm), ρ_1 is the density of the particle (g/cm^3), ρ_2 is the density of the vitreous (g/cm^3), and μ is the viscosity of the vitreous (dyne/ sec/cm^2). As Stoke's law is based on the settling of a particle in a stagnant liquid layer, and because the vitreous in vivo will have some mobility because of the eye movements and the fluid clearance, the nanoparticle sedimentation times in vivo could be shorter than those estimated using Stoke's law. Thus, besides the particle properties, the properties of the vitreous including viscosity, density, convection currents, and fluid clearance determine particle settling. The site of injection within the vitreous will also determine when and where the particles settle within the vitreous cavity. Following deposition on the inner limiting membrane of the retina, nanoparticle penetration into the various retinal layers has been observed (32,33,35). Once the particles gain access to the retina, they might gain partial access to the systemic circulation via the retinal or choroid vasculature (Fig. 1). In addition, as most of the intravitreal injections are given close to the limbus in the pars plana area where the retina is absent, settling and initial penetration of the particles into the iris and the ciliary body have been observed (Fig. 3) in some studies (32,33). Penetration into the ciliary body might be facilitated by the anatomical proximity of these structures to the site of intravitreal injection, the high porosity of these structures, and the elimination mechanism of the particles via the anterior segment.

Influence of the particle's size on disposition is best understood using nondegradable particles. Employing such a strategy in a rabbit study, Sakurai et al. (36) investigated the effect of particle size on the intravitreal disposition of nanoparticles, by selecting nonbiodegradable polystyrene particles of three sizes, 50, 200, and 2000 nm. The investigators observed a decrease in intravitreal half-life with an increase in particle size (Fig. 4) (36). The clearance of all these particles is much slower compared to sodium fluorescein, for which a half-life of 7.8 ± 0.7 hours was observed. The 50- and 200-nm particles penetrated the ocular tissues to a significantly greater extent compared to the 2000-nm particles, as assessed by fluorescence microscopy at one month postadministration. However, the nanoparticle concentrations in the ocular tissues were not quantified. It was suggested that the larger particles (2 μm) are mainly eliminated through the anterior chamber angle, whereas the smaller particles may be cleared via the retina as well as the anterior chamber angle. In a rat study, polylactide nanoparticles of 140 and 310 nm did not exhibit any differences in distribution following intravitreal administration (33). Both sizes of particles were seen to be penetrating the inner layers of the retina, concentrating in the RPE. The penetration pattern was similar for both sizes of particles. This is possibly because the range of sizes tested was very narrow compared to the study of Sakurai et al. (36).

Vitrectomy increases microparticle clearance from the posterior segment. The influence of vitrectomy on particle

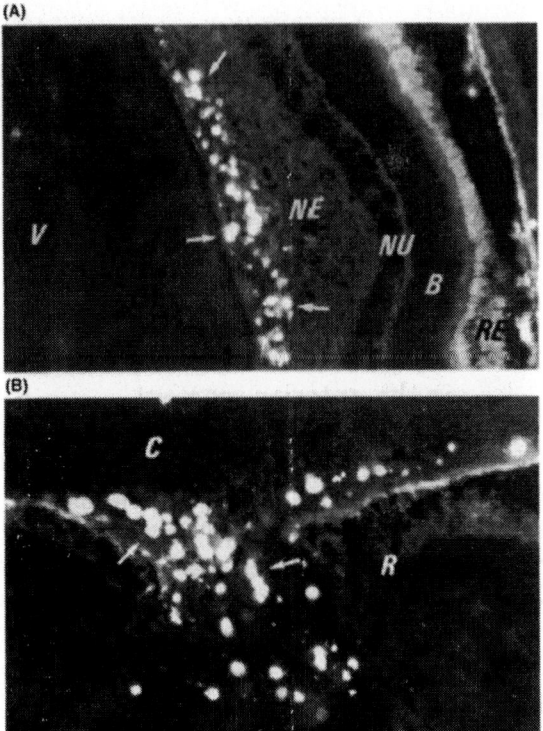

Figure 3 Ocular tissue distribution of ganciclovir-loaded albumin nanoparticles two weeks postintravitreal injection. Female Wistar rats (11–13 weeks old) were injected with 5 µL of suspension of albumin nanoparticles (304 ± 47 nm mean size and 200 mg/mL) intravitreally: (**A**) localization of the particles in the posterior structures, and (**B**) localization of the particles in the anterior structures. *Abbreviations:* B, photoreceptor layer; C, ciliary muscle; NE, neuronal interplay area; NU, outer and inner nuclear layers; R, retina; RE, retinal pigment epithelium; and V, vitreous cavity. *Source:* From Ref. 32.

clearance was assessed in rabbits using ~50 µm 5-fluorouracil–poly(lactic/glycolic acid) (PLGA) particles (37). No particles were observed in the vitreous cavity at the end of 48 ± 5.2 days after injection in normal animals. In the vitrectomized animals, particle clearance was more rapid with almost complete disappearance of the particles occurring at

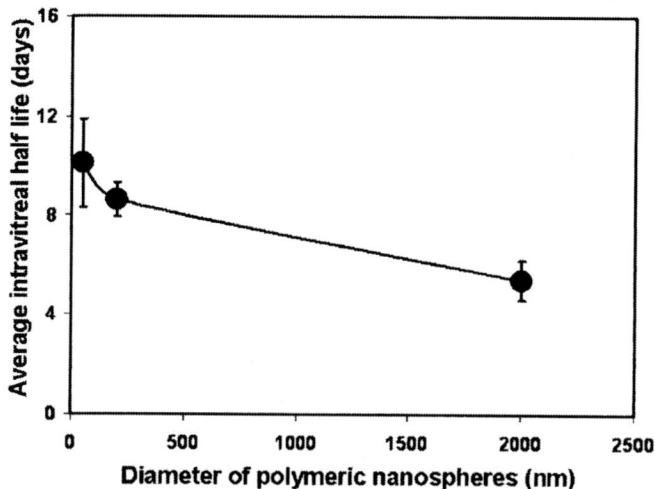

Figure 4 Influence of size on the intravitreal disposition of polymeric nanoparticles. Microparticles (2 μm) and nanoparticles (200
and 50 nm) of fluorescent polystyrene were injected into the vitreous cavity of pigmented rabbits. Data are expressed as mean -
± SEM of $n = 5$. *Source*: Based on data from Ref. 36.

the end of 14 ± 2.4 days. Fundus examination, a subjective
technique, was used in this study. The authors did not measure the penetration of the particles into the ocular tissues.
In another study, 60 μm particles could be observed in the
vitreous cavity in 73% of rats at two months postintravitreal
administration (15).

Bourges et al. (33) investigated intravitreal disposition of
polylactide nanoparticles in rats. The nanoparticles migrated
in all directions and initially adsorbed to the lens posterior
capsule and entered the iris and ciliary body. The particles
also penetrated through the retinal layers and a significant
accumulation was seen in the RPE. By four weeks the particles disappeared from the vitreous and preferentially accumulated in the RPE layer.

Inflammatory conditions within the vitreous can
enhance particle clearance. The influence of experimental
autoimmune uveitis, a vitreal inflammatory condition, on
the disposition of intravitreally administered nanoparticles

was assessed by DeKozak et al. (34) in a rat model. In normal eyes, the nanoparticles entered the intraocular tissues with uptake occurring within 24 hours in the cells of the iris and in the astrocytes within the inner limiting membrane of the retina. In about three days, the nanoparticles were found in the RPE, choroid, and ciliary body. However, under inflammatory conditions at the end of 24 hours, besides the astrocytes, the penetration was also observed in the RPE and anterior chamber of the eye. In addition, the nanoparticles were taken up by infiltrating macrophages at the end of one day. At three days postadministration, under inflammatory conditions, the nanoparticles were detected in the cervical lymph nodes and rare nanoparticles were detected in the spleen and the liver, which was not observed when ocular inflammation was absent.

Following intravitreal administration, 300 nm albumin nanoparticles were retained in the vitreous at the end of two weeks in rats (32). The nanoparticles were mainly seen overlaying the retina on the inner limiting membrane. Ploylactide–rhodamine nanoparticles of 140 ± 20 nm were shown to be retained to a small extent within the vitreous, for at least one month after administration in rats (33). Significant particle intensity could be detected in the intraocular tissues including the ganglion cell layer, rod outer segments, and, specifically, the RPE at one month postadministration. The presence of particles in the inner layers of the retina and the RPE has also been demonstrated by Merodio et al. (32). Penetration of the particles after two weeks was seen in the neural retina as well as the RPE (Fig. 3). Unfortunately, the particle levels were not quantified in these studies.

RPE uptake of the microparticles and nanoparticles has been demonstrated in vitro (33,38–41). The RPE has a natural mechanism of phagocytosing the rod outer segments, and investigators believe that it could be one of the reasons for increased uptake of nanoparticles by the RPE. The uptake of nanoparticles of different sizes has been investigated (39). Fluorescent carboxylate–modified polystyrene particles with sizes ranging from 20 to 2000 nm were incubated with the ARPE-19 cells for three hours (Fig. 5). The uptake of the

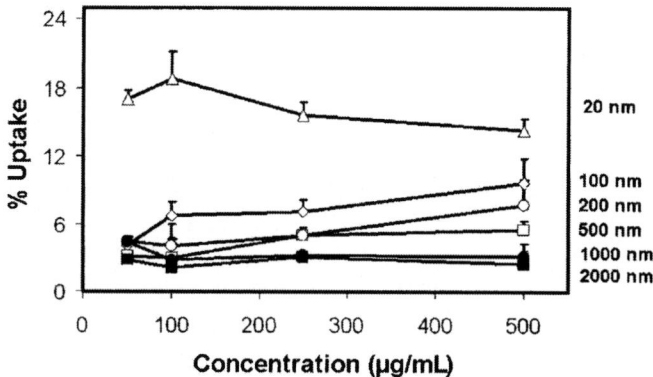

Figure 5 Size dependent in vitro uptake of fluorescent polystyrene nanoparticles and microparticles by ARPE-19 cells. Confluent ARPE-19 cells grown in 48-well plates were exposed to nanoparticle suspensions of various sized nanoparticles at concentrations ranging from 50 to 500 µg/mL for three hours. The data are expressed as mean ± SD for $n = 4$. *Source*: From Ref. 39.

particles into the cells was visualized using confocal microscopy and the particle uptake was quantified using spectrofluorometry. The authors observed that the percentage uptake of the nanoparticles increased with a decrease in particle size (Fig. 5) with the percentage uptake being as high as 20% by a 1-cm^2 monolayer of RPE cells in three hours. To elucidate the mechanism of uptake of the particles by the RPE cells, the authors determined the uptake of 20 and 2000 nm particles in the presence of colchicine, cytochalasin B, and sodium azide. These inhibitors have been previously shown to reduce particle uptake in cells (42–44). The cellular energy depletion using sodium azide did not decrease the uptake of the 20 nm particles by ARPE-19 cells but reduced it for the 2000-nm particles suggesting that uptake of larger particles can be energy dependent. Colchicine, a compound that depolymerizes microtubules, reduced the uptake of 20- and 2000-nm particles suggesting that microtubules are involved in the uptake of nano- and microparticles. Cytochalasin B, a compound that impairs actin gelation and microfilament assembly, reduced the uptake of 20 nm particles but not of 2000 nm particles suggesting that microfilaments might be more critical for the uptake of nanoparticles as compared to microparticles.

The degradation of biodegradable particles occurs in the eye. Some investigations reported smaller particles and partially degraded particles in their histological examinations. Four months after administration of rhodamine-loaded nanoparticles in the vitreous, the particles were seen in many ocular tissues including the inner retinal layers, the RPE, and even in the choroids (33). Along with intact nanoparticles, the investigators observed some partially hydrolyzed nanoparticles. Release of rhodamine from these particles gave a diffused red staining to the neural retina and the RPE (33).

Thus, intravitreally administered nanoparticles and microparticles are removed by the retinal as well as the anterior segment pathways, with the nanoparticles better persisting in the vitreous compared to microparticles based on a few limited studies. The half-lives of both microparticles and nanoparticles are much greater than those for a solution-dosage form (36). Nanoparticles are more permeable through the various layers of the retina compared to microparticles. Particulate systems have a tendency to accumulate in the RPE consistent with the phagocytotic nature of this cell layer. The more prolonged retention of nanoparticles in the vitreous compared to microparticles might be due to their extremely low settling velocities as opposed to reduced clearance by cells. Most of the studies performed in the vitreal disposition of particles to date are qualitatively based on tissue images. In the future, more quantitative studies should be undertaken.

Periocular Disposition of Nanoparticles

There have not been many attempts to investigate the use of periocular nanoparticulate systems. The value of this route of administration for retinal drug delivery has not been fully utilized. However, in recent years, there has been resurgence in the use of periocular routes for drug delivery to the posterior segment (45). To date, nanoparticle disposition from the periocular routes including subconjunctival, peribulbar, subtenon, and retrobulbar routes has not been compared. There have been a few studies with periocular microparticles, but no published studies with periocular nanoparticles. In a study with Adriamycin-loaded polylactide microparticles, significant retention and degradation of the microspheres was observed at the site of administration (46). The investigators observed infiltrating cells, which reduced with time. At the end of 12 weeks, only remnant pieces of microparticles could be seen in the conjunctival tissue. Following intravitreal particle administration of retinoic acid–loaded poly(lactide-*co*-glycolide) microparticles in rabbits, an accidental leakage under the conjunctiva on needle removal was observed by Giordano et al. (47). The small amount of microspheres that leaked out could not be seen seven days after the incident using slit-lamp examination. Indirect evidence of disposition can be obtained from studies that have investigated the use of micro- or nanoparticulate systems for sustained drug delivery. In one such study, Kompella et al. (48) investigated the use of nano- and microparticles of budesonide formulated using polylactide as the encapsulating polymer. There was a higher burst and lower subsequent release rate with the nanoparticles as compared to the microparticles (Fig. 6). Also, the ocular tissue levels of budesonide were several folds higher with microparticles at the end of seven days as compared to the nanoparticles of equivalent dose (Fig. 6). However, when compared to the solution of budesonide, the nanoparticles provided significantly higher budesonide levels in the retina and other ocular tissues at the end of seven days postadministration. Though the authors primarily consider the differences in drug release rates to be the reason for the higher tissue levels of budesonide following subconjunctival administration, the possibility of differential disposition of the particles from the subconjunctival space could not be ruled out from their study. In trying to investigate the disposition of particulate systems from the subconjunctival space, Amrite and Kompella (49), using nonbiodegradable fluorescent particles of 20-nm and 2-μm sizes, demonstrated that the larger particles were almost completely retained in the periocular space for up to 60 days postadministration. However, they observed that the 20 nm particles disappeared

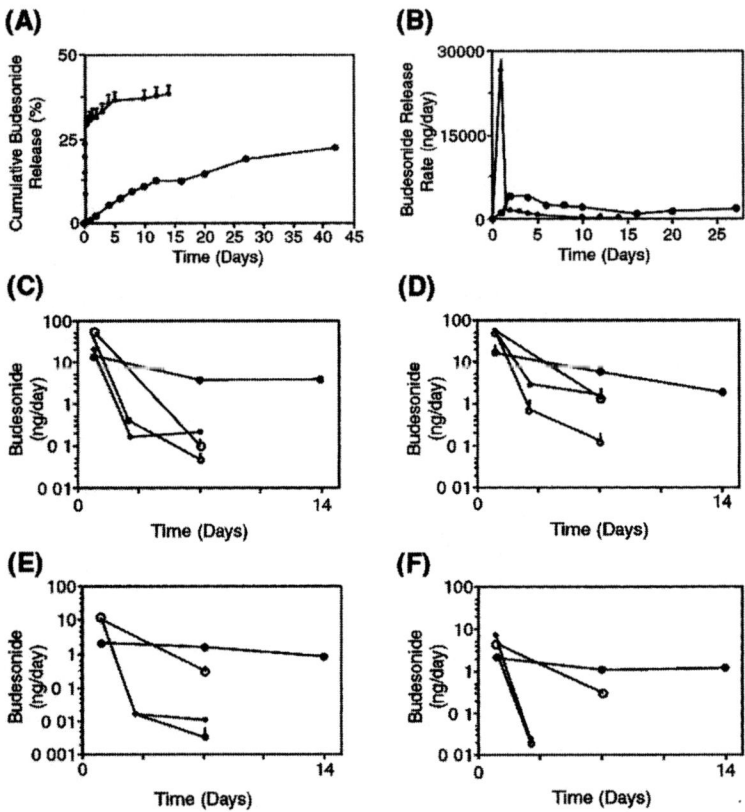

Figure 6 Ocular tissue distribution of budesonide after subconjunctival administration of budesonide–PLGA nanoparticles and microparticles. In vitro release profile of budesonide from PLA nanoparticles (*small filled circles*) and microparticles (*large filled circles*). (**A**) Cumulative budesonide release versus time. (**B**) Budesonide release rate (ng/day) versus time. Budesonide was administered in the eyes of rats, in the form of a solution (50 or 75 μg to one eye, *small and large open circles*, respectively), nanoparticles (50 μg to one eye, *small filled circle*), or microparticles (75 μg to one eye, *large filled circle*), and drug levels were estimated in (**C**) retina, (**D**) vitreous, (**E**) cornea, and (**F**) lens. Data are expressed as mean ± SD for $n = 4$. *Abbreviations*: PLGA, poly (lactic/glycolic acid); PLA, poly (L-lactide). *Source*: From Ref. 48.

rapidly from the site of injection with 25% of the dose remaining at the end of day one and < 15% at the end of seven days postadministration. The authors did not observe the presence of 20 nm nanoparticles at the site of administration at 60 days postadministration. There was no significant penetration of either the nanoparticles or the microparticles into the ocular tissues except the sclera for a period of up to 60 days (49–51). Further investigations by the authors suggested that particles 200 nm in diameter show a similar retention behavior in the subconjunctival space as the larger 2-μm particles (49). The authors have also shown that the ocular penetration of the different particles is < 0.1% of the administered dose for 20-nm particles with no particles detected in any of the ocular tissues for the 200- and 2-μm particles at the 1, 7, and 60 days time points they studied (50,51).

In a study with periocular microparticles of a PKC inhibitor, Saishin et al. (52) reported a significant presence of microparticles under the conjunctiva 10 days postadministration. Their gross pathological examination showed the presence of a large collection of microspheres beneath the conjunctiva that was similar for the drug-loaded and the placebo microparticles of PLGA. The authors reported that the microparticles occupied an entire quadrant outside of the eye extending up to the optic nerve in their pig model. The pigs were injected with 100 mg of microspheres containing PKC 412 into one eye, whereas the other eye received 100 mg of placebo microspheres.

In summary, periocularly administered particulate systems exhibit unique disposition behavior. Micro- and nanoparticles above 200 nm do not gain access to the intraocular tissues including choroid, retina, and vitreous. Particles of this size range can be completely retained at the site of injection for at least up to two months, indicating their potential usefulness as sustained drug delivery systems. The comparison of various periocular routes with respect to the differences in particle disposition has yet to be undertaken. A probable route for the disposition of nanoparticles from the periocular space is schematically presented in Figure 1. From periocularly administered particulate systems, the drug can be released to enter the conjunctiva, tear film, cornea, and aqueous humor or the drug might cross the sclera and choroid to reach the retina and the vitreous. Among the different periocular routes, viz., the subconjunctival, subtenon, and retrobulbar administrations, the disposition can differ and it has yet to be investigated.

OCULAR DRUG DELIVERY ENHANCEMENT USING NANOPARTICLES

One of the major reasons for research in developing systems other than solutions for ophthalmic drug delivery has been the need to reduce the dosing frequency by prolonging drug effects, especially for treating chronic ocular disorders. This can be achieved by sustained drug delivery or by providing greater dose delivery. Because of the limitations of topical solutions in delivering a high dose fraction to the anterior segment tissues and because of a limited prolongation of duration of effect with an increase in dose, approaches to sustain drug delivery have been widely investigated. Such an approach utilizes the drug better than a pulsatile bolus delivery of the drug. Compared to a solution form of the drug, a slower release can be obtained with a suspension form of the drug. Novel delivery systems like the nano- and microparticulate systems with suitable surface features provide newer means of sustaining and prolonging the drug delivery. The micro- and nanoparticulate systems have been investigated to sustain ocular drug delivery by both the topical and the injectable routes of administration. The microparticles, because of their lower surface area:volume ratio, can potentially prolong the drug release better as compared to nanoparticles.

Topically Applied Nanoparticles

Nanoparticles can be administered by the topical route in the form of drops. Noninvasive repeated dosing is possible in this case. As explained earlier, the residence time of an ophthalmic drug solution in the tears and cornea is only of the order of a few minutes, and hence the effective levels in the ocular tissues are not sustained for a long time. Therefore, the purpose of topically applied nanoparticles would be mainly to sustain the drug levels in the anterior tissues for a period of a few hours. The residence time can be increased by increasing the viscosity of the formulation. Alternatively, nanoparticles can be coated or prepared with mucoadhesive or bioadhesive polymers that can interact with the precorneal mucus or cells to increase the residence time, and hence the duration of drug action.

The first prolonged effect nanoparticle system utilized pilocarpine as the model drug (53). The particles were formulated using cellulose acetate phthalate (CAP) as the polymer. Compared to a solution-dosage form, CAP nanoparticle suspension increased the AUC of the drug by 50% in the aqueous humor and the miosis time from 4 to 10 hours. The enhancement in the AUC was due to decreased elimination rate of the drug in the CAP nanoparticles. The CAP dissolved at the tear pH of 7.2, forming a viscous polymer solution when the formulation (which had a pH of 4.5) was administered topically. Similarly, compared to the solution-dosage form, pilocarpine-loaded PBCA nanoparticles increased the miosis time from 3 to 4 hours and prolonged the IOP lowering effects from 4 to >9 hours (31). The polyalkyl cyanoacrylate colloidal systems are eliminated from the tears with a residence time of 15–20 minutes, which is significantly higher than the residence time of eye drops (5–10 minutes) (2,54). The increase in retention could be due to mucoadhesiveness of poly(alkyl cyanoacrylates) (PACAs), which leads to binding of these particles directly to the cornea and the conjunctiva (24). Chitosan, a mucoadhesive polymer, can be used to coat drug-loaded nanoparticles to prolong the contact time of the formulation with the ocular surface and to enhance particle uptake by the ocular tissues (55). Utilizing chitosan as an encapsulating polymer, a two-to sixfold increase in the corneal and conjunctival levels of cyclosporine A was demonstrated when compared to a cyclosporine A suspension (13). The levels of cyclosporine A in the conjunctiva and cornea were subtherapeutic at the end of 24 and 48 hours after administration of cyclosporine A suspension, but therapeutic with the chitosan nanoparticles. No difference was, however, observed in the corneal and conjunctival levels, when the cyclosporine A suspension was compared to a cyclosporine A suspension in a chitosan solution.

To understand the influence of a positive charge of chitosan on ocular drug delivery from nanoparticles, the effect of
coating PECL nanoparticles of indomethacin with either positively charged poly(L-lysine) or chitosan was assessed (56).
The AUC values were four times greater with the nanoparticle systems as compared to indomethacin solution. The
chitosan-coated nanoparticles had an eight times higher
AUC as compared to the indomethacin solution. When comparing uncoated PECL nanoparticles of indomethacin with
either poly(L-lysine)-coated or chitosan-coated nanoparticles,
it was observed that the AUC in the cornea or the aqueous
humor was not different between the uncoated and poly(L-
lysine)-coated nanoparticles. However, the AUC was twice
as much with the chitosan-coated nanoparticles. The authors
suggested that the higher levels with chitosan-coated nanoparticles were due to the mucoadhesive properties of chitosan
as both the chitosan-coated as well as the poly(L-lysine)-
coated nanoparticles had similar positive surface charge.
The effect was probably due to an enhanced uptake of the particles by the cornea as well as the possible opening of tight
junctions by chitosan in the corneal epithelium (9).

Polyethylene glycol (PEG) coating of nanoparticles is a
useful approach to enhance the ocular effects of drug-loaded
nanoparticles following topical application. A comparison of
PEG-coated acyclovir-loaded nanoparticles of polyehtyl-2-cya-
noacrylate (PECA) and a suspension of acyclovir or a physical
mixture of acyclovir and unloaded PECA nanoparticles indicated a 25-fold increase in the drug level in the aqueous humor
with the PEG-coated nanoparticles, when compared to the
drug suspension or the physical mixture (57). Giannavola compared the ocular bioavailability of acyclovir–poly(L-lactide)
(PLA) nanoparticles, acyclovir–PLA physical mixture suspension, acyclovir suspension, and acyclovir–PLA–PEG nanoparticles (15). There was a sustained delivery of acyclovir with the
nanoparticle formulations for a period of over six hours in the
aqueous humor. The drug AUC values in the aqueous humor
were several folds higher with the nanoparticle formulations
as compared to the control acyclovir suspension or the acyclovir nanoparticle physical mixture (Fig. 7). The aqueous humor

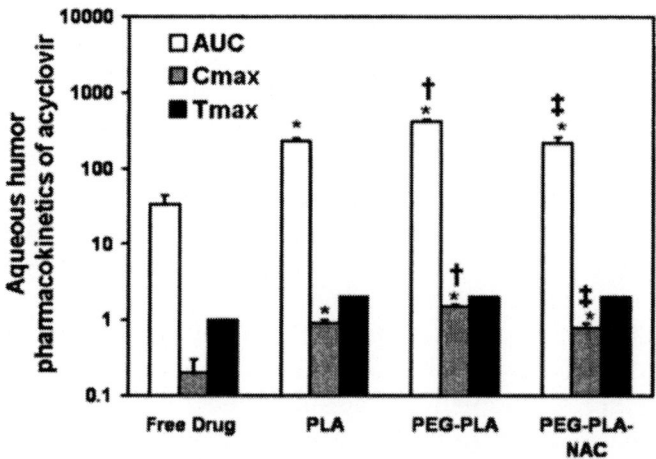

Figure 7 Aqueous humor pharmacokinetics of acyclovir after a single instillation of various formulations in the rabbit eye. Free drug refers to a dispersion of acyclovir in sterile isotonic phosphate buffer. PLA refers to a suspension of nanoparticles of acyclovir prepared using poly(L-lactide) as the encapsulating polymer and coated with poly(ethylene glycol). PLA–PEG–NAC refers to administration of the PLA–PEG nanoparticle suspension of acyclovir after pretreatment with NAC, which is a mucolytic. $*, p < 0.001$ versus free drug formulation; $\dagger, p < 0.001$ versus PLA nanospheres; $\ddagger, p < 0.001$ versus PEG-coated PLA nanospheres. The AUC refers to area under the curve from 0 to 6 hours (μg/hr/mL), $C_{\max}$ is the maximum concentration in the aqueous humor (μg/mL), and $T_{\max}$ is the time for maximum concentration (hr). Data are expressed as mean $\pm$ SD for $n = 4$. *Abbreviations:* PLA, poly(l-lactide); PEG, poly(ethylene glycol); NAC, N-acetyl cysteine. *Source*: Based on data from Ref. 15.

AUC values were significantly higher for the PEG-coated nanoparticles when compared to the uncoated nanoparticles. When the mucous was removed from the ocular surface by pretreatment with N-acetyl cysteine, there was no difference in the aqueous humor AUC of PEG-coated and uncoated nanoparticles. The increase in the AUC with PEG-coated nanoparticles, therefore, could be attributed to their enhanced interaction with the mucous layer. A comparison of the interaction of PEG-coated PECL–rhodamine nanoparticles and chitosan-coated PECL–rhodamine nanoparticles with the ocular mucosa revealed that these coatings enhanced the penetration of the encapsulated dye through the cornea (58). In addition, nanoparticle systems coated with either PEG or chitosan were able to penetrate the corneal surface by a transcellular pathway. The PEG coating enhanced the transport of the nanocapsules across the whole epithelium, whereas with the chitosan coating enhanced the transport only in the superficial layers of the corneal epithelium.

The beneficial effects of topically applied nanoparticles appear to be dependent on the type of drug chosen and the release rates. Early studies indicated no beneficial effects with nanoparticles encapsulating progesterone and hydrocortisone, possibly because these drugs are relatively well absorbed and their release rate from the particles was very slow (17,27). Subsequently developed nanoparticles were aimed at more rapid release rates, with a large fraction of drug release occurring within a few hours (57). Such formulations resulted in better drug delivery or effects, usually spanning < 1 day.

An interesting approach to prolong delivery from nanosystems in the precorneal area is to use contact lenses impregnated with nanoparticles that are optically acceptable. Such an approach was employed by Gulsen and Chauhan. (59). They have utilized a dispersion of microemulsion drops in hydroxyethylmethacrylate hydrogels (Fig. 8). The particles in the microemulsion have a very narrow size distribution and diameter below 50 nm. The lenses are clear and the particles do not interfere with vision. Their system has demonstrated sustained release of lidocaine in vitro for over eight days. In the absence of in vivo studies, no real conclusions about the benefits of this approach can be made.

The tissue uptake of nanoparticles and/or drug encapsulated in nanoparticles is elevated under inflammatory conditions. Zimmer et al. (27) using albumin nanoparticles entrapping radioactive hydrocortisone, measured the drug levels as opposed to the concentration of the particles in various tissues. The drug levels with the nanoparticles were lower in the aqueous as compared to the reference solution for the initial time points in both healthy and inflamed eyes.

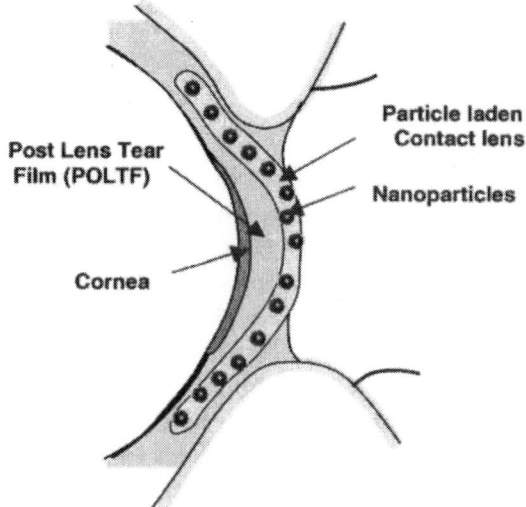

Figure 8 Schematic illustration of a nanoparticle-loaded ophthalmic contact lens for sustained topical ophthalmic drug delivery. *Source*: From Ref. 59.

However, at later time points (60–240 minutes) there was no difference in the drug levels. This could be due to slower release of the drug from the nanoparticles. In case of both the reference solution and the nanoparticles, the drug levels were higher in the inflamed eyes as compared to the normal eyes indicating increased uptake of nanoparticles and/or hydrocortisone under inflammation. The trend was similar in the cornea.

In summary, topically applied nanoparticles elevated drug uptake and sustained drug delivery in some instances. PEG containing nanoparticles or those coated with mucoadhesive polymers seem to provide a greater advantage. Nanoparticles in a topical suspension have prolonged the drug delivery for at most a few hours. New approaches such as nanoparticle or drug bound contact lenses might offer an alternative for prolonged drug delivery for at least a few days. The uptake and transport of the nanoparticles depends on the properties of the particle as well as the disease condition of the ocular surface. Mucoadhesive particles or particles with PEG coating seem to enter the cell layers better. The uptake of particles, and hence drug delivery is higher under inflammatory conditions.

Intravitreal Nanoparticles

Sustained drug delivery/enhanced uptake by the cells has been the mainstay for the investigations in the use of intravitreal nanoparticles. El-Samaligy et al. (60). have utilized poly(ethyl cyanoacrylate) (PECA) nanoparticles of acyclovir and ganciclovir for sustained delivery to the retina after intravitreal administration in rabbits. The drug tissue concentrations were significantly higher in the retina with the nanoparticles when compared to the drug solutions. The nanoparticles also provided therapeutically effective concentrations of ganciclovir for a period of over 10 days. Compared to the solution-dosage form, the nanosphere formulations resulted in lower plasma concentrations of ganciclovir. A special feature used with nanoparticles delivered by the intravitreal route has been to deliver macromolecules like DNA. In a recent study Bejjani et al. (41) have evaluated the use of nanoparticles for gene transfection of the RPE in vivo. The investigators utilized nanoparticles of PLGA encapsulating the DNA for nuclear red fluorescent protein. The gene expression was found in the inner retinal layers to a small extent with the gene expression mostly localized to the RPE. The gene expression for red fluorescent protein was sustained for a period of 14 days. This RPE localized gene expression is consistent with their previous studies on disposition of nanoparticles from the vitreous (33). There was also no toxicity observed with the particulate system. An interesting observation made by the authors was the higher in vivo gene expression when compared to in vitro expression in ARPE-19 cells. The possible reasons for this include the in vitro dilution of the plasmid due to the division of cells and the presence of more active phagocytic mechanism in the RPE cells in vivo. Sustained delivery using intravitreal particles is an area of intense preclinical and clinical research. In vitro studies with poly(lactide–co–glycolide) nanoparticles of a vascular endothelial growth factor (VEGF) antisense oligonucleotide demonstrated efficacy in inhibiting VEGF secretion from ARPE-19 cells, which was similar to standard lipofectin treatment and significantly higher than naked antisense oligonucleotide (Fig. 9) (40).

Table 1 lists polymeric nanoparticles administered by various routes and their outcomes. In summary, intravitreally administered nanoparticles sustain ocular drug delivery better than topically administered nanoparticles. This is due to the slower clearance of the particles from the intraocular sites compared to their clearance from the precorneal area. Intravitreally administered nanoparticles gain access to the RPE and facilitate gene expression therein.

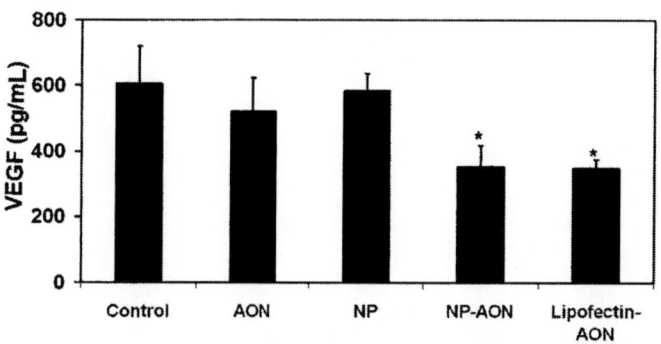

Figure 9 VEGF-antisense oligonucleotide nanoparticles inhibit VEGF secretion from ARPE-19 cells. A suspension of nanoparticles (105 ng/mL) containing 1 μM antisense oligonucleotide were incubated with ARPE-19 cells from day four to day six of seeding. Lipofectin (a commercially available transfection agent) treatment for four hours on day four was used as a positive control. The media were then replaced with serum-free medium on day six in all cases and the VEGF secretion over 12 hours was estimated using an ELISA. *, Significantly different from control ($p < 0.05$). The data are expressed as mean ± SD for $n = 4$. *Abbreviations*: VEGF, vascular endothelial growth factor; PLGA, Poly(lactic/glycolic acid); AON, naked antisenense oligonucleotide; NP, PLGA nanoparticles with no antisense; and NP-AON, PLGA nanoparticles encapsulating VEGF antisense oligonulcleotide; ELISA, enzyme-linked immunosorbent assay. *Source*: Based on data from Ref. 40.

Table 1 In Vivo Studies with Nanoparticulate Systems in the Eye

Drug	Polymer	Route	Species	Particle size (nm)/enhancement	References
Pilocarpine	PBCA	Topical	Rabbits	Miotic time increased from 3 to 4 hours. Miotic response was enhanced by 33% as compared to control solution	31
Betaxolol	PACA	Topical	Rabbits	240 nm/ocular bioavailability and effects improved	61
Amikacin sulfate	PBCA	Topical	Rabbits	300 nm/dextran 70,000 was used as stabilizer and greater concentration of nanoparticles was achieved in cornea and aqueous humor	18
Betaxolol	PECL	Topical	Rabbits	Intraocular pressure (IOP) reduction response was longer & greater as compared to solution	62
Pilocarpine	PBCA	Topical	Rabbits	190 nm/aqueous humor AUC of pilocarpine increased by 23%	26
Pilocarpine	CAP	Topical	Rabbits	300 nm/AUC of miosis versus time curve increased by 50%	53
Cartelol	PECL	Topical	Rabbits	> 8 hours of therapeutic effect. Systemic side effects were reduced	19
Metipranolol	Different polymers	Topical	Rabbits	Varying sizes from 100 to 300 nm/efficacy did not increase but systemic side effects were reduced	63

Ibuprofen	Eudragit® RS100	Topical	Rabbits	100 nm/bioavailability and pharmacological activity increased	21
Cloricromene	Eudragit® RL100	Topical	Rabbits	62.9 nm (86.5%) and 201.4 nm (11.7%)/ AUC increased by 80%. Drug effects increased	8
7-Hydroxy-2-dipropyl-aminotetra-lin	Calcium phosphate	Topical	Rabbits	600 nm/enhanced IOP lowering	11
Acyclovir	PLA and PEG–PLA	Topical	Rabbits	50–130 nm/aqueous humor AUC increased by sevenfold as compared to control suspension with PLA. AUC increased by 14-fold with PLA–PEG	15
Tobramycin	Solid lipids	Topical	Rabbits	<100 nm/ aqueous humor AUC increased by fourfold as compared to control commercial formulation. Duration of action extended from 4 hr to 6 hr	10
Betaxolol chlorhydrate	IBCA	Topical	Rabbits	240–280 nm/effect was equivalent to the commercial eyedrops	64
Ganciclovir	Albumin	Intravitreal	Rabbits	290 and 304 nm/well tolerated. Localized in surrounding tissues of the vitreous. Inflammatory or toxic response was not significant	32
Tamoxiphen	PEG–PHDCA copolymer	Intravitreal	Rats	112 ± 25 nm/better effect with decreased dose as compared to control solution	34

(Continued)

Table 1 In Vivo Studies with Nanoparticulate Systems in the Eye (*Continued*)

Drug	Polymer	Route	Species	Particle size (nm)/enhancement	References
Acyclovir and ganciclovir	PECA	Intravitre al	Rabbits	525 nm/retinal AUC of both acyclovir and ganciclovir increased. For over 10 days concentrations were therapeutically effective	60
Rhodamine/ Nile red	PLGA	Intravitreal	Rats	140 and 310 nm/up to four months, nanoparticles were present in the retina and specifically the RPE	33
Gene for red nuclear fluorescent protein	PLGA	Intravitreal	Rats	643 ± 74 nm/gene expression in RPE for over 14 days	41
Budesonide	PLA	Periocular	Rats	345 ± 2 nm/higher drug levels were present in the retina, vitreous, lens, and cornea with a 50-µg dose as compared to a 50- or even 75-µg solution of budesonide seven days postadministration	48

Abbreviations: PACA, poly(alkyl cyanoacrylate); PBCA, poly(butyl cyanoacrylate); PLGA, poly(L-lactide-*co*-glycolide); PLA, poly(L-lactide); PEG, poly(ethylene glycol); IBCA, isobutyl cyanoacrylate; CAP, cellulose acetate phthalate; PECA, poly(ethyl cyanoacrylate); PHDCA, poly(hexadecyl cyanoacrylate).

SAFETY AND TOLERABILITY OF
PARTICULATE SYSTEMS

The safety and tolerability of the nanoparticulate systems largely depends on the type of polymer used for formulating the system. Thus, only polymeric materials that are known to be biodegradable to relatively nontoxic products should be considered for the preparation of therapeutic nanoparticles. Other than the polymer, the safety and the tolerability will depend on the drug being encapsulated and also to some degree on the physicochemical characteristics of the nanoparticles including size, surface charge, and hydrophilic/lipophilic properties. The formulation of nanoparticles using several polymers involves the use of organic solvents for dissolving the polymers. The residual organic solvents in the particles can also lead to some toxic effects and decrease the tolerability (65,66).

Topically Applied Nanoparticles

Table 1 shows a list of polymers that have been used for nanoparticulate systems intended for the topical ocular route. Many studies have evaluated the relative safety and tolerability of nanoparticulate systems. PACAs are the most extensively studied carriers for topical ophthalmic applications. Among the various PACAs, the PBCA has been used extensively as an artificial tissue and bone glue (67). It is well tolerated and significantly less toxic than various other tissue glues. The toxicity of PBCA nanoparticles was assessed by Couvreur et al. (68,69). The polymer is relatively safe, as LD_{50} is as high as 500 mg/kg after intravenous administration in rabbits. In their subacute studies, no significant changes were observed in the histological pattern of the tissues and also no significant effects were observed on the body weight or blood parameters including blood pH, cell counts, and differential cell counts. The authors also reported that PBCA as well as poly(hexyl cyanoacrylate) did not show any toxicity in the tissues. No mutagenicity was observed with the nanoparticles or their degradation products as determined by the

Ames test. The particles were well tolerated by rabbit eyes with no signs of inflammation or reddening even after multiple dosing. The observations about tolerability have been similar in a more recent study evaluating the ocular tolerability of cyclophosphamide-loaded PBCA nanoparticles (23). The cytotoxicity of nanoparticles of poly(alkyl cyanoacrylate) with different alkyl chain lengths was evaluated by Lherm et al. (70) using cultured fibroblasts. These studies indicated that the toxicity was the most with ethyl and isobutyl derivatives, intermediate with the methyl derivative, and the lowest with isohexyl derivative. In fact, the cytotoxicity correlated well with the velocity of the polymer degradation (70).

Pignatello et al. (20) evaluated the ocular tolerability of Eudragit® nanosuspensions in rabbits. Only mild conjunctival hyperemia was observed 10 minutes after the end of treatment in 2 out of 10 eyes with the Eudragit® RS100 and in 3 out of 10 eyes with the Eudragit® RL100 nanosuspensions. The hyperemia disappeared at 6 and 24 hours posttreatment. The investigators support the use of these polymers because of the absence of any severe inflammation or discomfort in vivo.

The tolerability of acyclovir-PLA nanoparticles was evaluated using a Draize test in the rabbit eyes (15). The PLA nanospheres showed no signs of ocular inflammation or tissue alterations in the rabbit eyes. In addition, the particles did not cause conjunctival swelling or discharge, iris hyperemia, or corneal opacification.

Ocular tolerability of PEG-coated PECA nanospheres in rabbits indicated no severe ocular inflammation, with only a mild conjunctival hyperemia 10 minutes after the end of the treatment (57). In vitro toxicity studies for chitosan nanoparticles using a conjunctival cell line indicated no cytotoxicity (indicated by difference in survival) up to a concentration of 2 mg/mL (12,13).

Intravitreal Nanoparticles

Several studies have investigated the safety and the tolerability of intravitreally applied nanoparticles (33,34,41,71–73). There are some concerns about the safety of intravitreal particulate systems. Algvere and Martini demonstrated that intravitreal administration of colloidal carbon nanoparticles induce neovascularization (71–74). In their studies with cynomolgus monkeys, they observed that with the injection of intravitreal carbon nanoparticles (size 20–70 nm), there was conspicuous cyclitis one week after the administration. The cyclitis was characterized by exudative separation of the non-pigmented and pigmented ciliary epithelial layers, the presence of inflammatory cells, and premacular detachment of the vitreous. The pathological changes continued for weeks, and at the end of 10 weeks, all the injected eyes had extensive retinal detachment with pre- and subretinal collagenous cellular membranes. The authors concluded that these changes were the inflammatory responses to the particles.

From the studies mentioned earlier, it appears that intravitreal administration of nanoparticles could stimulate some adverse reactions. However, it is not evident as to whether the earlier inflammatory response was specific to carbon nanoparticles or if it was a generalized response to nanoparticles. In studies with intravitreal albumin nanoparticles in rats, Merodio et al. (32) reported less severe side effects. The authors observed that there was absence of any cellular infiltration following the association of nanoparticles with ocular cells. However, the authors also reported that the cryoarchitecture of the outer retina was distorted to a certain degree. Merodio et al. (32) reported similar changes in eyes that had received just plain solution of the drug and no nanoparticles suggesting that the observed distortions may be related to either the drug in question or the way in which the tissues were processed. No photoreceptor degeneration was observed in the eyes injected with the albumin nanoparticles and the neural retina showed no signs of alteration.

Giordano et al. (75) investigated the biodegradation and tissue tolerability of intravitreal biodegradable PLGA microspheres. The authors observed only a mild, localized, nonprogressive, foreign body reaction in response to microsphere injection. The retina and the choroid were normal and the ERG showed no abnormalities of the microspheres at one and six month postadministration. No clinical inflammatory signs were observed by slit-lamp at four days postinjection and thereafter.

The anatomy and tissue integrity was well preserved after intravitreal administration of PLA nanoparticles entrapping the fluorochrome rhodamine 6G (33). There was a nonspecific activation of glial cells and a mild transient inflammatory reaction.

De Kozak et al. (34) investigated the use of intravitreal nanoparticles made of poly[(hexadecyl cyanoacrylate)-co-ethylene glycol] (PHDCA–PEG) copolymer. In their histological analysis in normal rats, the investigators observed phagocytosis of nanoparticles by macrophages at eight hours postadministration. The investigators concluded that the injection of those nanoparticles could have a modest inflammatory reaction.

Acute ocular tolerability of PECA nanoparticles encapsulating acyclovir and ganciclovir were evaluated by El-Samaligy et al. (60) in rabbit eyes. At six days postadministration, the investigators observed lens opacification and vitreous turbidity, which was present throughout the 10-day study. The vitreous opacification was probably due to the opaque colloidal dispersion, which was administered.

There is a possibility of interference with vision if intravitreal particles, which might come in the path of light (76). As the diameter of the particles becomes greater than 50 nm, the light scattering by the particles can interfere with the vision. The scattering intensity can be calculated from the refractive index of the particles and the particle size. Based on such calculations, it was suggested that particles below 50 nm would be the most effective in avoiding this complication. However, even with <50 nm particles, as the mass injected increases, significant vision interference can be anticipated. A loss of vision by one line in the eye chart was estimated for 10 mg of 50-nm particles (76). Another problem that is commonly present with colloidal systems (both nano- and microparticles) is the aggregation of particles. Because of the high surface-free energy, the nanoparticles have a natural (thermodynamic) tendency to aggregate, which can result in an increase in particle size. Thus, a formulation that has particles that do not by themselves interfere with vision, might cause vision disturbances because of aggregation of particles within the vitreous.

The observations mentioned suggest that the intravitreal route of administration with nanoparticulate systems needs further evaluation. The long-term safety of this route needs to be investigated and also the influence of the particles on the vision and retinal function needs further evaluation. This route would at least produce a mild inflammatory reaction and it is essential to evaluate the risk/benefit ratio when utilizing intravitreal and other systems clinically.

Periocular Nanoparticles

The safety of nanoparticles after periocular administration has not been evaluated to a very large extent. As mentioned earlier, the type of the polymer and its degradation products are major determinants of the safety and acceptability of a polymeric formulation. The periocular route offers the advantage that the polymer burden on the sensitive tissues of the eye like the retina can be reduced after periocular administration.

Kimura et al. (46) have evaluated the effects of PLA microspheres on the ocular and periocular tissues after subconjunctival administration. They observed that one week postadministration, the site of administration contained a few inflammatory cells and a little fibrous tissue. At the injection site the microspheres were degraded over time and were phagocytosed by several multinucleated giant cells. The retina and the ciliary body were found to be normal in the histopathologic examinations. Similarly, other studies have not shown any significant toxicity to the ocular tissues following the administration of nano- or microparticles by the periocular route (48,52).

Local inflammation at the site of administration was observed after periocular injection. This involved an increased number of mast cells and polymorphonuclear (PMNs) neutrophils. This may be a xeno-response and also possibly an elimination mechanism of the particles. However, this was thought to be reversible and no extensive damage has been reported.

CONCLUSIONS

Nanoparticulate delivery systems have potential applications for ocular drug delivery. However, nanoparticles are not a universal solution for all problems associated with ocular drug delivery. Many of the problems associated with the delivery of ocular therapeutics including rapid clearance, short duration of action, and inefficient uptake can in part be addressed using nanoparticulate systems. Nanoparticles coated or prepared with mucoadhesive or bioadhesive polymers will likely prolong precorneal residence time of the drug in the tear film and help increase drug uptake into the cornea and conjunctiva following topical administration.

Following intravitreal administration, the nanoparticles can penetrate the retinal layers and accumulate in various retinal cells, especially the RPE cells. In addition, intravitreally administered small nanoparticles are less likely to interfere with vision compared to microparticles. There is also a potential for the use of nanoparticles for transscleral drug, gene, and protein delivery via the periocular routes of administration. Although nanoparticles can sustain drug delivery by various routes, microparticles are in general better for sustaining drug delivery because of their low surface:volume ratio. However, selection of a particular dosage form should be based on several factors including the disease condition, anatomical target, drug properties, and patient compliance.

In addition to the aforementioned advantages and delivery routes, nanoparticles can also be utilized for systemic delivery to the retina. This would be beneficial for approaches such as photodynamic therapy for the treatment of neovascular complications like diabetic retinopathy and age related macular degeneration (77–81). The systemic half-life of these particles can be increased by using PEGylated nanoparticles, which can reduce particle clearance by the reticuloendothelial system (82).

Thus, nanoparticles can be administered by topical, intravitreal, periocular, or systemic routes for the therapy of ocular complications. Their value can be further enhanced by understanding the influence of nanoparticle properties on in vitro and in vivo drug disposition. To better utilize nanoparticulate systems in the eye, approaches should be developed to better sustain drug delivery from nanoparticles and to reduce the aggregation of these particles in vitro and in vivo.

ACKNOWLEDGMENTS

This work was supported by NIH grants EY013842 and DK064172. The authors are thankful to the University of Nebraska Medical Center for a graduate student fellowship award to Aniruddha Amrite.

REFERENCES

1. Machaand SA, Mitra K. Overview of ocular drug delivery. In: Mitra AK, ed. Ophthalmic drug delivery systems. Vol. 130. New York: Marcel-Dekker, 2003:1–12.

2. Lee VHL. Precorneal, corneal, and postcoreneal factors. In: Mitra AK, ed. Ophthalmic drug delivery systems. Vol. 58. New York: Marcel-Dekker, 2003:59–81.

3. Kompellaand UB, Lee VHL. Barriers to drug transport in ocular epithelia. In: Lee VHL, ed. Transport processes in pharmaceutical systems. New York: Marcel Dekker, 1999:317–376.

4. Lingand TL, Combs DL. Ocular bioavailability and tissue distribution of [14C]ketorolac tromethamine in rabbits. J Pharm Sci 1987; 76:289–294.

5. Miller MH, Madu A, Samathanam G, Rush D, Madu CN, Mathisson K, Mayers M. Fleroxacin pharmacokinetics in aqueous and vitreous humors determined by using complete concentration-time data from individual rabbits. Antimicrob Agents Chemother 1992; 36:32–38.

6. Barbault-Foucher S, Gref R, Russo P, Guechot J, Bochot A. Design of poly-epsilon-caprolactone nanospheres coated with bioadhesive hyaluronic acid for ocular delivery. J Controlled Release 2002; 83:365–375.

7. Bucolo C, Maltese A, Puglisi G, Pignatello R. Enhanced ocular anti-inflammatory activity of ibuprofen carried by an Eudragit RS100 nanoparticle suspension. Ophthal Res 2004; 34:319–323.

8. Bucolo C, Maltese A, Maugeri F, Busa B, Puglisi G, Pignatello R. Eudragit RL100 nanoparticle system for the ophthalmic delivery of cloricromene. J Pharm Pharmacol 2004; 56: 841–846.

9. Calvo P, Vila-Jato JL, Alonso MJ. Comparative in vitro evaluation of several colloidal systems, nanoparticles, nanocapsules, and nanoemulsions as ocular drug carriers. J Pharm Sci 1996; 85:530–536.

10. Cavalli R, Gasco MR, Chetoni P, Burgalassi S, Saettone MF. Solid lipid nanoparticles (SLN) as ocular delivery system for tobramycin. Int J Pharm 2002; 238:241–245.

11. Chu TC, He Q, Potter DE. Biodegradable calcium phosphate nanoparticles as a new vehicle for delivery of a potential ocular hypotensive agent. J Ocul Pharmacol Ther 2002; 18:507–514.

12. de Campos AM, Diebold Y, Carvalho EL, Sanchez A, Alonso MJ. Chitosan nanoparticles as new ocular drug delivery systems: in vitro stability, in vivo fate, and cellular toxicity. Pharm Res 2004; 21:803–810.

13. De Campos AM, Sanchez A, Alonso MJ. Chitosan nanoparticles: a new vehicle for the improvement of the delivery of drugs to the ocular surface. Application to cyclosporin A. Int J Pharm 2001; 224:159–168.

14. Fitzgerald P, Hadgraft J, Kreuter J, Wilson CG. A scintigraphic evaluation of microparticulate ophthalmic delivery systems: liposomes and nanoparticles. Int J Pharm 1987; 40:81–84.

15. Giannavola C, Bucolo C, Maltese A, et al. Influence of preparation conditions on acyclovir-loaded poly-D,L-lactic acid nanospheres and effect of PEG coating on ocular drug bioavailability. Pharm Res 2003; 20:584–590.

16. Langer K, Mutschler E, Lambrecht G, et al. Methylmethacrylate sulfopropylmethacrylate copolymer nanoparticles for drug delivery. Part III: evaluation as drug delivery system for ophthalmic applications. Int J Pharm 1997; 158:219–231.

17. Li VH, Wood RW, Kreuter J, Harmia T, Robinson JR. Ocular drug delivery of progesterone using nanoparticles. J Microencapsul 1986; 3:213–218.

18. Losa C, Calvo P, Castro E, Vila-Jato JL, Alonso MJ. Improvement of ocular penetration of amikacin sulphate by association to poly(butylcyanoacrylate) nanoparticles. J Pharm Pharmacol 1991; 43:548–552.

19. Marchal-Heussler L, Sirbat D, Hoffman M, Maincent P. Poly(-epsilon-caprolactone) nanocapsules in carteolol ophthalmic delivery. Pharm Res 1993; 10:386–390.

20. Pignatello R, Bucolo C, Puglisi G. Ocular tolerability of Eudragit RS100 and RL100 nanosuspensions as carriers for ophthalmic controlled drug delivery. J Pharm Sci 2002; 91:2636–2641.

21. Pignatello R, Bucolo C, Ferrara P, Maltese A, Puleo A, Puglisi G. Eudragit RS100 nanosuspensions for the ophthalmic controlled delivery of ibuprofen. Eur J Pharm Sci 2002; 16:53–61.

22. Pignatello R, Bucolo C, Spedalieri G, Maltese A, Puglisi G. Flurbiprofen-loaded acrylate polymer nanosuspensions for ophthalmic application. Biomaterials 2002; 23:3247–3255.

23. Salgueiro A, Egea MA, Espina M, Valls O, Garcia ML. Stability and ocular tolerance of cyclophosphamide-loaded nanospheres. J Microencapsul 2004; 21:213–223.

24. Wood RW, Li VH, Kreuter J, Robinson JR. Ocular disposition of poly-hexyl-2-cyano[3-14C]acrylate nanoparticles in the albino rabbit. Int J Pharm 1985; 23:175–183.

25. Zimmer A, Kreuter J, Robinson JR. Studies on the transport pathway of PBCA nanoparticles in ocular tissues. J Microencapsul 1991; 8:497–504.

26. Zimmer A, Mutschler E, Lambrecht G, Mayer D, Kreuter J. Pharmacokinetic and pharmacodynamic aspects of an ophthalmic pilocarpine nanoparticle-delivery-system. Pharm Res 1994; 11:1435–1442.

27. Zimmer A, Maincent P, Thouvenot P, Kreuter J. Hydrocortisone delivery to healthy and inflamed eyes using a micellar polysorbate 80 solution or albumin nanoparticles. Int J Pharm 1994; 110:211–222.

28. Qaddoumi MG, Ueda H, Yang J, Davda J, Labhasetwar V, Lee VH. The characteristics and mechanisms of uptake of PLGA nanoparticles in rabbit conjunctival epithelial cell layers. Pharm Res 2004; 21:641–648.

29. Qaddoumi MG, Gukasyan HJ, Davda J, Labhasetwar V, Kim KJ, Lee VH. Clathrin and caveolin-1 expression in primary pigmented rabbit conjunctival epithelial cells: role in PLGA nanoparticle endocytosis. Mol Vis 2003; 9:559–568.

30. Hansen SH, Sandvig K, van Deurs B. Molecules internalized by clathrin-independent endocytosis are delivered to endosomes containing transferrin receptors. J Cell Biol 1993; 123:89–97.

31. Diepold R, Kreuter J, Himber J, et al. Comparison of different models for the testing of pilocarpine eyedrops using conventional eyedrops and a novel depot formulation (nanoparticles). Graefes Arch Clin Exp Ophthalmol 1989; 227:188–193.

32. Merodio M, Irache JM, Valamanesh F, Mirshahi M. Ocular disposition and tolerance of ganciclovir-loaded albumin nanoparticles after intravitreal injection in rats. Biomaterials 2002; 23:1587–1594.

33. Bourges JL, Gautier SE, Delie F, et al. Ocular drug delivery targeting the retina and retinal pigment epithelium using polylactide nanoparticles. Invest Ophthalmol Vis Sci 2003; 44:3562–3569.

34. De Kozak Y, Andrieux K, Villarroya H, et al. Intraocular injection of tamoxifen-loaded nanoparticles: a new treatment of experimental autoimmune uveoretinitis. Eur J Immunol 2004; 34:3702–3712.

35. Merodio M, Espuelas MS, Mirshahi M, Arnedo A, Irache JM. Efficacy of ganciclovir-loaded nanoparticles in human cytomegalovirus (HCMV)-infected cells. J Drug Target 2002; 10:231–238.

36. Sakurai E, Ozeki H, Kunou N, Ogura Y. Effect of particle size of polymeric nanospheres on intravitreal kinetics. Ophthal Res 2001; 33:31–36.

37. Moritera T, Ogura Y, Honda Y, Wada R, Hyon SH, Ikada Y. Intravitreal drug delivery by microspheres of biodegradable polymers. Nippon Ganka Gakkai Zasshi 1990; 94:508–513.

38. Kimura H, Ogura Y, Moritera T, Honda Y, Tabata Y, Ikada Y. In vitro phagocytosis of polylactide microspheres by retinal pigment epithelial cells and intracellular drug release. Curr Eye Res 1994; 13:353–360.

39. Aukunuruand JVU, Kompella B. In vitro delivery of nano- and micro-particles to human retinal pigment epithelial (ARPE-19) cells. Drug Deliv Technol 2002; 2:50–57.

40. Aukunuru JV, Ayalasomayajula SP, Kompella UB. Nanoparticle formulation enhances the delivery and activity of a vascular endothelial growth factor antisense oligonucleotide in human retinal pigment epithelial cells. J Pharm Pharmacol 2003; 55:1199–1206.

41. Bejjani RA, BenEzra D, Cohen H, et al. Nanoparticles for gene delivery to retinal pigment epithelial cells. Mol Vis 2005; 11:124–132.

42. Band H, Bhattacharya A, Talwar GP. Mechanism of phagocytosis by Schwann cells. J Neurol Sci 1986; 75:113–119.

43. Silverstein SC, Steinman RM, Cohn ZA. Endocytosis. Annu Rev Biochem 1977; 46:669–722.

44. Robinson BV. The pharmacology of phagocytosis. Rheumatol Rehabil Suppl 1978; 37–46.

45. Raghava S, Hammond M, Kompella UB. Periocular routes for retinal drug delivery. Expert Opin Drug Deliv 2004; 1:99–114.

46. Kimura H, Ogura Y, Moritera T, et al. Injectable microspheres with controlled drug release for glaucoma filtering surgery. Invest Ophthalmol Vis Sci 1992; 33:3436–3441.

47. Giordano GG, Refojo MF, Arroyo MH. Sustained delivery of retinoic acid from microspheres of biodegradable polymer in PVR. Invest Ophthalmol Vis Sci 1993; 34:2743–2751.

48. Kompella UB, Bandi N, Ayalasomayajula SP. Subconjunctival nano- and microparticles sustain retinal delivery of budesonide, a corticosteroid capable of inhibiting VEGF expression. Invest Ophthalmol Vis Sci 2003; 44:1192–1201.

49. Amrite AC, Kompella B. Microparticles but not nanoparticles are retained in the subconjunctival space at 2-months post-administration. Vol. 45. Association for Research in Vision and Ophthalmology: FL, 2004:E-Abstract 5067.

50. Amrite AC, Ayalasomayajula SP, Kompella UB. Ocular distribution of intact nano- & microparticles following subconjunctival & systemic routes of administration. Drug Deliv Technol 2003; 3:52–57.

51. Amrite AC, Ayalasomayajula SP, Kompella UB. Nano- and Micro-Particles Reach Retina Following Systemic but Not Subconjunctival Administration. Association for Research in Vision and Ophthalmology, Vol. 44, Ft. Lauderdale, FL, 2003, pp. E-Abstract 4449.

52. Saishin Y, Silva RL, Callahan K, et al. Periocular injection of microspheres containing PKC412 inhibits choroidal neovascularization in a porcine model. Invest Ophthalmol Vis Sci 2003; 44:4989–4993.

53. Gurny R. Preliminary study of prolonged acting drug delivery systems for the treatment of glaucoma. Pharm Acta Helv 1981; 56:130–132.

54. Diepold R, Kreuter J, Guggenbuhl P, Robinson JR. Distribution of poly-hexyl-2-cyano-[3-14C]acrylate nanoparticles in healthy and chronically inflamed rabbit eyes. Int J Pharm 1989; 54:149–153.

55. Alonsoand MJ, Sanchez A. The potential of chitosan in ocular drug delivery. J Pharm Pharmacol 2003; 55:1451–1463.

56. Calvo P, Alonso MJ, Vila-Jato JL, Robinson JR. Improved ocular bioavailability of indomethacin by novel ocular drug carriers. J Pharm Pharmacol 1996; 48:1147–1152.

57. Fresta M, Fontana G, Bucolo C, Cavallaro G, Giammona G, Puglisi G. Ocular tolerability and in vivo bioavailability of poly(-ethylene glycol) (PEG)-coated polyethyl-2-cyanoacrylate nanosphere-encapsulated acyclovir. J Pharm Sci 2001; 90:288–297.

58. De Campos AM, Sanchez A, Gref R, Calvo P, Alonso MJ. The effect of a PEG versus a chitosan coating on the interaction of drug colloidal carriers with the ocular mucosa. Eur J Pharm Sci 2003; 20:73–81.

59. Gulsen D, Chauhan A. Dispersion of microemulsion drops in HEMA hydrogel: a potential ophthalmic drug delivery vehicle. Int J Pharm 2005; 292:95–117.

60. El-Samaligy MS, Rojanasakul Y, Charlton JF, Weinstein GW, Lim JK. Ocular disposition of nanoencapsulated acyclovir and ganciclovir via intravitreal injection in rabbit's eye. Drug Deliv 1996; 3:93–97.

61. Marchal-Heussler L, Maincent P, Hoffman M, Sirbat D. Value of the new drug carriers in ophthalmology: liposomes and nanoparticles. J Fr Ophtalmol 1990; 13:575–582.

62. Marchal-Heussler L, Fessi H, Devissaguet JP, Hoffman M, Manicent P. Colloidal drug delivery systems for the eye: a comparison of efficacy of three different polymers: polyisobutylcyanoacrylate, polylactic-co-glycolic acid, polyepsilon–caprolacton. STP Pharma Sci 1992; 2:98–104.

63. Losa C, Marchal-Heussler L, Orallo F, Vila Jato JL, Alonso MJ. Design of new formulations for topical ocular administration: polymeric nanocapsules containing metipranolol. Pharm Res 1993; 10:80–87.

64. Marchal-Heussler L, Maincent P, Hoffman M, Spittler J, Couvreur P. Antiglaucomatous activity of betaxolol chlorhydrate sorbed onto different isobutylcyanoacrylate nanoparticle preparations. Int J Pharm 1990; 58:115–122.

65. Tham R, Bunnfors I, Eriksson B, Larsby B, Lindgren S, Odkvist LM. Vestibulo-ocular disturbances in rats exposed to organic solvents. Acta Pharmacol Toxicol (Copenh) 1984; 54:58–63.

66. Hempel-Jorgensen A, Kjaergaard SK, Molhave L. Cytological changes and conjunctival hyperemia in relation to sensory eye irritation. Int Arch Occup Environ Health 1998; 71:225–235.

67. Zimmerand A, Kreuter J. Microspheres and nanoparticles used in ocular delivery systems. Adv Drug Deliv Rev 1995; 16:61–73.

68. Couvreur P, Kante B, Grislain L, Roland M, Speiser P. Toxicity of polyalkylcyanoacrylate nanoparticles II: doxorubicin-loaded nanoparticles. J Pharm Sci 1982; 71:790–792.

69. Couvreur P. Polyalkylcyanoacrylates as colloidal drug carriers. Crit Rev Ther Drug Carrier Syst 1988; 5:1–20.

70. Lherm C, Muller RM, Puisieux F, Couvreur P. Alkylcyanoacrylate drug carriers: II. Cytotoxicity of cyanoacrylate nanoparticles with different alkylchain length. Int J Pharm 1991; 84:13–22.

71. Algvereand P, Martini B. Sequelae of intravitreal phagocytic activity in response to microparticles. Acta Ophthalmol Suppl 1985; 173:107–110.

72. Algvereand P, Martini B. Experimental intravitreal proliferation and neovascularization in the cynomolgus monkey. Graefes Arch Clin Exp Ophthalmol 1986; 224:69–75.

73. Algvere P, Wallow IH, Martini B. The development of vitreous membranes and retinal detachment induced by intravitreal carbon microparticles. Graefes Arch Clin Exp Ophthalmol 1988; 226:471–478.

74. Martini B. Proliferative vitreo-retinal disorders: experimental models in vivo and in vitro. Acta Ophthalmol Suppl 1992; 201: 1–63.

75. Giordano GG, Chevez-Barrios P, Refojo MF, Garcia CA. Biodegradation and tissue reaction to intravitreous biodegradable poly(D,L-lactic-*co*-glycolic)acid microspheres. Curr Eye Res 1995; 14:761–768.

76. Maurice D. Review: practical issues in intravitreal drug delivery. J Ocul Pharmacol Ther 2001; 17:393–401.

77. Oliveira CA, Machado AE, Pessine FB. Preparation of 100 nm diameter unilamellar vesicles containing zinc phthalocyanine and cholesterol for use in photodynamic therapy. Chem Phys Lipids 2005; 133:69–78.

78. Ladd BS, Solomon SD, Bressler NM, Bressler SB. Photodynamic therapy with verteporfin for choroidal neovascularization in patients with diabetic retinopathy. Am J Ophthalmol 2001; 132:659–667.

79. Hooperand CYR, Guymer H. New treatments in age-related macular degeneration. Clin Exp Ophthalmol 2003; 31:376–391.

80. Wormald R, Evans J, Smeeth L, Henshaw K. Photodynamic therapy for neovascular age-related macular degeneration. Cochrane Database Syst Rev 2003; 2:CD002030.

81. Vargas A, Pegaz B, Debefve E, et al. Improved photodynamic activity of porphyrin loaded into nanoparticles: an in vivo evaluation using chick embryos. Int J Pharm 2004; 286:131–145.

82. Otsuka H, Nagasaki Y, Kataoka K. PEGylated nanoparticles for biological and pharmaceutical applications. Adv Drug Deliv Rev 2003; 55:403–419.

12

DNA Nanoparticle Gene Delivery Systems

MOSES O. OYEWUMI and KEVIN G. RICE

Division of Medicinal and Natural Products
Chemistry, College of Pharmacy, University of Iowa,
Iowa City, Iowa, U.S.A.

GENE DELIVERY VECTORS

Gene therapy involves the introduction of DNA or ribonucleic acid (RNA) into target cells to either express or suppress the biosynthesis of proteins (1,2). The ability to manipulate protein expression in humans could provide a cure or treatment for many diseases that are currently untreatable by conventional drug therapy. The potential therapeutic utility could be greatest for inherited diseases, such as cystic fibrosis and hemophilia, since the genetic basis of these diseases is well known. In principle, replacement of a single defective gene in the affected cells could permanently halt the symptoms. Gene therapy may also have a unique impact on certain acquired diseases such as cancer and AIDS by virtue of the ability to use the genetics of the affected cell to mediate its own destruction (3–6). The therapeutic success of gene therapy is largely dependent on the development of efficient delivery systems for DNA.

A great deal of attention has been placed on viral gene delivery vectors. Retrovirus, adenovirus, and adeno-associated virus are nanoparticulate gene transfer agents capable of mediating high levels of gene expression (7–9). However, to be suitable as gene delivery agents for use in humans, a number of host immune responses must be overcome (4,10). Nonviral delivery systems have been increasingly proposed as alternatives to viral vectors because of potential advantages since they are amenable to synthetic manipulations, cell/tissue targeting, low immune response, and unrestricted plasmid size.

Nonviral gene delivery systems are typically composed of plasmid DNA condensed into nanoparticles by a cationic polymer (11). As such, they are incapable of replication in the host and because their chemical composition is known, they can be designed to minimize host immune responses. However, unlike viral gene delivery vectors, nonviral gene delivery systems mediate moderate to high gene expression levels in cell culture, but often fail to produce significant levels of gene expression in vivo (12).

This difference in gene transfer efficiency between viral and nonviral gene delivery systems is most likely the result of numerous complementary mechanisms that the virus has evolved over millions of years to maximize transfection of the host. These mechanisms include the ability to circulate in the blood, bind to cell surface receptors, gain entry into the cell, avoid lysosomal destruction, survive degradation in the cytosol, and deliver genetic material to the nucleus. In contrast, most nonviral gene delivery systems depend on the properties of a single polymer, selected primarily for its ability to mediate gene transfer in rapidly dividing cells in culture, with little regard to overcoming biological barriers in the circulation or inside the target cell, which normally are quiescent.

The mechanism of nonviral gene transfer in cell cultures is primarily pinocytosis facilitated by electrostatic or hydrophobic interactions between the gene vector and the cell surface. Not surprisingly, many gene transfer nanoparticles are electropositively charged and thereby bind ionically to the electronegative surface of the cells composed of proteoglycans or sialylated glycoproteins. Several studies have cited a correlation between the size of nanoparticles and their ability to transfect cells in culture; however, this relationship is not clear-cut (13,14). Logically, the smaller nanoparticles of less than 100 nm would be able to enter cells more easily through pinocytosis; however, on occasion larger particles of 200–300 nm are found to be as efficient or more efficient in gene transfer in cell cultures (15). This is partly the influence of sedimentation that occurs during in vitro gene transfer (14,16).

Depending on the in vitro gene transfer protocol, gene delivery nanoparticles may be allowed to sediment for anywhere from 4 to 48 hours onto cells prior to determining gene transfer efficiency. Likewise, most in vitro gene transfer protocols are performed in the presence of complex buffers containing serum proteins that bind to polycationic DNA nanoparticles that alter their size during the course of transfection.

Despite this ambiguity, the need to maintain a small particle size to mediate gene targeting in vivo is not under dispute. The physiological barriers that block the extravasation of liposome particles greater than 200 nm in diameter are the same that block the targeting of gene delivery particles.

The following summarizes some of the fundamentals involved in preparing and using DNA nanoparticles to mediate gene transfer. A comparison of DNA nanoparticles prepared using different polymers reveals a remarkable similarity in their physical properties. The major problems confronting the use of nanoparticles for gene delivery will be discussed with a focus on approaches to optimize gene transfer nanoparticles for use in vivo.

POLYMERS USED TO PREPARE DNA NANOPARTICLES

Polyethylenimine (PEI) has the highest cationic charge density of any macromolecule and is thus effective in DNA condensation (16–18). Every third atom on the backbone of the polymer is a nitrogen atom. All of the nitrogen atoms on linear PEI are protonatable, but in branched PEI, only two-thirds of nitrogen atoms can be charged (19,20). PEI has a buffering capacity at pH 4–6, and as such, it possesses the ability to destabilize lysosomal membranes and facilitate endosomal escape of gene transfection agents (Fig. 1) (20).

Low molecular weight (10 kDa) PEI results in efficient gene transfer with lower toxicity in comparison to high molecular weight PEI; however, direct intravenous dosing of linear PEI (22 kDa) can be lethal (21,22).

Polyamidoamine dendrimers are a class of polymers in which an amine starting material is repeatedly substituted at its amino termini to provide a branched structure. This class of highly branched spherical polymers has unique surface topology of primary amino groups resulting in high positive charge densities with low cytotoxicity (17,23).

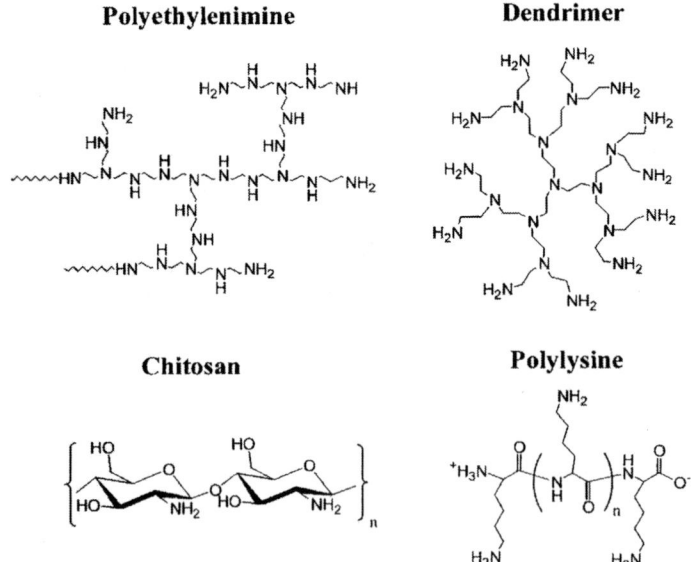

Figure 1 Chemical structure of DNA condensing polymers. The chemical structures of polyethylenimine, dedrimers, chitosan, and polylysine are compared.

Polylysine was one of the first polycation polymers to be employed for gene delivery (24). Polylysine is a biodegradable linear polypeptide of varying length of 20–1000 amino acids. The level of gene transfer mediated by polylysine DNA is significantly boosted when combined with a lysomotropic agent like chloroquine (25). The transfection efficiency of polylysine DNA condensates increases with increasing molecular weight but the associated toxicity also increases (26–29). Consequently, low molecular weight peptide carriers offer the advantage of lower toxicity and defined chemical structure and purity (30,31). Fully functionalized peptides have been developed that possess targeting ligands and polyethylene glycol (PEG) to produce DNA nanoparticles that are stable in the blood and target to hepatocytes and Kupffer cells in vivo (32,33).

Chitosan is a polysaccharide copolymer of *N*-acetyl-D-glucosamine and D-glucosamine obtained by partial alkaline deacetylation of chitin. Chitosan effectively condenses DNA and protects it from nuclease degradation (17,34). It has the advantage of being a nontoxic cationic polymer with low immunogenicity (34).

Cationic lipids have been used extensively as gene delivery vectors (35). The head group on cationic lipids is protonated at physiological pH affording binding to DNA (12,36,37). Some examples of the most prevalent cationic lipids used in gene transfer are dioleoylpropyl trimethylammonium chloride (DOTMA), dioleoyl trimethylammonium propane (DOTAP), or dimethylaminoethane carbamoyl cholesterol (DC-Chol).

PHYSICAL PROPERTIES OF DNA NANOPARTICLES

When complexed with a polycation, plasmid DNA undergoes a conformational change from a hydrodynamic size of 200–300 nm to particles of less than 100 nm (Fig. 2). Thus, condensed DNA occupies only 10^{-3}–10^{-4} of the volume of plasmid DNA (11,38). A focus on the conformation of DNA

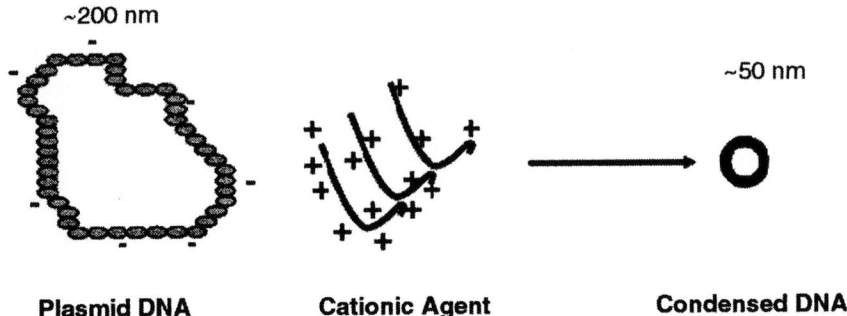

~200 nm

~50 nm

Plasmid DNA   **Cationic Agent**   **Condensed DNA**

Figure 2 DNA condensation process. A schematic illustration of the condensation of DNA resulting from the electrostatic interaction between plasmid DNA (polyanionic; approximate size of 200 nm) and a cationic agent (polymer). Electropositive DNA particles are obtained when cationic agents are used in excess of a charge ratio of 1:1.

may be insightful toward understanding DNA condensation. Plasmid DNA has a highly organized chemical structure. The volume occupied by a random DNA coil is dependent on its molecular weight, size, and persistence length. The flexibility of DNA is characterized by the persistent length and the distance between its ends. The persistent length of DNA has two components: the electrostatic contributions due to the repulsion between two strands and the intrinsic stiffness of the uncharged helix. Mechanistic investigations have concluded that polycationic polymers cause DNA condensation through a number of events such as localized bending or distortion of DNA and decreasing the net charge on DNA coupled with decreasing the unfavorable DNA segment–segment interactions (39–41).

Measurement of submicrometer particles can be carried out by photon correlation spectroscopy (PCS) or dynamic light scattering (DLS). PCS determines the hydrodynamic diameter of nanoparticles via Brownian motion. As such, accurate size determination is dependent on certain properties of the liquid medium such as absence of dust contaminants, viscosity, refractive index of particles, and temperature. Other methods of determining the sizes and morphology of nanoparticles include: transmission electron microscopy, scanning electron microscopy, and atomic force spectroscopy (42,43). The electrophoretic mobility based on the zeta potential measurements will provide information on nanoparticle surface charge.

The most important variables to control when forming DNA condensates are the concentration of DNA, the charge ratio of DNA to cationic polymer, and the buffer (44). The concentration of DNA is kept relatively dilute (20–100 µg/mL) since the particle size increases as the concentration of DNA condensate increases (29). DNA condensates tend to remain small (<100 nm in diameter) below a concentration of 100 µg/mL but then increase sharply and become visible flocculates at a concentration of >200–300 µg/mL (45). This phenomenon occurs irrespective of the polymer used or charge ratio, and is one of the most difficult problems confronting the use of DNA condensates. Typically, DNA condensates are prepared by mixing plasmid DNA with a cationic polymer in which the order of mixing and vortex speed of mixing play more subtle roles in influencing the size of DNA nanoparticles. After their formation, dilution of DNA nanoparticles will tend not to decrease their particle size (45,46). However, attempts to concentrate DNA condensates, either by evaporation under vacuum or by freeze drying will result in a dramatic increase in particle size. The simple act of freeze/thaw can also influence the particle size of DNA nanoparticles (47).

The charge ratio of DNA nanoparticles is the calculated ratio of amines on the polymer relative to the phosphates on DNA at a given stoichiometry of polymer to DNA. One generally assumes that all amines (1', 2', and 3') are equally protonated and carry a single positive charge and that each phosphodiester carries a negative charge. The calculated charge ratio is a simple way to compare the stoichiometry of different polymers binding to plasmid DNA, especially since most of the polymers under study are heterogenous, such that their molecular weight cannot be accurately known.

When a cationic polymer binds to plasmid DNA, sodium ions are displaced and the electronegative charge is partially satisfied. At a charge ratio of approximately 1:1, DNA nanoparticles will sharply grow in particle size and exhibit neutral zeta potential. Adding polymers in a stepwise titration in excess of a charge ratio of 1:1 results in a decrease in the DNA particle size to less than 100 nm and a conversion to electropositive particles, generally at a charge ratio (N:P) of 2:1 (48). The magnitude of the electropositive zeta potential is somewhat dependent on the ionic strength and pH of the buffer; however, at pH 7 in 5 mM N-2-hydroxyethylpiperazine-N'-2-ethanesulfonic acid (HEPES) the zeta potential of fully condensed DNA is about +30 mV (48). Further titration of polymer into DNA condensates neither decreases particle size nor increases zeta potential, suggesting that all of the accessible phosphate groups have been fully titrated with cationic polymer and that residual polymer remains unbound to the nanoparticles in solution.

DNA condensates are normally prepared at near-neutral pH in low ionic strength buffer such as HEPES, avoiding the use of sodium phosphate. This is based on the observation that even dilute DNA particles have a great propensity to flocculate in sodium containing buffers, especially in normal saline. In addition to the type of buffer used, the counterion of the cationic polymer will have an influence over the size of DNA nanoparticles. Generally, a bromide or chloride counterion will be substituted with an acetate or trifluoroacetate to decrease the size or polydispersity of DNA nanoparticles.

The size of DNA nanoparticles can be determined by quasielastic light scattering (QELS) in the concentration range of 20–50 μg/mL of DNA. Deconvolution of the light scattering using a multimodel analysis generally leads to the identification of two populations of DNA condensates (45). The major population representing >90% of the mass of DNA possesses a diameter of <100 nm, whereas the minor population of approximately 5–10% represents larger particles of typically two to three times the diameter of that of the major population (45). Systematically increasing the concentration of DNA condensates leads to an increase in the minor population and a proportional decrease in the major population, eventually leading to the formation of very large particles.

The size of DNA nanoparticles will also be somewhat dependent on the number of plasmids in each particle. DNA condensates may include one or multiple plasmids in a single particle; however, the observed size of the particle is less influenced by the number of plasmids per particle compared to the aggregation of particles (26,49). Some DNA nanoparticles are reportedly very small, presumably arising from the condensation of a single plasmid per particle (50).

Using both electron microscopy and light scattering measurements, several studies have shown that varying the DNA lengths (400–50,000 bp) over a wide range has little influence on the mean particle size of DNA nanoparticles. Furthermore, these studies establish that particle size is independent of the DNA sequence (38,51–53). The independence of particle size using different sized DNA has also been observed with poly-L-lysine (PLL) DNA particles (54). Likewise, the mean particle size of condensed DNA was not significantly different using linear, supercoiled, and circular forms of DNA (Table 1) (55).

Although few studies have directly addressed the issue, it is also notable that the size and polydispersity of DNA nanoparticles is very dependent on the size of polymer used. For example, very short polylysine peptides (8–13 residues) are able to weakly bind and condense plasmid DNA, but the resultant particle size is > 300–3000 nm. Increasing the length of polylysine peptides to 18 or greater leads to maximal condensation of plasmid DNA to particles of <100 nm in diameter (56). Conversely, as discussed above, decreasing the size of plasmid DNA, even by sonication to form short oligonucleotides (dp 100 or less), still results in the formation of DNA condensates that are <100 nm.

Table 1 Particle Sizes and Shapes of DNA Nanoparticles

Condensing polymer	Shape of NPs	Size (nm)	References
Polylysine	Spherical	10–100	31,41
Peptide	Toroids, rods	20–100	29
PEI	Toroids	20–80	57
Chitosan	Spherical, toroids	40–300	34,58
Polyamidoamine dendrimer	Spherical	80–100	49

Abbreviation: PEI, polyethyleneimine.

The precise shape of DNA nanoparticles has been the focus of several studies. This is usually determined by electron microscopic analysis of immobilized DNA condensates. The type of polymer used and its counterion will most often have an influence on the degree to which DNA nanoparticles appear as spherical, toroids, or rods (Fig. 3).

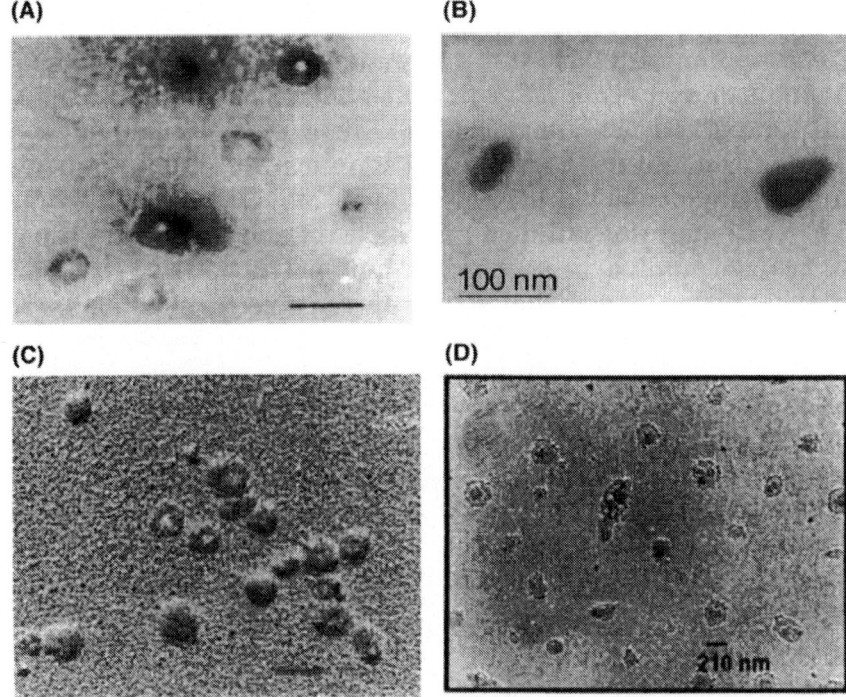

Figure 3 Electron microscopy of DNA nanoparticles. (**A**) TEM of PEI–DNA complexes (N/P = 10), bar = 100 nm. (**B**) PEG-peptide DNA condensates (5 mg/mL) following freeze drying and rehydration. TEM of 5 mg/mL in water. (**C**) Electron micrograph of dendrimer–DNA complex made using fractured dendrimer: DNA at a weight ratio of 4:1, bar = 100 nm. (**D**) Scanning electron micrograph of chitosan–DNA nanoparticles. *Abbreviations*: TEM, transmission electron micrograph; PEG, polyethylene glycol. *Source*: From Refs. 34, 45, 57, 59.

BIODISTRIBUTION AND TRAFFICKING OF DNA NANOPARTICLES

The rate and extent of clearance of gene vectors from systemic circulation will greatly impair the efficiency of gene delivery in vivo. Following intravenous (i.v.) dosing, DNA nanoparticles (unmodified) are rapidly cleared from circulation by the reticuloendothelial system (RES). The RES (mononuclear phagocyte system) are phagocytic cells that function as the body disposal mechanism for foreign particles and macromolecules. These cells are found throughout the body, either free in blood circulation or in fixed sites such as liver (Kupffer cells), spleen, and bone marrow. Therefore, the primary site of biodistribution is the liver and spleen. Within the liver, it is the Kupffer cells that are mostly responsible for capturing circulating DNA nanoparticles. The rate and extent of RES uptake may be influenced by particle size, hydrophobicity, and charge. With increasing particle sizes of greater than 200 nm in diameter, biodistribution to the spleen becomes appreciable and can even become the primary site of DNA nanoparticle biodistribution.

Colloidal instability in blood circulation can also result in the distribution of DNA nanoparticles mainly in the lung after i.v. dosing in mice (30,60). This is generally true for electropositive DNA particles prepared at a charge ratio 2:1 or higher in which the surface of the particle is not sterically stabilized, such as by the covalent linkage of PEG. Under this condition the positively charged particle will rapidly attract and bind electronegative protein in the serum, such as albumin, and quickly grow in particle size with physical entrapment of DNA particles in the capillary beds of the lung. Negatively or neutrally charged DNA nanoparticles are able to circumvent this effect; however, this requires the use of long cationic polymers to form condensates that are sufficiently stable in the circulation and avoid dissociation and metabolism of DNA (16).

When used in vivo, cationic lipid DNA complexes form aggregates due to binding to anionic serum proteins (61). This results in nonspecific biodistribution and entrapment in the lung. Cationic lipid DNA complexes are also weakly formed and thereby tend to dissociate at physiological concentrations of salt. This leads to the premature release of plasmid DNA, which is susceptible to metabolism by DNAse. These properties contribute to the well-known phenomenon of cationic lipid DNA complexes mediating maximal in vitro expression in serum-free media. Certain cationic lipids also have been shown to be toxic, especially at high concentrations due to protein kinase C inhibition (62,63).

Upon internalization into the target cells, it is important that the DNA be released from the endosomes to avoid transport to lysosomes, which is a major site of DNA metabolism. Achieving endosomal escape is one of the most difficult barriers for nanoparticulate gene delivery systems. PEI is believed to increase gene transfer by buffering the endosome, thereby causing osmotic lysis (64). Alternatively, fusogenic peptides undergo a conformational change at a lower pH allowing their binding and lysis of endosomal membranes (65). Upon release into the cytosol, DNA must be internalized in the nucleus to effect gene expression (66). Whether this process is an active or passive diffusion and involves uncoating of DNA are primary subjects of debate (12,66).

CONCLUSIONS

The continued interest in developing DNA particles as nonviral gene delivery agents for use in humans has driven scientists to increasingly investigate polymers that mediate potent in vitro transfection even though there is no direct correlation with their in vivo efficacy. The only common criterion appears to be that DNA nanoparticles remain small, less than 100 nm in diameter. Although the cationic charge of DNA nanoparticles facilitates their cell surface binding and internalization in vitro, it is also responsible for nonspecific binding and aberrant biodistribution in vivo. The ease of preparation and physical properties of DNA nanoparticles, being sufficiently stable when prepared fresh, support their use as nonviral gene transfer agents that mediate moderate levels of reporter genes in cell culture. However, as drug delivery systems for i.v. dosing in animals, the inability to overcome a limited DNA concentration while maintaining a small particle size and the inability to prepare and store DNA particles are significant barriers to their use.

However, considering the options, many scientists still believe that the physical problems related to aggregation and the biological barriers that diminish transfection efficiency can, in time, be solved. If so, chemically well-defined DNA particles that mediate sufficient targeting and gene expression would have a significant role in molecular medicine. The next decade will be a critical period in pursuit of these goals.

REFERENCES

1. Durland RH, Eastman EM. Manufacturing and quality control of plasmid-based gene expression systems. Adv Drug Deliv Rev 1998; 30:33–48.

2. Labhasetwar V, Chen B, Muller DWM, et al. Gene-based therapies for restenosis. Adv Drug Deliv Rev 1997; 24:109–20.

3. Hashida M, Nishikawa M, Yamashita F, Takakura Y. Cell-specific delivery of genes with glycosylated carriers. Adv Drug Deliv Rev 2001; 52:187–196.

4. Mansouri S, Lavigne P, Corsi K, Benderdour M, Beaumont E, Fernandes JC. Chitosan-DNA nanoparticles as nonviral vectors in gene therapy: strategies to improve transfection efficacy. Eur J Pharm Biopharm 2004; 57:1–8.

5. Engelhardt JF, Yang Y, Statford-Perricaudet LD, et al. Direct gene transfer of human CFTR into human bronchial epithelia of xenografts with E1-deleted adenovirus. Nat Gene 1993; 4:27–34.

6. Cristiano RJ, Smith LC, Woo SL. Hepatic gene therapy: adenovirus enhancement of receptor-mediated gene expression in primary hepatocytes. Proc Natl Acad Sci USA 1993; 90:2122–2166.

7. Lewis PF, Emerman M. Passage through mitosis is required for oncoretroviruses but not the human immunodeficiency virus. J Virol 1994; 68:510–16.

8. Oligino TJ, Yao Q, Ghivizzani SC, Robbins P. Vector systems for gene transfer to joints. Clin Orthop 2000; 379:S17–S30.

9. El-Aneed A. An overview of current delivery systems in cancer gene therapy. J Controlled Release 2004; 94:1–14.

10. Yang Y, Li Q, Erte HCJ, Wilson JM. Cellular and humoral immune responses to viral antigens create barriers to lung-directed gene therapy with recombinant adenoviruses. J Virol 1995; 69:2004–2015.

11. De Smedt SC, Demeester J, Hennink WE. Cationic polymer based gene delivery systems. Pharm Res 2000; 17:113–126.

12. Brown MD, Schatzlein AG, Uchegbu IF. Gene delivery with synthetic (non viral) carriers. Int J Pharm 2001; 229:1–21.

13. Wagner E, Cotten M, Mechtler K, Kirlappos H, Birnstiel, ML. DNA-binding transferrin conjugates as functional gene-delivery agents: synthesis by linkage of polylysine or ethidium homodimer to the transferrin carbohydrate moiety. Bioconjug Chem 1991; 2:226–231.

14. Ogris M, Steinlein P, Kursa M, Mechtler K, Kircheis R, Wagner E. The size of DNA/transferrin-PEI complexes is an important factor for gene expression in cultured cells. Gene Ther 1998; 5:1425–1433.

15. Kim JS, Maruyama A, Akaike T, Kim, SW. In vitro gene expression on smooth muscle cells using a terplex delivery system. J. Controlled Release 1997; 47:51–59.

16. Boussif O, Lezoualc'h F, Zanta MA, et al. A versatile vector for gene and oligonucleotide transfer into cells in culture and in vivo: polyethylenimine. Proc Nat Acad Sci USA 1995; 92:7297–7301.

17. Garnett MC. Gene-delivery systems using cationic polymers. In: Critical reviews in therapeutic drug carrier systems. Begell House, Inc., 1999:147–207.

18. Tiyaboonchai W, Woiszwillo J, Middaugh CR. Formulation and characterization of DNA-polyethylenimine-dextran sulfate nanoparticles. Eur J Pharm Sci 2003; 19:191–202.

19. Klemm AR, Young D, Lloyd JB. Effects of polyethylenimine on endocytosis and lysosome stability. Biochem Pharmacol 1998; 56:41–46.

20. Kichler A, Leborgne C, Coeytaux E, Danos O. Polyethylinimine-mediated gene delivery: a mechanistic study. J Gene Med 2001; 3:135–144.

21. Fischer D, Bieber T, Li Y, Elsasser HP, Kissel T. A novel non-viral vector for DNA delivery based on low molecular weight, branched polyethylenimine: effect of molecular weight on transfection efficiency and cytotoxicity. Pharm Res 1999; 16:1273–1279.

22. Chollet P, Favrot, MC, Hurbin A, Coll JL. Side-effects of a systemic injection of linear polyethylenimine-DNA complexes. J Gene Med 2002; 4:84–91.

23. Mamede M, Saga T, Ishimori T, et al. Hepatocyte targeting of 111In-labeled oligo-DNA with avidin or avidin-dendrimer complex. J Controlled Release 2004; 95:133–141.

24. Wu GY, Wu CH. Receptor-mediated in vitro gene transformation by a soluble DNA carrier system. J Biol Chem 1987; 262:4429–32 [published erratum appears in J Biol Chem 1988; 263(1):588].

25. Pouton CW, Lucas P, Thomas BJ, Uduehi, AN, Milroy DA, Moss SH. Polycation-DNA complexes for gene delivery: a comparison of the biopharmaceutical properties of cationic polypeptides and cationic lipids. J Controlled Release 1998; 53: 289–299.

26. Wolfert, MA, Seymour LW. Atomic force microscopic analysis of the influence of the molecular weight of poly(L)lysine on the size of polyelectrolyte complexes formed with DNA. Gene Ther 1996; 3:269–273.

27. Ledley F. Pharmaceutical approach to somatic gene therapy. Pharm Res 1996; 13:1595–1614.

28. McKenzie DL, Kwok KY, Rice KG. A potent new class of reductively activated peptide gene delivery agents. J Biol Chem 2000; 275:9970–9977.

29. Duguid JG, Li C, Shi M, et al. A physicochemical approach for predicting the effectiveness of peptide-based gene delivery systems for use in plasmid-based gene therapy. Biophys J 1998; 74:2802–2814.

30. Collard WT, Yang Y, Kwok KY, Park Y, Rice KG. Biodistribution, metabolism, and in vivo gene expression of low molecular weight glycopeptide polyethylene glycol peptide DNA co-condensates. J Pharm Sci 2000; 89:499–512.

31. Smith LC, Duguid J, Wadhwa MS, et al. Synthetic peptide-based DNA complexes for nonviral gene delivery. Adv Drug Del Rev 1998; 30:115–131.

32. Park Y, Kwok KY, Boukarim C, Rice KG. Synthesis of sulfhy-dryl crosslinking poly (ethylene glycol) peptides and glycopep-tides as carriers for gene delivery. Bioconjug Chem 2002; 13:232–239.

33. Kwok KY, Park Y, Yongsheng Y, McKenzie DL, Rice KG. In vivo gene transfer using sulfhydryl crosslinked PEG-peptide/glycopeptide DNA co-condensates. J Pharm Sci 2003; 92: 1174–1185.

34. Mao HQ, Roy K, Troung-Le VL, et al. Chitosan-DNA nanopar-ticles as gene carriers: synthesis, characterization and trans-fection efficiency. J Controlled Release 2001; 70:399–421.

35. Felgner PL, Ringold GM. Cationic liposome-mediated transfec-tion. Nature 1989; 337:387–388.

36. Remy JS, Sirlin C, Vierling P, Behr JP. Gene transfer with a series of lipophilic DNA-binding molecules. Bioconjug Chem 1994; 5:647–654.

37. Wheeler CJ, Felgner PL, Tsai YJ, et al. A novel cationic lipid greatly enhances plasmid DNA delivery and expression in mouse lung. Proc Nat Acad Sci USA 1996; 93:11454–11459.

38. Bloomfield, VA. DNA condensation. Curr Opin Struct Biol 1996; 6:334–341.

39. Kabanov AV, Kabanov VA. DNA complexes with polycations for the delivery of genetic material into cells. Bioconjug Chem 1995; 6:7–20.

40. Ledley FD. Pharmaceutical approach to somatic gene therapy. Pharm Res 1996; 13:1595–1614.

41. Vijayanatham V, Thomas T, Thomas TJ. DNA nanoparticles and development of DNA delivery vehicles for gene therapy. Biochemistry 2002; 41:14085–14094.

42. Luo D, Woodrow-Mumford K, Belcheva N, Saltzman WM. Controlled DNA delivery systems. Pharm Res 1999; 16:1300–1308.

43. Danielsen S, Varum KM, Stokke BT. Structural analysis of chitosan mediated DNA condensation by AFM: influence of chitosan molecular parameters. Biomacromolecules 2004; 5:928–936.

44. Kabanov AV, Kabanov VA. Interpolyelectrolyte and block ionomer complexes for gene delivery: physico-chemical aspects. Adv Drug Deliv Rev 1998; 30:49–60.

45. Kwok KY, Adami RC, Hester KC, Park Y, Thomas S, Rice KG. Strategies for maintaining the particle size of peptide DNA condensates following freeze-drying. Int J Pharm 2000; 203:81–88.

46. Anchordoquy TJ, Carpenter JF, Kroll DJ. Maintenance of transfection rates and physical characterization of lipid/DNA complexes after freeze-drying and rehydration. Arch Biochem Biophys 1997; 348:199–206.

47. Armstrong, TKC, Girouard LG, Anchordoquy TJ. Effects of PEGylation on preservation of cationic lipid/DNA complexes during freeze-thawing and lyophilization. J Pharm Sci 2002; 91:2549–2558.

48. Tomlinson E, Rolland AP. Controllable gene therapy. Pharmaceutics of non-viral gene delivery systems. J Controlled Release 1996; 39:357–372.

49. Tang MX, Szoka FC. The influence of polymer structure on the interactions of cationic polymers with DNA and morphology of the resulting complexes. Gene Ther 1997; 4:823–832.

50. Liu G, Li, D, Pasumarthy MK, et al. Nanoparticles of compacted DNA transfect postmitotic cells. J Biol Chem 2003; 278:32578–32586.

51. Bloomfield VA, He S, Li AH, Arscott PB. Light scattering studies on DNA condensation. Biochem Soc Trans 1991; 19:496.

52. Pouton CW. Nuclear import of polypeptides, polynucleotides and supramolecular complexes. Adv Drug Deliv Rev 1998; 34:51–64.

53. Park SY, Harriers D, Gelbart WM. Topological defects and optimum size of DNA condensates. Biophys J 1998; 75: 714–720.

54. Wagner E, Cotten M, Foisner R, Birnstiel ML. Transferrin-polycation-DNA complexes: the effect of polycations on the structure of the complex and DNA delivery to cells. Proc Natl Acad Sci USA 1991; 88:4255–4259.

55. Adami RC, Collard WT, Gupta, SA, Kwok KY, Bonadio J, Rice KG. Stability of peptide-condensed plasmid DNA formulations. J Pharm Sci 1998; 87:678–683.

56. Wadhwa MS, Collard WT, Adami RC, McKenzie DL, Rice, KG. Peptide-mediated gene delivery: influence of peptide structure on gene expression. Bioconjug Chem 1997; 8:81–88.

57. Bettinger T, Remy JS, Erbacher, P. Size reduction of galactosy-lated PEI/DNA complexes improves lectin-mediated gene transfer into hepatocytes. Bioconjug Chem 1999; 10:558–561.

58. MacLaughlin FC, Mumper RJ, Wang J, et al. Chitosan and depolymerized chitosan oligomers as condensing carriers for in vivo plasmid delivery. J Controlled Release 1998; 56: 259–272.

59. Tang MX, Redemann CT, Szoka FC Jr. In vitro gene delivery by degraded polyamidoamine dendrimers. Bioconjug Chem 1996; 7:703–714.

60. Li S, Rizzo MA, Bhattacharya S, Huang L. Characterization of cationic lipid-protamin-DNA (LPD) complexes for intravenous gene delivery. Gene Ther 1998; 5:930–937.

61. Liu F, Huang L. Development of non-viral vectors for systemic gene delivery. J Controlled Release 2002; 78:259–266.

62. Rando RR. Regulation of protein kinase C activity by lipids. FASEB J 1988; 2:2348–2355.

63. Zhao XB, Lee RJ. Tumor-selective targeted delivery of genes and antisense oligodeoxyribonucleotides via the folate receptor. Adv Drug Deliv Rev 2004; 56:1193–1204.

64. Wightman L, Kircheis R, Rossler V, et al. Different behavior of branched and linear polyethylenimine for gene delivery in vitro and in vivo. J Gene Med 2001; 3:362–372.

65. Plank C, Oberhauser B, Mechtler K, Koch C, Wagner E. The influence of endosome-disruptive peptides on gene transfer using synthetic. J Biol Chem 1994; 269:12918–12924.

66. Vacik J, Dean B, Zimmer W, Dean D. Cell-specific nuclear import of plasmid DNA. Gene Ther 1999; 6:1006–1014.

13

Nanotechnology and Nanoparticles: Clinical, Ethical, and Regulatory Issues

MAKENA HAMMOND

College of Pharmacy, University of Nebraska Medical Center, Omaha, Nebraska, and Virginia State University, Petersburg, Virginia, U.S.A.

UDAY B. KOMPELLA

Department of Pharmaceutical Sciences, College of Pharmacy, University of Nebraska Medical Center, Omaha, Nebraska, U.S.A.

INTRODUCTION

Unequivocally, nanotechnology is quickly becoming a vanguard with respect to drug delivery systems. This results from the fact that products of this technology such as nanoparticles can be used to treat a wide variety of challenging diseases including diabetes, thromboses, heart disease, neurodegenerative disorders, and cancer, for which therapeutic alternatives are limited. However, as with any new technology, the risks of nanoparticulate systems must be heavily researched to ensure that the advantages of therapeutic treatment far outweigh any possible side effects. The pharmaceutical manufacturers are primarily responsible for ensuring the safety and efficacy of nanoparticles for clinical use. The United States Food and Drug Administration (FDA) is the authority that ensures that the nanoparticle-based products meet the regulatory standards for approval. The purpose of this chapter is to briefly summarize the clinical progress to date with nanoparticles, the most tangible therapeutic systems of nanotechnology, and the ethical issues and regulatory challenges associated with the products of nanotechnology (1).

CLINICAL ASPECTS

Several nanoparticle technologies are currently in clinical trials and a few have progressed to clinical use. NanoCrystal™ technology from Elan Pharmaceuticals International, Ltd. is one breakthrough technology that is being licensed to pharmaceutical companies for specialized drug delivery systems. Currently, there are some FDA approved drug products employing this technology. Rapamune® (Wyeth-Ayerst Laboratories), an oral tablet dosage form containing nanoparticles of the immunosuppressant drug rapamycin, was approved by the U.S. FDA during the year 2000. Prior to the development of this product, rapamycin was only available as a solution dosage form, which required refrigeration storage and mixing with water or orange juice prior to administration. The tablet dosage form employing nanocrystals is a more convenient dosage form. Emend®, an antiemetic oral capsule dosage form of aprepitant used in conjunction with cancer therapy, also utilizes nanocrystal technology. Emend, developed by Merck & Co., Inc., as a new chemical entity, was approved by the U.S. FDA on March 26, 2003. Probably, the potential trendsetter of the approved nanoparticles products is Abraxane™, developed by American Biosciences, Inc. and American Pharmaceutical Partners, Inc. The FDA approved this product on January 7, 2005. This product contains placlitaxel albumin-bound particles and allows the delivery of a higher dose of paclitaxel over 30 minutes without steroid premedication and without the toxicities associated with solvents (cremophor) in the previously approved paclitaxel injection. Abraxane has been approved for treatment of breast cancer after failure of combination chemotherapy for metastatic disease or relapse within six months of adjuvant chemotherapy. A similar advance is with Rexin-G^{TM} (Epeius Biotechnology Corporation), a nanoparticle medicine for pancreatic cancer, which received orphan drug status on August 15, 2003, from the FDA for treating metastatic or nonresectable stage IV pancreatic cancer (2,3). Rexin-G comprises a pathotropic retroviral vector carrying a cytocidal gene construct. The pharmaceutical products discussed above are summarized in Table 1 .

One area of nanotechnology that holds tremendous promise is medical imaging for diagnostic purposes. An imaging approach based on ferumoxtran-10, comprising ultrasmall particles of iron oxide in the nanometer range, has been shown to enhance the sensitivity of MRI in predicting lymph node metastasis without losing the specificity in humans (4). Similar observations were also made in the diagnosis of lymph node metastases in urinary bladder cancer, breast cancer, and head and neck cancer (5–7).

Another emerging imaging approach is based on quantum dots, referred to as Q-dots, which emit photons when

Table 1 Examples of Pharmaceutical Products Based on Nanosize Materials

Brand name	Description	Advantage
Emend	Nanocrystallized aprepitant (antiemetic) in a capsule	Enhanced dissolution rate and bioavailability
Rapamune	Nanocrystallized rapamycin (immunosuppressant) in a tablet	Enhanced dissolution rate and bioavailability
Abraxane	Paclitaxel (anticancer drug) bound albumin particles	Enhanced dose tolerance and hence, effect. Elimination of solvent associated toxicity
Rexin-G	A retroviral vector carrying cytotoxic genes	Effective in pancreatic cancer treatment

stimulated by UV light. Currently, preclinical studies are exploring their potential application in the diagnosis of cancer. Investigators at Emory University employed Q-dots as a means to identify human prostate cancer cells growing in mice (8,9). The investigators encapsulated Q-dots in triblock copolymers and further functionalized these systems with tumor cell targeting ligands. Following injection of the Q-dots, the investigators were able to identify the location of cancer cells by the colors emitted by the Q-dots under a mercury lamp. Q-dots have also been used in tracking metastatic tumor cell extravasation and to tag biomarkers such as Her-2 (10,11). Currently, Evident Technologies commercially offers EviTag^{TM}, a synthetically coated Q-dot developed specifically for use in a biological setting for cell identification purposes. Polymer coated Q-dots are safer. Because of their small size, the Q-dots will likely be eliminated rapidly from the system, similar to small dendrimer nanoparticles (12). However, their long-term toxicity and biodegradation has yet to be assessed. If the safety of such systems is established, Q-dots could prove to be valuable tools in diagnosis and treatment.

Nanospectra Biosciences has created ~130 nm nanoshells, which consist of a silica core surrounded by gold that can be further coated with targeting ligands or polymers (13). Six hours following intravenous injection of nanoshells coated with polyethylene glycol in immunocompetent mice bearing subcutaneous colon carcinomas, tumors could be illuminated with a diode laser (808 nm, 4 W/cm^2, 3 min). The nanoshells used in this study exhibit tunable optical absorptivities, with strong absorption of near-infrared light. Following extravasation of nanoshells at tumor locations, the phototherapy protocol used results in the heating of nanoshells, and hence the surrounding tissue, leading to cell death. According to this principle, the above phototherapy with nanoshells resulted in tumor-free mice for > 90 days compared to untreated animals, which developed significant tumors within 6–19 days. This procedure has the projected use of targeting micrometastases, which are cancer cells that are too small for surgeons to locate and remove. Similar thermal therapy of tumors can also be accomplished with magnetic nanoparticles (14,15). Following the administration of biocompatible magnetic nanoparticles and the exposure of AC magnetic fields of 18 kA/m to prostate cancer bearing rat models, intratumoral temperatures as high as 70°C could be attained. This approach known as magnetic fluid hyperthermia resulted in tumor growth inhibition by about 50%.

In addition to locating and eliminating tumor cells, nanoparticles are being employed for the treatment of ischemic stroke. ImaRx Therapeutics recently completed a phase II study in April 2005 for SonoLysis™ in the treatment of acute ischemic stroke (16). SonoLysis uses an external ultrasound and nanobubbles or microbubbles designed to efficiently eliminate blood clots without using potentially harmful lytic drugs. The bubbles are injected intravenously and allowed to accumulate at the site of the clot. The ultrasound causes the bubbles to pulsate and eventually break apart, resulting in removal of the clot.

Nanoparticles are also of value in treating diabetes. Flamel Technologies, Inc. (www.flamel.com) recently released phase II data on Basulin®, a long-acting insulin, which utilizes a nano-particulate system named Medusa™. This system is based on self-assembling poly(amino acid) polymers containing leucine and glutamic acid. These carriers assemble the protein of interest into nanocarriers that are 20–50 nm in diameter. This technology is being extended by Flamel Technologies to other proteins including erythropoietin, human growth hormone, interferons, and interleukins. These nanocarriers are capable of releasing proteins in a sustained manner. Diasome Pharmaceuticals in the United States is developing an orally administered liposomal nanoparticulate system of insulin. This system has shown blood glucose reductions comparable to injected regular insulin in humans (17).

ENVIRONMENTAL, SOCIAL, AND ETHICAL ISSUES

Though the advent of nanotechnology has caused quite a stir in the scientific community, little has been done to learn the long-term effects of nanotechnology on the environment. Before nanotechnology is largely used to enhance the quality of life, it must be made certain that health risks are considered, social and ethical issues are addressed, public opinions are gathered, and regulatory matters are assessed from all aspects.

As of now, sufficient knowledge regarding the environmental and societal effects of pharmaceutical nanoparticles does not exist. On these counts, knowledge from the effects of other types of particles, e.g., air pollutants, can be helpful. Several air pollutants are in the nanosize range and they never fully settle. Thus, one can assume that therapeutic nanoparticles will accumulate in the air, water, and soil, resulting in lasting effects in biological systems. In addition, the advantageous surface area of nanoparticles may pose an added risk of explosion and fire hazard because of enhanced reactivity. As particle size decreases, the propensity toward violent dust cloud explosions increases leaving those who work in such industries susceptible to endangerment (18–20). Thus, it is exceedingly important that the environmental effects of nanoparticles be thoroughly considered.

When complications arise in the environment, it is inevitable that there are adverse health effects as well. The three primary routes of penetration of nanoparticles into the human body, as discussed in a previous chapter are through the skin, intestinal tract, and lungs (21). Some nanoparticles can penetrate deep into the dermis layer of the skin, thereby entering the systemic circulation. Following ingestion, nanoparticles might cross the intestinal barriers to enter the systemic circulation. Following inhalation, besides exerting respiratory effects, nanoparticles might translocate from the highly vascular lungs into the systemic circulation, potentially resulting in cardiovascular effects. Inhalation is by far the most likely route for systemic entry of environmental nanoparticles. Furthermore, some nanoparticles tend to clump in the air or within the body, which might block respiratory passages and vasculature.

In addition to adverse environmental and health risks, ethical issues related to nanotechnology have to be considered.

Although no fundamentally new ethical dilemmas are anticipated, the relevant ethical issues must not be ignored and a
conscientious approach to the development of nanotechnology
products is recommended (22,23). There are concerns that
nanotechnology will increase the gap between developed and
underdeveloped countries because of different abilities to utilize it, creating a so-called nano-divide. Therefore, opportunities to apply nanotechnology to benefit poor or developing
countries should be sought. Civil liberties should also be considered. Even though this technology may improve security,
safety, and individualized healthcare, there are concerns that
it might reduce privacy by increasing surveillance and release
of personal information without consent.

Disability rights groups might oppose to proposed nanotechnology-based interventions that enhance human capacities, on the grounds that this might lead to stigmatization
of those without enhanced capacities. This leads to many
uncertainties because it is still relatively unknown what is
and what will be available and when (24). Little is known
as to what the future of nanotechnology holds. Any technology
whose benefits are even slightly obscure and those that cause
discordance of major political interest groups are usually lit-
tered with controversy; as was seen with the advent of
nuclear energy, reproductive technologies, and biotechnology.
Much care must be taken to ensure that nanotechnology does
not overstate its boundaries.

Currently, awareness of nanotechnologies is very
low among the general public. In one survey, 51.8% of 1536
American respondents indicated that they had heard nothing
about nanotechnology (25). About 39.8% of the respondents
believed that the benefits of nanotechnology outweigh the
risks and 38.3% perceived that benefits equal risks. Among
the most important potential benefits of nanotechnology,
new ways to detect and treat diseases was identified by most
(57.2%), followed by new ways to clean up the environment
(15.8%), increased national security and defense (11.8%), physical and mental improvements for humans (11.5%), and
cheaper, better consumer products (3.8%). In general, science
and technology are not the top priorities of the government

and a lack of public education will make it harder for legislation concerning nanotechnology to be passed. Not only does the public need to know that nanotechnology exists but they also need to have accurate awareness of both the benefits and risks so that they can make informed decisions. It could be years before complete studies of the health and environmental impacts of nanotechnology come together, but public opinion could become established long before then.

REGULATORY CHALLENGES

Nanotechnology is an exciting new field with hopes for improvements in a wide variety of uses. Nanotechnology is new, and there is much research needed before sufficient regulation can be determined. Precisely for this reason, the regulation of nanotechnology products including nanoparticles is nonspecific. The Royal Society and the Royal Academy of Engineering took a proactive approach and prepared a comprehensive report on the opportunities and challenges of nanotechnology and made recommendations to ensure that regulations reflect the fact that nanoparticulate material may have greater toxicity than material in the larger size range (26). This report recommends voluntary disclosures on all products containing nanomaterials. These organizations also recommend that all relevant regulators review regulations within their area and ensure that they keep pace with future developments. The United States regulatory agencies are also taking a cautious approach as discussed in the following.

The mission of the U.S. Food and Drug Administration is to protect public health by ensuring safe and effective medical products and safe foods for humans and animals. It is envisioned that nanotechnologies and nanomaterials may be utilized in some of the wide range of products regulated by the U.S. FDA, including foods, cosmetics, drugs, devices, and veterinary products (Table 2). There are also a wide range of products involving nanotechnologies, which are regulated by other federal agencies. According to the FDA, nanotechnology

Table 2 Products Regulated by the U.S. FDA

Foods
    All interstate domestic and imported, including produce, fish, shellfish,
    shell eggs, and milk (not meat or poultry)
    Bottled water
    Wine, < 7% alcohol
    Infant formula
Food additives
    Colors
    Food containers
Cosmetics
Dietary supplements
Animal feeds
Pharmaceuticals
    Human
    Animal
    Tamper resistant packaging
Medical devices
Radiation emitting electronic products
Vaccines
Blood products
Tissues
Sterilants

---

is research and technology or the development of products regulated by the FDA that involve all of the following:

1. The existence of materials or products at the atomic, molecular, or macromolecular levels, where at least one dimension that affects the functional behavior of the drug/device product is in the length scale range of approximately 1–100 nm.
2. The creation and use of structures, devices, and systems that have novel properties and functions because of their small size.
3. The ability to control or manipulate the product on the atomic scale.

The FDA website (www.fda.gov) is informative regarding how different products are regulated, and how one should proceed to get the product approved for marketing. Although there are no nanotech-specific guidance documents at this time, all existing guidance documents would apply to nano-tech products.

The FDA paradigm for regulation of new products is based on the concepts of risk management, which includes identification, analysis, and control of risk. Nanomaterials may be a part or a consequence in several products approved by the FDA. For instance every approved biodegradable device and dosage form potentially contributes nanoparticulates to the biological system during its degradation process. In addition, several approved protein–polymer conjugates can be considered as nanomaterials. However, FDA has not experienced an adverse reaction related to the "nano" size of resorbable drug or medical device products.

The regulation and approval by the FDA is on a "product-by-product" basis, with the overall regulation process falling into three stages: premarket approval, premarket acceptance, and postmarket surveillance.

- *Premarket approval*: Prior to market introduction of any new pharmaceuticals, high-risk medical devices, food additives, colors, and biologicals, FDA approval is required. The producer/sponsor of the product is responsible for identifying and assessing the risks presented by the product. This party will also be responsible for indicating means to minimize the risks in a product application. These documents are reviewed by the FDA staff, with the assistance of an advisory committee. Often, a preapproval inspection of the manufacturing plant is required.
- *Premarket acceptance*: This category refers to products that are often copies of similar products that were approved previously or are products prepared to approved specifications. Examples include pharmaceuticals that are manufactured to existing USP Monographs and medical devices marketed with 510(k) premarket notifications. For these products, the FDA receives and reviews some form of notice that the products will be marketed and the products undergo a more rapid review process than premarket approval.

- *Postmarket surveillance*: In this third category, FDA manages the risks of "generally recognized as safe" (GRAS) products like foods, cosmetics, radiation emitting electronic products, and materials such as food additives and food packaging. For products in this category, market entry, and distribution are at the discretion of the manufacturer/producer. These products are generally regulated by the application of good manufacturing practices. FDA takes regulatory action if adverse events that threaten public or individual health occur.

The FDA coordinates policies within itself and with other government agencies. Regular nanotechnology discussions within the FDA are coordinated by the Office of Science and Health Coordination (OSHC). Centers within also organize similar meetings to share experiences with the review of the products, ensure that each center is aware of product guidance that may be developing elsewhere within the agency, and generally educate staff and policy makers about nanotechnology. Safety issues are identified and studied. As a member of the Nanoscale Science and Engineering Technology (NSET) Subcommittee of the National Science and Technology Council (NSTC) Committee on Technology, the FDA coordinates knowledge and policy with the other U.S. Government agencies. To define new test methods/protocols to ensure the safety of these products, the FDA and National Institute for Occupational Safety and Health (NIOSH) cochair the NSET Working Group on Nanomaterials Environmental and Health Implications (NEHI). In addition, FDA is a direct contributor to the evaluations of the toxicity of materials supported by the National Institute of Environmental Health Sciences (NIEHS) and the National Toxicology Program (NTP). To further facilitate the regulation of nanotechnology products, the FDA has formed an interactive Nanotechnology Interest Group (NTIG), which is made up of representatives from all of the centers.

The FDA expects that many nanotechnology products span the regulatory boundaries between pharmaceuticals, medical devices, and biologicals. These can be regulated as

"Combination Products" through established regulatory pathways. For combination products, the FDA will determine the "primary" mode of action of the product, i.e., a drug, medical device, or biological product. After this classification, the product will be managed by the appropriate FDA center with consultations from the other centers. Thus, more than one the FDA center might regulate nanoparticles and related systems. Because the FDA has traditionally regulated many products with particulate materials in the nanosize range, the FDA believes that the existing battery of pharmacotoxicity tests is probably adequate for most nanotechnology products. Particle size is not the issue. As and when new toxicological risks that derive from the new materials and/or new conformations of existing materials are identified, the FDA will require new tests.

The FDA regulations are for products, not technologies. In addition, the FDA regulates only the claims made by the product sponsor. If the manufacturer makes no nanotechnology claims regarding the manufacture or performance of the product, the FDA may be unaware at the time that the product under review employed nanotechnology.

Finally, the FDA has only limited authority over some potentially high-risk products, such as cosmetics. Many products are regulated only if they cause adverse health-related events in use. To date there have been few resources available to assess the risks of these products. Other government agencies have different missions with respect to nanotechnology. These include the need to solve environmental problems, improve technology to address disease, etc. Few resources currently exist to assess the risks that would result from the wide-scale deployment of nanotechnology products.

CONCLUSIONS

Nanoparticulate systems can potentially be used for enhanced drug dissolution and bioavailability, enhanced cellular uptake, improved tumor targeting, and diagnostic purposes. While some nanoparticle-based products are already approved by the U.S. FDA, several others are currently under development and clinical assessment, with anticipated FDA approval in the not so distant future. The public, although moderately knowledgeable about nanotechnology, perceive health benefits as the primary advantage of nanotechnology-based products. While there are ethical issues with respect to the accessibility of products of nanotechnology to the common man, the emergence of new ethical issues is unlikely with this technology. Toxicity of airborne nanoparticles, although a concern, given the prior development of nanoparticle-based products for pharmaceutical purposes, is unlikely to be a major issue. Also, the public is of the opinion that the benefits of the technology will outweigh the risks. The FDA takes a risk-based management approach for products of new technologies. The manufacturers are ultimately responsible for developing safe and efficacious pharmaceutical products based on nanoparticles.

ACKNOWLEDGMENTS

This work was supported by NIH grants DK064172, DK064172–02S1, and EY013842.

REFERENCES

1. Kayser O, Olbrich C, Yardley V, Kiderlen AF, Croft SL. Formulation of amphotericin B as nanosuspension for oral administration. Int J Pharm 2003; 254:73–75.

2. Gordonand EM, Hall FL. Nanotechnology blooms, at last (Review). Oncol Rep 2005; 13:1003–1007.

3. Gordon EM, Cornelio GH, Lorenzo CC III, et al. First clinical experience using a 'pathotropic' injectable retroviral vector (Rexin-G) as intervention for stage IV pancreatic cancer. Int J Oncol 2004; 24:177–185.

4. Rockall AG, Sohaib SA, Harisinghani MG, et al. Diagnostic performance of nanoparticle-enhanced magnetic resonance imaging in the diagnosis of lymph node metastases in patients with endometrial and cervical cancer. J Clin Oncol 2005; 23:2813–2821.

5. Deserno WM, Harisinghani MG, Taupitz M, et al. Urinary bladder cancer: preoperative nodal staging with ferumoxtran-10-enhanced MR imaging. Radiology 2004; 233:449–456.

6. Michel SC, Keller TM, Frohlich JM, et al. Preoperative breast cancer staging: MR imaging of the axilla with ultrasmall superparamagnetic iron oxide enhancement. Radiology 2002; 225:527–536.

7. Anzai Y, Piccoli CW, Outwater EK, et al. Evaluation of neck and body metastases to nodes with ferumoxtran 10-enhanced MR imaging: phase III safety and efficacy study. Radiology 2003; 228:777–788.

8. Gao X, Cui Y, Levenson RM, Chung LW, Nie S. In vivo cancer targeting and imaging with semiconductor quantum dots. Nat Biotechnol 2004; 22:969–976.

9. Gao X, Yang L, Petros JA, Marshall FF, Simons JW, Nie S. In vivo molecular and cellular imaging with quantum dots. Curr Opin Biotechnol 2005; 16:63–72.

10. Voura EB, Jaiswal JK, Mattoussi H, Simon SM. Tracking metastatic tumor cell extravasation with quantum dot nanocrystals and fluorescence emission-scanning microscopy. Nat Med 2004; 10:993–998.

11. Wu X, Liu H, Liu J, et al. Immunofluorescent labeling of cancer marker Her2 and other cellular targets with semiconductor quantum dots. Nat Biotechnol 2003; 21:41–46.

12. Nigavekar SS, Sung LY, Llanes M, et al. 3H dendrimer nanoparticle organ/tumor distribution. Pharm Res 2004; 21:476–483.

13. O'Neal DP, Hirsch LR, Halas NJ, Payne JD, West JL. Photothermal tumor ablation in mice using near infrared-absorbing nanoparticles. Cancer Lett 2004; 209:171–176.

14. Johannsen M, Jordan A, Scholz R, et al. Evaluation of magnetic fluid hyperthermia in a standard rat model of prostate cancer. J Endourol 2004; 18:495–500.

15. Johannsen M, Thiesen B, Jordan A, et al. Magnetic fluid hyperthermia (MFH) reduces prostate cancer growth in the orthotopic Dunning R3327 rat model. Prostate 2005; 64:283–292.

16. Unger EC, Porter T, Culp W, Labell R, Matsunaga T, Zutshi R. Therapeutic applications of lipid-coated microbubbles. Adv Drug Deliv Rev 2004; 56:1291–1314.

17. Davis SN, Geho B, Tate D, et al. The effects of HDV-insulin on carbohydrate metabolism in Type 1 diabetic patients. J Diabetes Complications 2001; 15:227–233.

18. Abrahamsonand J, Dinniss J. Ball lightning caused by oxidation of nanoparticle networks from normal lightning strikes on soil. Nature 2000; 403:519–521.

19. Sikkeland T, Skuterud L, Goltsova NI, Lindmo T. Reconstruction of doses and deposition in the western trace from the Chernobyl accident. Health Phys 1997; 72:750–758.

20. Chesser RK, Bondarkov M, Baker RJ, Wickliffe, JK, Rodgers BE. Reconstruction of radioactive plume characteristics along Chernobyl's Western Trace. J Environ Radioact 2004; 71:147–157.

21. Hoet PH, Bruske-Hohlfeld I, Salata OV. Nanoparticles—known and unknown health risks. J Nanobiotechnol 2004; 2:12.

22. Grunwald A. Nanotechnology—a new field of ethical inquiry? Sci Eng Ethics 2005; 11:187–201.

23. Berne RW. Towards the conscientious development of ethical nanotechnology. Sci Eng Ethics 2004; 10:627–638.

24. Grunwald A. The case of nanobiotechnology. EMBO Rep 2004; 5:S32–S36.

25. Cobband MC, Macoubrie J. Public perceptions about nanotechnology: risks, benefits and trust. J Nanoparticle Res 2004; 6:395–405.

26. Report of The Royal Academy of Engineering, The Royal Society. Nanoscience and nanotechnologies: opportunities and uncertainties. 2004.

Index